ⓘnquire

ⓘnteract

ⓘnspire

ⓘnvent

This ⓘScience Interactive Student Textbook Belongs to:

_____

Name

_____

Teacher/Class

Where am I located?

The dot on the map shows where my school is.

FLORIDA
COURSE 2

SCIENCE

Glencoe

D1535947

Leopard, *Panthera pardus*

Once common across southern Asia and most of Africa, most leopards exist today in sub-Saharan Africa in rain forests and deserts. They are the smallest of the big cats—tigers, lions, jaguars, and leopards. Leopards are known for their ability to climb trees while carrying prey.

Education

Copyright © 2012 by The McGraw-Hill Companies, Inc.

All rights reserved. No part of this publication may be reproduced or distributed in any form or by any means, or stored in a database or retrieval system, without the prior written consent of The McGraw-Hill Companies, Inc., including, but not limited to, network storage or transmission, or broadcast for distance learning.

Send all inquiries to:
McGraw-Hill Education
8787 Orion Place
Columbus, OH 43240

ISBN: 978-0-07-660221-6
MHID: 0-07-660221-4

Printed in the United States of America.

2 3 4 5 6 7 8 9 WDQ 15 14 13 12 11 10

The *McGraw·Hill* Companies

Authors and Contributors

Consulting Authors

Alton L. Biggs
Biggs Educational Consulting
Commerce, TX

Ralph M. Feather, Jr., PhD
Assistant Professor
Department of Educational
Studies and Secondary Education
Bloomsburg University
Bloomsburg, PA

Douglas Fisher, PhD
Professor of Teacher Education
San Diego State University
San Diego, CA

Edward P. Ortleb
Science/Safety Consultant
St. Louis, MO

Series Consultants

Science

Solomon Bililign, PhD
Professor
Department of Physics
North Carolina Agricultural and
Technical State University
Greensboro, NC

John Choinski
Professor
Department of Biology
University of Central Arkansas
Conway, AR

Anastasia Chopelas, PhD
Research Professor
Department of Earth and Space
Sciences
UCLA
Los Angeles, CA

David T. Crowther, PhD
Professor of Science Education
University of Nevada, Reno
Reno, NV

A. John Gatz
Professor of Zoology
Ohio Wesleyan University
Delaware, OH

Sarah Gille, PhD
Professor
University of California San
Diego
La Jolla, CA

David G. Haase, PhD
Professor of Physics
North Carolina State University
Raleigh, NC

Janet S. Herman, PhD
Professor
Department of Environmental
Sciences
University of Virginia
Charlottesville, VA

David T. Ho, PhD
Associate Professor
Department of Oceanography
University of Hawaii
Honolulu, HI

Ruth Howes, PhD
Professor of Physics
Marquette University
Milwaukee, WI

Jose Miguel Hurtado, Jr., PhD
Associate Professor
Department of Geological
Sciences
University of Texas at El Paso
El Paso, TX

Monika Kress, PhD
Assistant Professor
San Jose State University
San Jose, CA

Mark E. Lee, PhD
Associate Chair & Assistant
Professor
Department of Biology
Spelman College
Atlanta, GA

Linda Lundgren
Science writer
Lakewood, CO

Keith O. Mann, PhD
Ohio Wesleyan University
Delaware, OH

Charles W. McLaughlin, PhD
Adjunct Professor of Chemistry
Montana State University
Bozeman, MT

Katharina Pahnke, PhD
Research Professor
Department of Geology and
Geophysics
University of Hawaii
Honolulu, HI

Jesús Pando, PhD
Associate Professor
DePaul University
Chicago, IL

Hay-Oak Park, PhD
Associate Professor
Department of Molecular
Genetics
Ohio State University
Columbus, OH

David A. Rubin, PhD
Associate Professor of Physiology
School of Biological Sciences
Illinois State University
Normal, IL

Toni D. Saucy
Assistant Professor of Physics
Department of Physics
Angelo State University
San Angelo, TX

Authors and Contributors

Authors

American Museum of Natural History
New York, NY

Michelle Anderson, MS
Lecturer
The Ohio State University
Columbus, OH

Juli Berwald, PhD
Science Writer
Austin, TX

John F. Bolzan, PhD
Science Writer
Columbus, OH

Rachel Clark, MS
Science Writer
Moscow, ID

Patricia Craig, MS
Science Writer
Bozeman, MT

Randall Frost, PhD
Science Writer
Pleasanton, CA

Lisa S. Gardiner, PhD
Science Writer
Denver, CO

Jennifer Gonya, PhD
The Ohio State University
Columbus, OH

Mary Ann Grobbel, MD
Science Writer
Grand Rapids, MI

Whitney Crispen Hagins, MA, MAT
Biology Teacher
Lexington High School
Lexington, MA

Carole Holmberg, BS
Planetarium Director
Calusa Nature Center and
Planetarium, Inc.
Fort Myers, FL

Tina C. Hopper
Science Writer
Rockwall, TX

Jonathan D. W. Kahl, PhD
Professor of Atmospheric Science
University of Wisconsin-Milwaukee
Milwaukee, WI

Nanette Kalis
Science Writer
Athens, OH

S. Page Keeley, MEd
Maine Mathematics and Science Alliance
Augusta, ME

Cindy Klevickis, PhD
Professor of Integrated Science and Technology
James Madison University
Harrisonburg, VA

Kimberly Fekany Lee, PhD
Science Writer
La Grange, IL

Michael Manga, PhD
Professor
University of California, Berkeley
Berkeley, CA

Devi Ried Mathieu
Science Writer
Sebastopol, CA

Elizabeth A. Nagy-Shadman, PhD
Geology Professor
Pasadena City College
Pasadena, CA

William D. Rogers, DA
Professor of Biology
Ball State University
Muncie, IN

Donna L. Ross, PhD
Associate Professor
San Diego State University
San Diego, CA

Marion B. Sewer, PhD
Assistant Professor
School of Biology
Georgia Institute of Technology
Atlanta, GA

Julia Meyer Sheets, PhD
Lecturer
School of Earth Sciences
The Ohio State University
Columbus, OH

Michael J. Singer, PhD
Professor of Soil Science
Department of Land, Air and Water Resources
University of California
Davis, CA

Karen S. Sottosanti, MA
Science Writer
Pickerington, Ohio

Paul K. Strode, PhD
I.B. Biology Teacher
Fairview High School
Boulder, CO

Jan M. Vermilye, PhD
Research Geologist
Seismo-Tectonic Reservoir
Monitoring (STRM)
Boulder, CO

Judith A. Yero, MA
Director
Teacher's Mind Resources
Hamilton, MT

Dinah Zike, MEd
Author, Consultant, Inventor of Foldables
Dinah Zike Academy; Dinah-Might Adventures, LP
San Antonio, TX

Margaret Zorn, MS
Science Writer
Yorktown, VA

Series Consultants, continued

Malathi Srivatsan, PhD
Associate Professor of
Neurobiology
College of Sciences and
Mathematics
Arkansas State University
Jonesboro, AR

Cheryl Wistrom, PhD
Associate Professor of Chemistry
Saint Joseph's College
Rensselaer, IN

Reading

ReLeah Cossett Lent
Author/Educational Consultant
Blue Ridge, GA

Math

Vik Hovsepian
Professor of Mathematics
Rio Hondo College
Whittier, CA

Series Reviewers

Thad Boggs
Mandarin High School
Jacksonville, FL

Catherine Butcher
Webster Junior High School
Minden, LA

Erin Darichuk
West Frederick Middle School
Frederick, MD

Joanne Hedrick Davis
Murphy High School
Murphy, NC

Anthony J. DiSipio, Jr.
Octorara Middle School
Atglen, PA

Adrienne Elder
Tulsa Public Schools
Tulsa, OK

Carolyn Elliott
Iredell-Statesville Schools
Statesville, NC

Christine M. Jacobs
Ranger Middle School
Murphy, NC

Jason O. L. Johnson
Thurmont Middle School
Thurmont, MD

Felecia Joiner
Stony Point Ninth Grade Center
Round Rock, TX

Joseph L. Kowalski, MS
Lamar Academy
McAllen, TX

Brian McClain
Amos P. Godby High School
Tallahassee, FL

Von W. Mosser
Thurmont Middle School
Thurmont, MD

Ashlea Peterson
Heritage Intermediate Grade
Center
Coweta, OK

Nicole Lenihan Rhoades
Walkersville Middle School
Walkersvillle, MD

Maria A. Rozenberg
Indian Ridge Middle School
Davie, FL

Barb Seymour
Westridge Middle School
Overland Park, KS

Ginger Shirley
Our Lady of Providence Junior-
Senior High School
Clarksville, IN

Curtis Smith
Elmwood Middle School
Rogers, AR

Sheila Smith
Jackson Public School
Jackson, MS

Sabra Soileau
Moss Bluff Middle School
Lake Charles, LA

Tony Spoores
Switzerland County Middle
School
Vevay, IN

Nancy A. Stearns
Switzerland County Middle
School
Vevay, IN

Kari Vogel
Princeton Middle School
Princeton, MN

Alison Welch
Wm. D. Slider Middle School
El Paso, TX

Linda Workman
Parkway Northeast Middle
School
Creve Coeur, MO

With your book!

Answer questions, record data, and interact with images directly in your book!

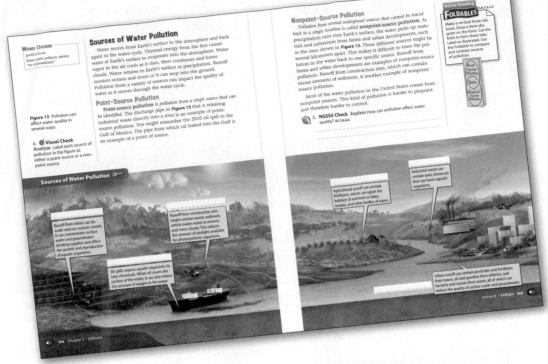

Online!

Log on to **Connect ED** for a digital version of this book that includes

- audio
- animations
- virtual labs

The Florida Teacher Advisory Board provided valuable input in the development of the © 2012 Florida student textbooks.

Ray Amil
Union Park Middle School
Orlando, FL

Maria Swain Kearns
Venice Middle School
Venice, FL

Ivette M. Acevedo Santiago, MEd
Resource Teacher
Lake Nona High School
Orlando, FL

Christy Bowman
Montford Middle School
Tallahassee, FL

Susan Leeds
Department Chair
Howard Middle School
Orlando, FL

Rachel Cassandra Scott
Bair Middle School
Sunrise, FL

Get Connected to

ConnectED

connectED.mcgraw-hill.com

Your online portal to everything you need!
- One-Stop Shop, One Personalized Password
- Easy Intuitive Navigation
- Resources, Resources, Resources

For Students
Leave your books at school. Now you can go online and interact with your StudentWorks™ Plus digital Student Edition from any place, any time!

For Teachers
ConnectED is your one-stop online center for everything you need to teach, including: digital eTeacherEdition, lesson planning and scheduling tools, pacing, and assessment.

For Parents
Get homework help, help your student prepare for testing, and review science topics.

Log on today and get ConnectED!

connectED.mcgraw-hill.com

Username: flscistudent
Password: science2012

Username and password allows you to experience all of the helpful features of ConnectED. For details about how to get your personal password for the life of the adoption, contact your Glencoe sales representative.

Authors and Contributors

Authors

American Museum of Natural History
New York, NY

Michelle Anderson, MS
Lecturer
The Ohio State University
Columbus, OH

Juli Berwald, PhD
Science Writer
Austin, TX

John F. Bolzan, PhD
Science Writer
Columbus, OH

Rachel Clark, MS
Science Writer
Moscow, ID

Patricia Craig, MS
Science Writer
Bozeman, MT

Randall Frost, PhD
Science Writer
Pleasanton, CA

Lisa S. Gardiner, PhD
Science Writer
Denver, CO

Jennifer Gonya, PhD
The Ohio State University
Columbus, OH

Mary Ann Grobbel, MD
Science Writer
Grand Rapids, MI

Whitney Crispen Hagins, MA, MAT
Biology Teacher
Lexington High School
Lexington, MA

Carole Holmberg, BS
Planetarium Director
Calusa Nature Center and Planetarium, Inc.
Fort Myers, FL

Tina C. Hopper
Science Writer
Rockwall, TX

Jonathan D. W. Kahl, PhD
Professor of Atmospheric Science
University of Wisconsin-Milwaukee
Milwaukee, WI

Nanette Kalis
Science Writer
Athens, OH

S. Page Keeley, MEd
Maine Mathematics and Science Alliance
Augusta, ME

Cindy Klevickis, PhD
Professor of Integrated Science and Technology
James Madison University
Harrisonburg, VA

Kimberly Fekany Lee, PhD
Science Writer
La Grange, IL

Michael Manga, PhD
Professor
University of California, Berkeley
Berkeley, CA

Devi Ried Mathieu
Science Writer
Sebastopol, CA

Elizabeth A. Nagy-Shadman, PhD
Geology Professor
Pasadena City College
Pasadena, CA

William D. Rogers, DA
Professor of Biology
Ball State University
Muncie, IN

Donna L. Ross, PhD
Associate Professor
San Diego State University
San Diego, CA

Marion B. Sewer, PhD
Assistant Professor
School of Biology
Georgia Institute of Technology
Atlanta, GA

Julia Meyer Sheets, PhD
Lecturer
School of Earth Sciences
The Ohio State University
Columbus, OH

Michael J. Singer, PhD
Professor of Soil Science
Department of Land, Air and Water Resources
University of California
Davis, CA

Karen S. Sottosanti, MA
Science Writer
Pickerington, Ohio

Paul K. Strode, PhD
I.B. Biology Teacher
Fairview High School
Boulder, CO

Jan M. Vermilye, PhD
Research Geologist
Seismo-Tectonic Reservoir Monitoring (STRM)
Boulder, CO

Judith A. Yero, MA
Director
Teacher's Mind Resources
Hamilton, MT

Dinah Zike, MEd
Author, Consultant, Inventor of Foldables
Dinah Zike Academy; Dinah-Might Adventures, LP
San Antonio, TX

Margaret Zorn, MS
Science Writer
Yorktown, VA

Consulting Authors

Alton L. Biggs
Biggs Educational Consulting
Commerce, TX

Ralph M. Feather, Jr., PhD
Assistant Professor
Department of Educational
Studies and Secondary Education
Bloomsburg University
Bloomsburg, PA

Douglas Fisher, PhD
Professor of Teacher Education
San Diego State University
San Diego, CA

Edward P. Ortleb
Science/Safety Consultant
St. Louis, MO

Series Consultants

Science

Solomon Bililign, PhD
Professor
Department of Physics
North Carolina Agricultural and
Technical State University
Greensboro, NC

John Choinski
Professor
Department of Biology
University of Central Arkansas
Conway, AR

Anastasia Chopelas, PhD
Research Professor
Department of Earth and Space
Sciences
UCLA
Los Angeles, CA

David T. Crowther, PhD
Professor of Science Education
University of Nevada, Reno
Reno, NV

A. John Gatz
Professor of Zoology
Ohio Wesleyan University
Delaware, OH

Sarah Gille, PhD
Professor
University of California San
Diego
La Jolla, CA

David G. Haase, PhD
Professor of Physics
North Carolina State University
Raleigh, NC

Janet S. Herman, PhD
Professor
Department of Environmental
Sciences
University of Virginia
Charlottesville, VA

David T. Ho, PhD
Associate Professor
Department of Oceanography
University of Hawaii
Honolulu, HI

Ruth Howes, PhD
Professor of Physics
Marquette University
Milwaukee, WI

Jose Miguel Hurtado, Jr., PhD
Associate Professor
Department of Geological
Sciences
University of Texas at El Paso
El Paso, TX

Monika Kress, PhD
Assistant Professor
San Jose State University
San Jose, CA

Mark E. Lee, PhD
Associate Chair & Assistant
Professor
Department of Biology
Spelman College
Atlanta, GA

Linda Lundgren
Science writer
Lakewood, CO

Keith O. Mann, PhD
Ohio Wesleyan University
Delaware, OH

Charles W. McLaughlin, PhD
Adjunct Professor of Chemistry
Montana State University
Bozeman, MT

Katharina Pahnke, PhD
Research Professor
Department of Geology and
Geophysics
University of Hawaii
Honolulu, HI

Jesús Pando, PhD
Associate Professor
DePaul University
Chicago, IL

Hay-Oak Park, PhD
Associate Professor
Department of Molecular
Genetics
Ohio State University
Columbus, OH

David A. Rubin, PhD
Associate Professor of Physiology
School of Biological Sciences
Illinois State University
Normal, IL

Toni D. Sauncy
Assistant Professor of Physics
Department of Physics
Angelo State University
San Angelo, TX

Inquiry LAB STATION

Labs, Labs, Labs

Launch Labs at the beginning of every lesson let you be the scientist! The ⚙LAB Station on **ConnectED** has all the labs for each chapter.

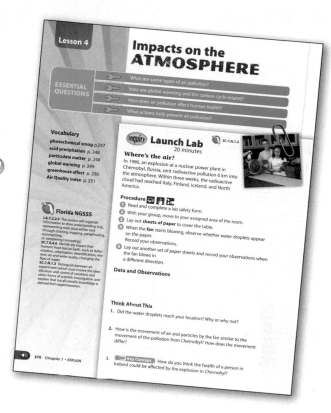

Lesson 4

Impacts on the ATMOSPHERE

ESSENTIAL QUESTIONS
- What are some types of air pollution?
- How are global warming and the carbon cycle related?
- How does air pollution affect human health?
- What actions help prevent air pollution?

Vocabulary
photochemical smog p.247
acid precipitation p. 248
particulate matter p. 248
global warming p. 249
greenhouse effect p. 250
Air Quality Index p. 251

Florida NGSSS
LA.7.2.2.3 The student will organize information to show understanding (e.g., representing main ideas within text through charting, mapping, paraphrasing, summarizing, or comparing/contrasting).
SC.7.E.6.6 Identify the impact that humans have had on Earth, such as deforestation, urbanization, desertification, erosion, air and water quality, changing the flow of water.
SC.7.N.1.3 Distinguish between an experiment (which must involve the identification and control of variables) and other forms of scientific investigation and explain that not all scientific knowledge is derived from experimentation.

Launch Lab
20 minutes
SC.7.N.1.3

Where's the air?
In 1986, an explosion at a nuclear power plant in Chernobyl, Russia, sent radioactive pollution 6 km into the atmosphere. Within three weeks, the radioactive cloud had reached Italy, Finland, Iceland, and North America.

Procedure
1. Read and complete a lab safety form.
2. With your group, move to your assigned area of the room.
3. Lay out **sheets of paper** to cover the table.
4. When the **fan** starts blowing, observe whether water droplets appear on the paper. Record your observations.
5. Lay out another set of paper sheets and record your observations when the fan blows in a different direction.

Data and Observations

Think About This
1. Did the water droplets reach your location? Why or why not?

2. How is the movement of air and particles by the fan similar to the movement of the pollution from Chernobyl? How does the movement differ?

3. **Key Concept** How do you think the health of a person in Iceland could be affected by the explosion in Chernobyl?

270 Chapter 7 • EXPLAIN

Virtual Labs

Virtual Labs provide a highly interactive lab experience.

i Read i Science

Sequence Words

While you read, watch for words that show the order events happen.

- first
- next
- last
- begins
- second
- later

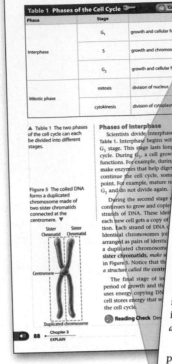

Vocabulary Help

Science terms are highlighted and reviewed to check your understanding.

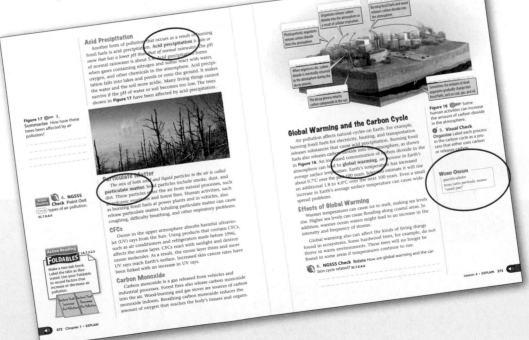

Write the answers to questions right in your book!

Concept Map

Each chapter's **Concept Map** gives you a place to show all the science connections you've learned.

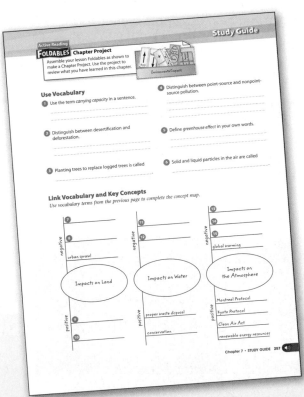

ⓘLAB STATION   Go online to access the labs for this chapter.

Skill Practice: How can you build your own scientific instrument?

Inquiry Lab: How can you help develop a bioreactor?

SCIENCE IN ACTION

Here are some of the exciting activities in this chapter!

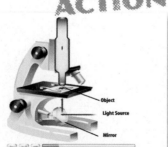

Concepts in Motion: Simple Microscope

BrainPOP: Scientific Methods

Page Keeley Science Probe

 **ConnectED**

There's More Online!
Video • Audio • Review • Inquiry • WebQuest • Assessment • Concepts in Motion • Multilingual eGlossary

FLORIDA
ⓘScience
Course 2

— Try It!

☐ **LAUNCH LABS**
Inquiry Apply It!
At the beginning of every lesson.

ⓘLAB STATION Go online to access the labs for this chapter.

MiniLabs: LESSON 1: Which materials will sink?
LESSON 2: Which liquid is densest?
LESSON 3: How do landforms compare?

Skill Practice: How can you find the density of a liquid?

Inquiry Lab: Modeling Earth and its Layers

SCIENCE IN ACTION

Here are some of the exciting activities in this chapter!

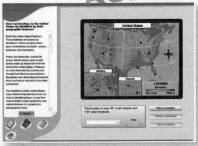

Virtual Lab: How can locations in the United States be identified by their geographic features?

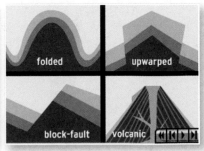

BrainPOP: Mountain Types

Concepts in Motion: Earth's Interior

 **There's More Online!**
Video • Audio • Review • Inquiry • WebQuest • Assessment • Concepts in Motion • Multilingual eGlossary

LAB STATION   Go online to access the labs for this chapter.

MiniLabs: LESSON 1: How does your garden grow?

LESSON 2: How do heat and pressure change rocks?

LESSON 3: How can you turn one sedimentary rock into another?

Skill Practice: How are rocks similar and different?

Inquiry Lab: Design a Forensic Investigation

SCIENCE IN ACTION

Here are some of the exciting activities in this chapter!

Virtual Lab: How are rocks classified?

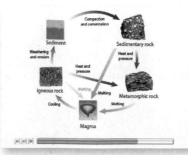

Concepts in Motion: Rock Cycle

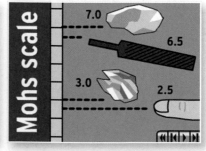

BrainPOP: Mineral Identification

 **Connect ED**   **There's More Online!**

Video • Audio • Review • Inquiry • WebQuest • Assessment • Concepts in Motion • Multilingual eGlossary

FLORIDA
iScience
Course 2

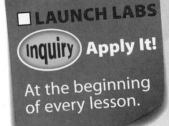

Try It!

☐ **LAUNCH LABS**

inquiry Apply It!

At the beginning of every lesson.

LAB STATION Go online to access the labs for this chapter.

MiniLabs: LESSON 1: What was that?
LESSON 2: How might Earth's early atmosphere formed?
LESSON 3: How do glaciers affect sea level?

Skill Practice: What do these numbers tell you about a rock's age?

Inquiry Lab: Design Your Own Grand Canyon

SCIENCE IN ACTION   Here are some of the exciting activities in this chapter!

Virtual Lab: What was Earth like throughout geologic time?

Concepts in Motion: Continental Drift

BrainPOP: Fossils

 **There's More Online!**
Video • Audio • Review • Inquiry • WebQuest • Assessment • Concepts in Motion • Multilingual eGlossary

LAB STATION   Go online to access the labs for this chapter.

<u>MiniLabs:</u>   LESSON 1: How do you use clues to put puzzle pieces together?

LESSON 2: How old is the Atlantic Ocean?

LESSON 3: How do changes in density cause motion?

<u>Skill Practice:</u> How do rocks on the seafloor vary with age away from the mid-ocean ridge?

<u>Inquiry Lab:</u> Movement of Plate Boundaries

SCIENCE IN ACTION

Here are some of the exciting activities in this chapter!

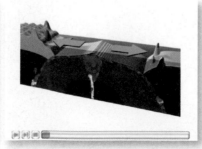

Virtual Lab: Where do most earthquake epicenters and volcanoes occur?

Concepts in Motion: Seafloor Spreading

Page Keeley Science Probe

ConnectED   **There's More Online!**
Video • Audio • Review • Inquiry • WebQuest • Assessment • Concepts in Motion • Multilingual eGlossary

FLORIDA
iScience
Course 2

Try It!

☐ **LAUNCH LABS**

Inquiry Apply It!

At the beginning of every lesson.

iLAB STATION Go online to access the labs for this chapter.

MiniLabs: LESSON 1: What will happen?

LESSON 2: What is the relationship between plate motion and landforms?

LESSON 3: How do folded mountains form?

LESSON 4: Can you analyze a continent?

Skill Practice: Can you measure how stress deforms putty?

Inquiry Lab: Design Landforms

SCIENCE IN ACTION

Here are some of the exciting activities in this chapter!

Virtual Lab: Why do things float?

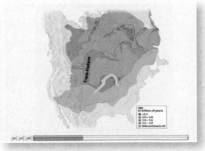

Concepts in Motion: Continent Formation

Page Keeley Science Probes

There's More Online!
Video • Audio • Review • Inquiry • WebQuest • Assessment • Concepts in Motion • Multilingual eGlossary

ⓘLAB
STATION   Go online to access the labs for this chapter.

<u>MiniLabs:</u>   LESSON 1: Can you use the Mercalli scale
to locate an epicenter?

LESSON 2: Can you model the movement
of magma?

<u>Skill Practice:</u> Can you locate an earthquake's epicenter?

<u>Inquiry Lab:</u>   The Dangers of Mount Rainier

SCIENCEIN
ACTION

Here are some of the exciting activities in this chapter!

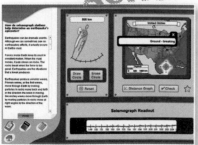

Virtual Lab: How do seismograph stations
help determine an earthquake's epicenter?

Concepts in Motion: Seismic Waves

What's Science Got to do With It?
Location, Location, Location

 **There's More Online!**

Video • Audio • Review • Inquiry • WebQuest • Assessment • Concepts in Motion • Multilingual eGlossary

FLORIDA
iScience
Course 2

Try It!

☐ **LAUNCH LABS**
inquiry **Apply It!**
At the beginning of every lesson.

iLAB STATION Go online to access the labs for this chapter.

MiniLabs: LESSON 2: What happens when you mine?
LESSON 3: What's in well water?
LESSON 4: What's in the air?

Skill Practice: What amount of Earth's resources do you use in a day?

Skill Practice: How will you design an environmentally safe landfill?

Inquiry Lab: Design a Green City

SCIENCE IN ACTION

Here are some of the exciting activities in this chapter!

What's Science Got to do With It?
Tire Tracks

Concepts in Motion: The Nitrogen Cycle

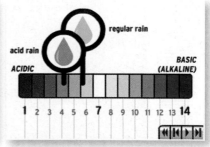

BrainPOP: Air Pollution

 **ConnectED**   **There's More Online!**
Video • Audio • Review • Inquiry • WebQuest • Assessment • Concepts in Motion • Multilingual eGlossary

Unit 2 # ENERGY AND ENERGY TRANSFORMATIONS

LAB STATION Go online to access the labs for this chapter.

MiniLabs: LESSON 1: How do waves transfer energy?

LESSON 2: How can you transfer energy?

LESSON 3: What affects the transfer of thermal energy?

Skill Practice: How can you classify different types of energy?

Inquiry Lab: Power a Device with a Potato

SCIENCE IN ACTION

Here are some of the exciting activities in this chapter!

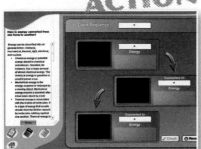

Virtual Lab: How is energy converted from one form to another?

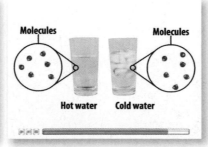

Concepts in Motion: Temperature and Molecular Motion

BrainPOP: Forms of Energy

 **ConnectED**

There's More Online!
Video • Audio • Review • Inquiry • WebQuest • Assessment • Concepts in Motion • Multilingual eGlossary

FLORIDA
iScience
Course 2

Try It!

☐ **LAUNCH LABS**

Inquiry **Apply It!**

At the beginning of every lesson.

LAB STATION   Go online to access the labs for this chapter.

MiniLabs: LESSON 1: How do waves travel through matter?

LESSON 2: How are wavelength and frequency related?

LESSON 3: How can reflection be used?

Skill Practice: How are the properties of waves related?

Inquiry Lab: Measuring Wave Speed

SCIENCE IN ACTION

Here are some of the exciting activities in this chapter!

Virtual Lab: What are some characteristics of waves?

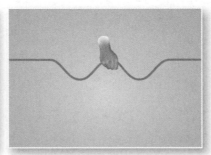

Concepts in Motion: Transverse Waves

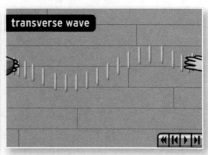

BrainPOP: Waves

 **There's More Online!**

Video • Audio • Review • Inquiry • WebQuest • Assessment • Concepts in Motion • Multilingual eGlossary

LAB STATION Go online to access the labs for this chapter.

<u>**MiniLabs:**</u> LESSON 1: Can you model a sound wave?

LESSON 2: Can you see a light beam in water?

LESSON 3: How does the size of an image change?

<u>**Skill Practice:**</u> How are light rays reflected from a plane mirror?

<u>**Inquiry Lab:**</u> The Images Formed by a Lens

SCIENCE IN ACTION

Here are some of the exciting activities in this chapter!

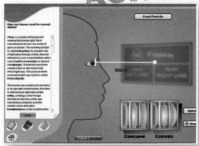

Virtual Lab: How are lenses used to correct vision?

What's Science Got to do With It? Crime Scene Investigation

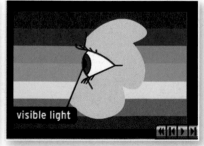

BrainPOP: Color

 **ConnectED**

There's More Online!
Video • Audio • Review • Inquiry • WebQuest • Assessment • Concepts in Motion • Multilingual eGlossary

Unit 3 HEREDITY AND REPRODUCTION

⚙️LAB STATION Go online to access the labs for this chapter.

<u>**MiniLabs:**</u>   LESSON 1: How does one cell produce four cells?

LESSON 2: What parts of plants can grow?

<u>**Inquiry Lab:**</u> Mitosis and Meiosis

-Try It!-

☐ **LAUNCH LABS**

Inquiry **Apply It!**

At the beginning of every lesson.

SCIENCE IN ACTION

Here are some of the exciting activities in this chapter!

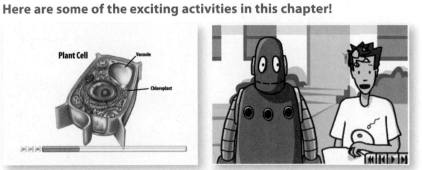

Virtual Lab: What is the life cycle of a simple plant?

Concepts in Motion: Meiosis

BrainPOP: Fertilization and Birth

 **There's More Online!**

Video • Audio • Review • Inquiry • WebQuest • Assessment • Concepts in Motion • Multilingual eGlossary

LAB STATION Go online to access the labs for this chapter.

MiniLabs: LESSON 1: Which is the dominant trait?
LESSON 2: Can you infer genotype?
LESSON 3: How can you model DNA?

Skill Practice: How can you use Punnett squares to model inheritance?

Inquiry Lab: Gummy Bear Genetics

SCIENCE IN ACTION

Here are some of the exciting activities in this chapter!

Virtual Lab: How are traits passed from parent to offspring?

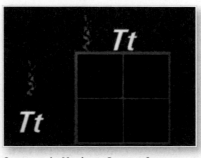

Concepts in Motion: Punnett Square

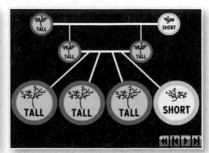

BrainPOP: Heredity

There's More Online!
Video • Audio • Review • Inquiry • WebQuest • Assessment • Concepts in Motion • Multilingual eGlossary

FLORIDA iScience Course 2

─Try It!

☐ **LAUNCH LABS**

inquiry Apply It!

At the beginning of every lesson.

LAB STATION Go online to access the labs for this chapter.

MiniLabs: LESSON 1: How do species change over time?
LESSON 2: Who survives?
LESSON 3: How related are organisms?

Skill Practice: Can you observe changes through time in collections of everyday objects?

Inquiry Lab: Model Adaptations in an Organism

SCIENCE IN ACTION

Here are some of the exciting activities in this chapter!

Virtual Lab: How can fossil and rock data determine when an organism lived?

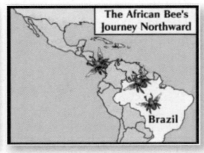

What's Science Got to do With It?
Killer Bees

Page Keeley Science Probe

 **ConnectED**

There's More Online!
Video • Audio • Review • Inquiry • WebQuest • Assessment • Concepts in Motion • Multilingual eGlossary

☐ Lesson 1   ☐ Lesson 2   ☐ Lesson 3

LAB STATION
Go online to access the labs for this chapter.

MiniLabs: LESSON 1: How many living and nonliving things can you find?

LESSON 2: How does a fish population change?

LESSON 3: How is energy transferred in a food chain?

Skill Practice: Can you make predictions about a population size?

Inquiry Lab: Can you observe part of the carbon cycle?

SCIENCE IN ACTION

Here are some of the exciting activities in this chapter!

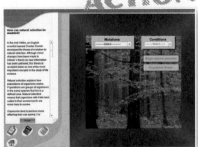

Virtual Lab: How can natural selection be modeled?

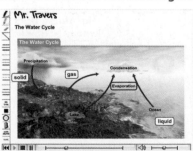

Concepts in Motion: The Water Cycle

BrainPOP: Ecosystems

 **There's More Online!**

Video • Audio • Review • Inquiry • WebQuest • Assessment • Concepts in Motion • Multilingual eGlossary

Scientific Explanations

An explanation helps provide answers to a question a scientist might be wondering about. Which of the following do you think involves providing a scientific explanation? Select the best response.

A. hypothesis

B. scientific theory

C. scientific law

D. hypothesis and scientific theory

E. scientific theory and scientific law

F. hypothesis, scientific theory, and scientific law

G. None of the above. An explanation is something else.

Explain your thinking. Describe how explanations are used in science.

FLORIDA
Nature of Science

Scientific
EXPLANATIONS

Methods of Science

This chapter begins your study of the nature of science, but there is even more information about the nature of science in this book. Each unit begins by exploring an important topic that is fundamental to scientific study. As you read these topics, you will learn even more about the nature of science.

FLORIDA BIG IDEAS

1 The Practice of Science

2 The Characteristics of Scientific Knowledge

3 The Role of Theories, Laws, Hypotheses, and Models

Think About It!

How can science provide answers to your questions about the world around you?

These two divers are collecting data about corals. They are marine biologists, scientists who study living things in oceans and other saltwater environments.

1 What information about corals do you think these scientists are collecting?

_____

_____

_____

2 What questions do you think they hope to answer?

_____

_____

_____

3 How do you think science can provide answers to their questions and your questions?

_____

_____

_____

_____

Florida NGSSS

LA.7.2.2.3 The student will organize information to show understanding (e.g., representing main ideas within text through charting, mapping, paraphrasing, summarizing, or comparing/contrasting);

MA.6.A.3.6 Construct and analyze tables, graphs, and equations to describe linear functions and other simple relations using both common language and algebraic notation.

SC.7.E.6.5 Explore the scientific theory of plate tectonics by describing how the movement of Earth's crustal plates causes both slow and rapid changes in Earth's surface, including volcanic eruptions, earthquakes, and mountain building.

SC.7.E.6.7 Recognize that heat flow and movement of material within Earth causes earthquakes and volcanic eruptions, and creates mountains and ocean basins.

SC.7.N.1.1 Define a problem from the seventh grade curriculum, use appropriate reference materials to support scientific understanding, plan and carry out scientific investigation of various types, such as systematic observations or experiments, identify variables, collect and organize data, interpret data in charts, tables, and graphics, analyze information, make predictions, and defend conclusions.

SC.7.N.1.3 Distinguish between an experiment (which must involve the identification and control of variables) and other forms of scientific investigation and explain that not all scientific knowledge is derived from experimentation.

SC.7.N.1.5 Describe the methods used in the pursuit of a scientific explanation as seen in different fields of science such as biology, geology, and physics.

 **Connect ED** **There's More Online!**
Video • Audio • Review • ⓘLab Station • WebQuest • Assessment • Concepts in Motion • Multilingual eGlossary

Understanding SCIENCE

ESSENTIAL QUESTIONS

🔑 What is scientific inquiry?

🔑 What are the results of scientific investigations?

🔑 How can a scientist prevent bias in a scientific investigation?

Vocabulary

science p. NOS 4
observation p. NOS 6
inference p. NOS 6
hypothesis p. NOS 6
prediction p. NOS 7
technology p. NOS 8
scientific theory p. NOS 9
scientific law p. NOS 9
critical thinking p. NOS 10

What is science?

The last time that you watched squirrels play in a park or in your yard, did you realize that you were practicing science? Every time you observe the natural world, you are practicing science. **Science** *is the investigation and exploration of natural events and of the new information that results from those investigations.*

When you observe the natural world, you might form questions about what you see. While you are exploring those questions, you probably use reasoning, creativity, and skepticism to help you find answers to your questions. People use these behaviors in their daily lives to solve problems, such as keeping a squirrel from eating birdseed, as shown in **Figure 1.** Similarly, scientists use these behaviors in their work.

Figure 1 The photos show two different bird feeder designs.

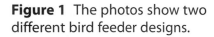 1. **Express** Someone used

and _____
to design each of these squirrel-proof bird feeders. These skills can be used to solve other problems. However, some solutions don't work.

Active Reading 2. **Recall** Describe a time when you used science to gain new information.

_____

_____

Branches of Science

No one person can study the entire natural world. Therefore, people tend to focus their efforts on one of the three fields or branches of science—life science, Earth science, or physical science, as described below. Then people or scientists can seek answers to specific problems within one field of science.

WORD ORIGIN

biology
from Greek *bios*, means "life";
and *logia*, means "study of"

Active Reading **3. Analyze** In the boxes below, write examples of the types of questions scientists might ask.

Life Science

Biology, or life science, is the study of all living things. This forest ecologist studies interactions in forest ecosystems, investigating lichens growing on Douglas firs.

Life Science Questions:

- [ ]
- [ ]
- [ ]

Earth Science

The study of Earth and space, including landforms, rocks, soil, and forces that shape Earth's surface, are Earth science. These Earth scientists are studying soil samples in Africa.

Earth Science Questions:

- [ ]
- [ ]
- [ ]

Physical Science

The study of chemistry and physics is physical science. Physical scientists study the interactions of matter and energy. This chemist is preparing antibiotic solutions.

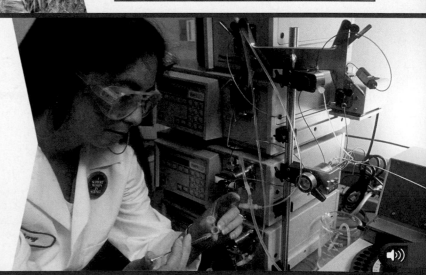

Physical Science Questions:

- [ ]
- [ ]
- [ ]

The Design Process

Science & Engineering

Create a Solution

Scientists investigate and explore natural events and then interpret data and share information learned from those investigations. How do engineers differ from scientists? Engineers design, construct, and maintain things that do not occur naturally. Roads, submarines, toys and games, microscopes, medical equipment, and amusement park rides all are the results of engineering. Science involves the practice of scientific inquiry, but engineering involves The Design Process—a set of methods used to find and create a solution to a problem or need.

Engineers have developed tools, such as submersibles and microscopes, that enable scientists to better explore the biological, physical, and chemical world, no matter where studies take place—under water, in a lab, or in a rain forest.

Alvin, a deep-sea submersible, has been in operation since 1964. It makes about 200 deep-sea dives each year and has helped scientists discover human artifacts, deep-sea organisms, and seafloor processes. Microscopes enable scientists to examine closely things that are not visible to the naked eye.

The Design Process

1. Identify a Problem or Need
- Determine a problem or need
- Document all questions, research, and procedures throughout the process

2. Research and Development
- Brainstorm solutions
- Research any existing solutions that address the problem or need
- Suggest limitations

3. Construct a Prototype
- Develop possible solutions
- Estimate materials, costs, resources, and time to develop the solutions
- Select the best possible solution
- Construct a prototype

4. Test and Evaluate Solutions
- Use models to test the solutions
- Use graphs, charts, and tables to evaluate results
- Analyze the solutions' strengths and weaknesses

5. Communicate Results and Redesign
- Communicate your design process and results to others
- Redesign and modify solution
- Construct final solution

It's Your Turn

DESIGN a magnifying tool using the design process.

Case STUDY

ESSENTIAL QUESTIONS

🔑 How do independent and dependent variables differ?

🔑 How is scientific inquiry used in a real-life scientific investigation?

Vocabulary

variable p. NOS 20

dependent variable p. NOS 20

independent variable p. NOS 20

constants p. NOS 20

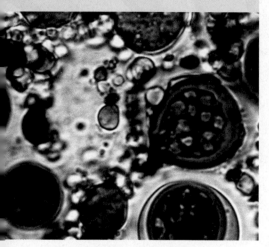

Figure 8 Florida companies experiment to develop larger, more cost-effective methods to produce biodiesel fuel using a variety of microalgae. Microalgae are plantlike organisms that can make oils.

Active Reading **1. Analyze** Summarize the relationship between independent and dependent variables.

Biodiesel from Microalgae

For the last few centuries, fossil fuels have been the main sources of energy for industry and transportation. But, studies have shown that burning fossil fuels negatively affects the environment. In addition, there are concerns about using up the world's reserves of fossil fuels.

Florida scientists have explored using protists to produce biodiesel. Biodiesel is fuel made primarily from living organisms. Protists, shown in **Figure 8**, are a group of microscopic organisms that usually live in water or moist environments. Microalgae are plantlike protists that make their own food using sunlight, water, and carbohydrates in a process called photosynthesis.

Designing a Controlled Experiment

Scientists use scientific inquiry to investigate the use of protists to make biodiesel. They design controlled experiments to test their hypotheses. Scientists conducting the studies practice scientific inquiry. The notebook pages contain information that a scientist might write in a science journal.

A controlled experiment is a scientific investigation that tests how one variable affects another. A **variable** *is any factor in an experiment that can have more than one value.* In controlled experiments, there are two types of variables. The **dependent variable** *is the factor measured or observed during an experiment.* The **independent variable** *is the factor that you want to test. It is changed by the investigator to observe how it affects a dependent variable.* **Constants** *are the factors in an experiment that remain the same.*

A controlled experiment has two groups—an experimental group and a control group. The experimental group is used to study how a change in the independent variable changes the dependent variable. The control group contains the same factors as the experimental group, but the independent variable is not changed. Without a control, it is difficult to know whether your experimental observations result from the variable you are testing or from another factor.

Biodiesel

The idea of engines running on fuel made from plant or plantlike sources is not entirely new. Rudolph Diesel, shown in **Figure 9,** invented the diesel engine. He used peanut oil to demonstrate how his engine worked. However, when petroleum was introduced as a diesel fuel source, it was preferred over peanut oil because it was cheaper.

Oil-rich food crops, such as soybeans, can be used as a source of biodiesel. However, some people are concerned that crops grown for fuel sources will replace crops grown for food. If farmers grow more crops for fuel, then the amount of food available worldwide will be reduced. Because of food shortages in many parts of the world, replacing food crops with fuel crops is not a good solution.

 Active Reading **2. Identify** (Circle) the product Rudolph Diesel used as fuel to demonstrate how his new engine worked.

Aquatic Species Program

In the late 1970s, the U.S. Department of Energy began funding its Aquatic Species Program (ASP) to investigate ways to remove air pollutants. Coal-fueled power plants produce carbon dioxide (CO_2), a pollutant, as a by-product. In the beginning, the study examined all aquatic organisms that use CO_2 during photosynthesis—their food-making process. During the studies, the project leaders noticed that some microalgae produced large amounts of oil. The program's focus soon shifted to using microalgae to produce oils that could be processed into biodiesel.

Active Reading **3. Analyze** Highlight the original purpose for which scientists studied microalgae.

Figure 9 Rudolph Diesel invented the first diesel engine in the early 1900s.

Scientific investigations often begin when someone observes an event in nature and wonders why or how it occurs.

A hypothesis is a tentative explanation that can be tested by scientific investigations. A prediction is a statement of what someone expects to happen next in a sequence of events.

Observation A:
While testing microalgae to discover if they would absorb carbon pollutants, ASP project leaders noticed that some species of microalgae had high oil content.

Hypothesis A:
Some microalgae species can be used as a source of biodiesel fuel because the microalgae produce oil.

Prediction A:
If the growing conditions are correct, then large amounts of oils will be produced.

One way to test a hypothesis is to design an experiment, collect data, and test predictions.

Design an Experiment and Collect Data:
Develop a rapid screening test to discover which species produce the most oil.
Independent Variable: amount of nitrogen available
Dependent Variable: amount of oil produced
Constants: the growing conditions of algae (temperature, water quality, exposure to the Sun, etc.)

During an investigation, observations, hypotheses, and predictions are often made when new information is discovered.

Observation B:
Based on previous studies, starving microalgae of nutrients could produce more oil.
Hypothesis B:
Microalgae grown with inadequate amounts of nitrogen alter their growth processes and produce more oil.
Prediction B:
If microalgae receive inadequate amounts of nitrogen, then they will produce more oil.

Figure 10 Green microalgae and diatoms showed the most promise during testing for biodiesel production.

Which Microalgae?

Microalgae are microscopic organisms that live in marine (salty) or freshwater environments. Like many plants and other plantlike organisms, they use photosynthesis and make sugar. The process requires light energy. Microalgae make more sugar than they can use as food. They convert excess sugar to oil. Scientists focused on these microalgae because their oil then could be processed into biodiesel.

Research began by collecting and identifying microalgae species. Focusing on microalgae in shallow, inland, saltwater ponds, scientists predicted that these microalgae were more resistant to changes in temperature and salt content in the water.

A test was developed to identify microalgae with high oil content. Soon, 3,000 microalgae species had been collected. Scientists checked these samples for tolerance to acidity, salt levels, and temperature and selected 300 species. Green microalgae and diatoms, as shown in **Figure 10,** showed the most promise. However, it was obvious that no one species was going to be perfect for all climates and water types.

Active Reading **4. Confirm** Highlight three factors for microalgae production that scientists considered in their research.

Oil Production in Microalgae

Scientists also began researching how microalgae produce oil. Some studies suggested that starving microalgae of nutrients, such as nitrogen, could increase the amount of oil they produced. However, starving microalgae also caused them to be smaller, resulting in no overall increase in oil production.

Outdoor Testing v. Bioreactors

Scientists first grew microalgae in outdoor ponds in New Mexico. However, outdoor conditions were very different from those in the laboratory. Cooler temperatures in the outdoor ponds resulted in smaller microalgae. Native algae species invaded the ponds, forcing out the experimental, high-oil-producing microalgae species.

Florida scientists continue to focus on growing microalgae in open ponds, similar to the one shown in **Figure 11.** Many scientists believe these open ponds are better for producing large quantities of biodiesel from microalgae. Some researchers are now growing microalgae in closed glass containers called bioreactors, also shown in **Figure 11.** Inside these bioreactors, organisms live and grow under controlled conditions. However, bioreactors are more expensive than open ponds.

Biofuel companies have been experimenting with low-cost bioreactors. Scientists hypothesize they could use long plastic bags, shown in **Figure 11,** instead of closed glass containers.

Active Reading

5. **Identify** In the boxes provided, state an advantage and a disadvantage for each microalgae environment shown.

Open Ponds	
Advantage:	
Disadvantage:	

Open ponds are less expensive than bioreactors for growing microalgae.

Figure 11
These three methods of growing microalgae are examples of three different hypotheses that are being tested in controlled experiments.

Microalgae grown in plastic bags are very expensive to harvest.

Microalgae grow under controlled conditions in glass bioreactors.

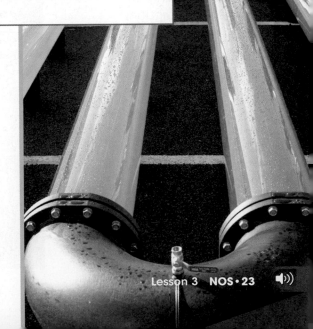

Glass Bioreactor Tubes	
Advantage:	
Disadvantage:	

Plastic Bags	
Advantage:	
Disadvantage:	

Inquiry
Visit ConnectED for this unit's **STEM** activity.

Patterns

It's a bird! It's a plane! No, it's Venus! Besides the Sun, Venus is brighter than any other star or planet in the sky. It is often seen from Earth without the aid of a telescope, as shown in **Figure 1.**

Astronomers study the pattern of each planet's orbit. A **pattern** is a consistent plan or model used as a guide for understanding and predicting things. Studying the orbital patterns of planets allows scientists to predict the future position of each planet. By studying the pattern of Venus's orbit, astronomers can predict when Venus will travel between Earth and the Sun and be visible from Earth, as shown in **Figure 2.** Using patterns, scientists are able to predict the date when you will be able to see this event in the future.

▲ **Figure 1** Venus is often so bright in the morning sky that it has been nicknamed the morning star.

Active Reading

1. Analyze Explain how the pattern of Venus's orbit is used to determine its position in the sky.

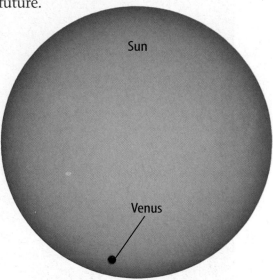

Figure 2 On June 8, 2004, observers around the world watched Venus pass in front of the Sun. This was the first time this event took place since 1882. ▶

Types of Patterns

Physical Patterns

A pattern that you can see and touch is a physical pattern. The crystalline structures of minerals are examples of physical patterns. When atoms form crystals, they produce structural, or physical, patterns. The crystal structure of the Star of India sapphire creates a pattern that reflects light in a stunning star shape.

Cyclic Patterns

An event that repeats many times again in a predictable order has a cyclic pattern. Since Earth's axis is tilted, the angle of the Sun's rays on your location on Earth changes as Earth orbits the Sun. This causes the seasons— winter, spring, summer, and fall— to occur in the same pattern every year.

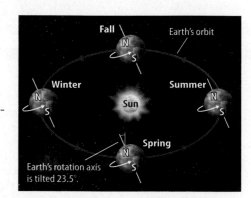

Patterns in Engineering

Engineers study patterns for many reasons, including to understand the physical properties of materials or to optimize the performance of their designs. Have you ever seen bricks with a pattern of holes through them? Clay bricks used in construction are fired, or baked, to make them stronger. Ceramic engineers understand that a regular pattern of holes in a brick assures that the brick is evenly fired and will not easily break.

Maybe you have seen a bridge constructed with a repeating pattern of large, steel triangles. Civil engineers, who design roads and bridges, know that the triangle is one of the strongest shapes in geometry. Engineers often use patterns of triangles in the structure of bridges to make them withstand heavy traffic and high winds.

Patterns in Physical Science

Scientists use patterns to explain past events or predict future events. At one time, only a few chemical elements were known. Chemists arranged the information they knew about these elements in a pattern according to the elements' properties. Scientists predicted the atomic numbers and the properties of elements that had yet to be discovered. These predictions made the discovery of new elements easier because scientists knew what properties to look for.

Look around. There are patterns everywhere—in art and nature, in the motion of the universe, in vehicles traveling on the roads, and in the processes of plant and animal growth. Analyzing patterns helps to understand the universe.

Patterns in Graphs

Scientists often graph their data to help identify patterns. For example, scientists might plot data from experiments on parachute nylon in graphs, such as the one below. Analyzing patterns on graphs then gives engineers information about how to design the strongest parachutes.

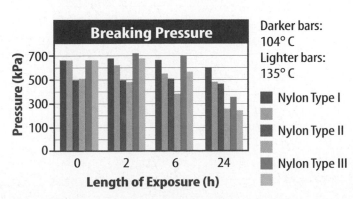

Breaking Pressure

Pressure (kPa)

Length of Exposure (h)

Darker bars: 104° C

Lighter bars: 135° C

Nylon Type I

Nylon Type II

Nylon Type III

Inquiry

LAB STATION

Try It!

SC.7.N.1.4

MiniLab *How strong is your parachute?* at connectED.mcgraw-hill.com

Apply It! After you complete the lab, answer these questions.

1. **Describe** Think of a pattern you use in your daily life. Describe how this pattern is useful to you to predict future events.

2. **Analyze** List possible uses of graphs. These might include personal time in sporting events or use of electric power. Analyze how such graphic patterns reflect a change that can be useful information.

Notes

PAGE KEELEY
**SCIENCE
PROBES**

Meteorite Impact

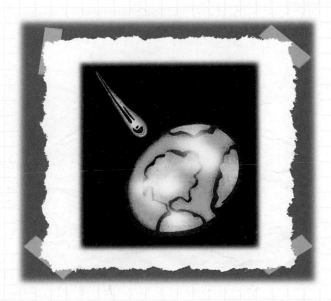

A meteorite is a solid object that comes from space, passes through our atmosphere, and lands on Earth. If a large meteorite were to strike Earth today, where would it most likely fall? Circle your prediction.

A. land

B. ice

C. ocean

D. freshwater

Explain your thinking. What reasoning did you use to make your prediction?

Earth's
LAYERS

The Big Idea

Think About It!

How is Earth structured?

These dancing lights in the night sky are called an aurora. Interactions between Earth's atmosphere and charged particles from the Sun cause an aurora. Conditions deep in Earth's interior structure create a magnetic field that attracts the charged particles to Earth's North Pole and South Pole.

1 How do you think Earth is structured?

_____
_____
_____
_____

2 How do you think Earth's core creates Earth's magnetic field?

_____
_____
_____
_____
_____

Get Ready to Read

What do you think about Earth's layers?

Do you agree or disagree with each of these statements? As you read this chapter, see if you change your mind about any of the statements.

AGREE DISAGREE

1 People have always known that Earth is round. ☐ ☐

2 Earth's hydrosphere is made of hydrogen gas. ☐ ☐

3 Earth's interior is made of distinct layers. ☐ ☐

4 Scientists discovered that Earth's outer core is liquid by drilling deep wells. ☐ ☐

5 All ocean floors are flat. ☐ ☐

6 Most of Earth's surface is covered by water. ☐ ☐

There's More Online!
Video • Audio • Review • ⒾLab Station • WebQuest • Assessment • Concepts in Motion • Multilingual eGlossary

9

Spherical EARTH

ESSENTIAL
QUESTIONS

 What are Earth's major systems and how do they interact?

 Why does Earth have a spherical shape?

Vocabulary

sphere p. 11

geosphere p. 12

gravity p. 13

density p. 15

inquiry Launch Lab

SC.7.N.1.3
SC.7.N.1.6

10 minutes

How can you model Earth's spheres?

Earth has different spheres made of water, solid materials, air, and life. Each sphere has unique characteristics.

Procedure

1. Read and complete a lab safety form.

2. Set a **clear plastic container** on your table, and add **gravel** to a depth of about 2 cm.

3. Pour equal volumes of **corn syrup** and **colored water** into the container.

4. Observe the container for 2 minutes. Record your observations below.

Data and Observations

Think About This

1. What happened to the materials?

2. 🔑 **Key Concept** Which Earth sphere did each material represent?

Florida NGSSS

LA.7.2.2.3 The student will organize information to show understanding (e.g., representing main ideas within text through charting, mapping, paraphrasing, summarizing, or comparing/contrasting);

MA.6.A.3.6 Construct and analyze tables, graphs, and equations to describe linear functions and other simple relations using both common language and algebraic notation.

SC.7.E.6.4 Explain and give examples of how physical evidence supports scientific theories that Earth has evolved over geologic time due to natural processes.

SC.7.N.1.3 Distinguish between an experiment (which must involve the identification and control of variables) and other forms of scientific investigation and explain that not all scientific knowledge is derived from experimentation.

SC.7.N.1.6 Explain that empirical evidence is the cumulative body of observations of a natural phenomenon on which scientific explanations are based.

inquiry **Why do you think Earth is spherical?**

1. Why do you think Earth is spherical? This image of Earth was taken from space. Notice Earth's shape and the wispy clouds that surround part of the planet. What else do you notice about Earth?

_____

_____

_____

_____

_____

_____

_____

Describing Earth

Imagine standing on a mountaintop. You can probably see that the land stretches out beneath you for miles. But you cannot see all of Earth—it is far too large. People have tried to determine the shape and size of Earth for centuries. They have done so by examining the parts they can see.

Many years ago, people believed that Earth was a flat disk with land in the center and water at the edges. Later they used clues to determine Earth's true shape, such as studying Earth's shadow on the Moon during an eclipse.

The Size and Shape of Earth

Now there are better ways to get a view of our planet. Using satellites and other technology, scientists know that Earth is a sphere. _A_ **sphere** _is shaped like a ball, with all points on the surface at an equal distance from the center._ But Earth is not a perfect sphere. As illustrated in **Figure 1,** Earth is somewhat flattened at the poles with a slight bulge around the equator. Earth has a diameter of almost 13,000 km. It is the largest of the four rocky planets closest to the Sun.

Figure 1 Earth is shaped like a sphere that is somewhat flattened.

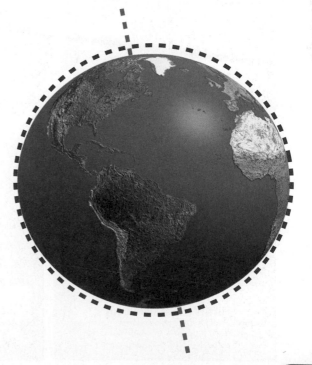

Earth Systems

Earth is large and complex. To simplify the task of studying Earth, scientists describe Earth systems, as shown in **Figure 2**. All of these systems interact by exchanging matter and energy. For example, water from the ocean evaporates and enters the atmosphere. Later, the water precipitates onto land and washes salts into the ocean.

The hydrosphere is water on Earth's surface, underground, and liquid water in the atmosphere. Most of the water in the hydrosphere is in salty oceans. Freshwater is in most rivers and lakes and underground. Frozen water, such as glaciers, is part of the cryosphere which overlaps with the hydrosphere.

The Atmosphere, the Hydrosphere, and the Cryosphere The atmosphere is the layer of gases surrounding Earth. It is Earth's outermost system. This layer is about 100 km thick. It is a mixture of nitrogen, oxygen, carbon dioxide, and traces of other gases.

The Geosphere and the Biosphere *The* geosphere *is Earth's entire solid body.* It contains a thin layer of soil and sediments covering a rocky center. It is the largest Earth system. The biosphere includes all living things on Earth. Organisms in the biosphere live within and interact with the atmosphere, hydrosphere, and even the geosphere.

Active Reading 2. **Identify** Highlight each of Earth's major systems and what they contain.

Earth's Systems 🔑

Figure 2 Earth's systems interact. A change in one Earth system affects all other Earth systems. They exchange energy and matter, making Earth suitable for life.

Active Reading 3. **Label** Fill in the name of each Earth system below based on the image and description provided.

layer of gases surrounding Earth

liquid water on Earth

Earth's entire solid body

all living organisms on Earth

frozen water on Earth

How did Earth form?

Earth formed about 4.6 billion years ago (bya), along with the Sun and the rest of our solar system. Materials from a large cloud of gas and dust came together, forming the Sun and all the planets. In order to understand how this happened, you first need to know how gravity works.

The Influence of Gravity

Gravity *is the force that every object exerts on every other object because of their masses.* The force of gravity between two objects depends on the objects' masses and the distance between them. The more mass either object has, or the closer together they are, the stronger the gravitational force. You can see an example of this in **Figure 3.**

Active Reading

FOLDABLES® LA.7.2.2.3

Make a half book from a sheet of paper. Label it as shown. Use it to organize your notes about Earth's formation.

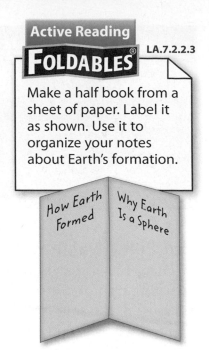

How Earth Formed   Why Earth Is a Sphere

Force of Gravity

The two objects in row A are the same distance apart as the two objects in row B. One of the objects in row B has more mass, creating a stronger gravitational force between the two objects in row B.

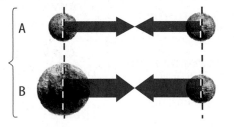

A

B

All four objects have the same mass. The two objects in row C are closer to each other than the two objects in row D and, therefore, have a stronger gravitational force between them.

C

D

Figure 3 Mass and distance affect the strength of the gravitational force between objects.

Active Reading **4. State** Why does Earth exert a greater gravitational force on you than other objects do?

_____

_____

_____

_____

The force of gravity is strongest between the objects in row B. Even though the objects in row A are the same distance apart as those in row B, the force of gravity between them is weaker because they have less mass. The force of gravity is weakest between the objects in row D.

As illustrated in **Figure 4,** all objects on or near Earth are pulled toward Earth's center by gravity. Earth's gravity holds us to Earth's surface. Since Earth has more mass than any object near you, it exerts a greater gravitational force on you than other objects do. You don't notice the gravitational force between less massive objects.

Figure 4 Earth's gravity pulls objects toward the center of Earth.

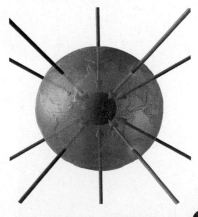

Active Reading **5. Name** Highlight the factors that affect the strength of the gravitational force between objects.

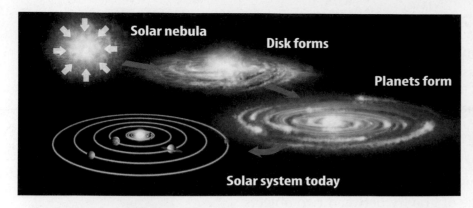

Solar nebula

Disk forms

Planets form

Solar system today

Figure 5 Gravity helped change a cloud of dust, gas, and ice, called a nebula, into our solar system. The Sun formed first, and the planets formed from the swirling disk of particles that remained.

Active Reading

6. Recall Our solar system formed from what type of cloud?

The Solar Nebula

The force of gravity played a major role in the formation of our solar system. As shown in **Figure 5,** the solar system formed from a cloud of gas, ice, and dust called a nebula. Gravity pulled the materials closer together. The nebula shrank and flattened into a disk. The disk began to rotate. The materials in the center of the disk became denser, forming a star—the Sun.

Next, the planets began to take shape from the remaining bits of material. Earth formed as gravity pulled these small particles together. As they collided, they stuck to each other and formed larger, unevenly shaped objects. These larger objects had more mass and attracted more particles. Eventually enough matter collected and formed Earth. But how did the unevenly shaped, young planet become spherical?

Early Earth

Eventually the newly formed Earth grew massive and generated thermal energy, commonly called heat, in its interior. The rocks of the planet softened and began to flow.

Gravity pulled in the irregular bumps on the surface of the newly formed planet. As a result, Earth developed a relatively even spherical surface.

 7. NGSSS Check Formulate How did Earth develop its spherical shape? SC.7.E.6.4

Inquiry SC.7.N.1.3

LAB STATION Try It!

MiniLab *Which materials will sink?* at connectED.mcgraw-hill.com

Apply It! After you complete the lab, answer these questions.

1. Would you expect to find more dense Earth layers on the surface of Earth or below?

2. Why?

The Formation of Earth's Layers

Thermal energy from Earth's interior affected Earth in other ways, as well. Before heating up, Earth was a mixture of solid particles. The thermal energy melted some of this material and it began to flow. As it flowed, Earth developed distinct layers of different materials.

The different materials formed layers according to their densities. **Density** *is the amount of mass in a material per unit volume.* Density can be described mathematically as

$$D = m/V$$

where D is the density of the material, m is the material's mass, and V is its volume. If two materials have the same volume, the denser material will have more mass.

 8. Decide Can a small object have more mass than a larger object? Explain your answer.

_____

_____

_____

There is a stronger gravitational force between Earth and a denser object than there is between Earth and a less dense object. You can see this if you put an iron block and a pinewood block with the same volumes in a pan of water. The wooden block, which is less dense than water, will float on the water's surface. The iron block, which is denser than water, will be pulled through the water to the bottom of the pan.

When ancient Earth started melting, something much like this happened. The densest materials sank and formed the innermost layer of Earth. The least dense materials stayed at the surface, and formed a separate layer. The materials with intermediate densities formed layers in between the top layer and the bottom layer. Earth's three major layers are shown in **Figure 6.**

MA.6.A.3.6

Math Skills

Solve One-Step Equations
Comparing the masses of substances is useful only if the same volume of each substance is used. To calculate density, divide the mass by the volume. The unit for density is a unit of mass, such as g, divided by a unit of volume, such as cm^3. For example, an aluminum cube has a mass of 27 g and a volume of 10 cm^3. The density of aluminum is 27 g / 10 cm^3 = 2.7 g/cm^3.

9. Practice
A chunk of gold with a volume of 5.00 cm^3 has a mass of 96.5 g. What is the density of gold?

Figure 6 Earth's geosphere is divided into three major layers.

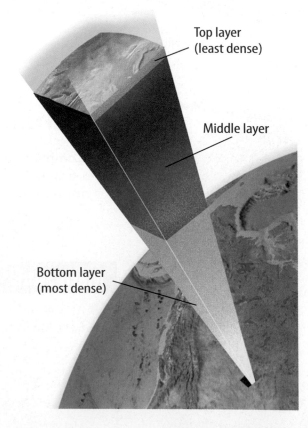

Top layer (least dense)

Middle layer

Bottom layer (most dense)

Earth's systems, including the atmosphere, hydrosphere, biosphere, and geosphere, interact with one another.

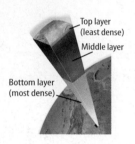

Top layer
(least dense)

Middle layer

Bottom layer
(most dense)

The geosphere is the solid body of Earth.

The solar system, including Earth, formed about 4.6 bya. Gravity caused particles to come together and formed a spherical Earth.

Use Vocabulary

1 The Earth system made mainly of surface water is called the

_____.

2 **Use the term** *density* in a sentence.

_____

Understand Key Concepts

3 Which is part of the atmosphere?

  (A) a rock   (C) oxygen gas

  (B) a tree   (D) the ocean

4 **Describe** how gravity affected Earth's shape during Earth's formation. SC.7.E.6.4

_____

_____

Interpret Graphics

5 **Organize** Complete the graphic organizer below. In each oval, list one of the following terms: *geosphere, hydrosphere, cryosphere, Earth, atmosphere,* and *biosphere.*

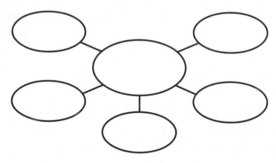

Critical Thinking

6 **Combine** your understanding of how Earth became spherical and observations of the Moon. Then form a hypothesis about the formation of the Moon. SC.7.E.6.4

_____

_____

Math Skills

7 At a given temperature, 3.00 m³ of carbon dioxide has a mass of 5.94 kg. What is the density of carbon dioxide at this temperature?

Time Capsules

◀ George Harlow studies diamonds to learn more about Earth's interior.

Formed billions of years ago in Earth's mantle, diamonds hold important clues about our planet's mysterious interior.

George Harlow is fascinated by diamonds. Not because of their dazzling shine or their value, but because of what they can reveal about Earth. He considers diamonds to be tiny time capsules that capture a picture of the ancient mantle, where they became crystals.

Most diamonds we find today formed billions of years ago deep within Earth's mantle, over 161 km below Earth's surface. Tiny bits of mantle, called inclusions, were trapped inside these extremely hard crystals as they formed. Millions of years later, the inclusions' diamond cases still protect them.

Harlow collects these diamonds from places such as Australia, Africa, and Thailand. Back in the lab, Harlow and his colleagues remove inclusions from diamonds. First, they break open a diamond with a tool similar to a nutcracker. Then they use a microscope and a pinlike tool to sift through the diamond rubble. They look for an inclusion, which is about the size of a grain of sand. When they find one, they use an electron microprobe and a laser to analyze the inclusion's composition, or chemical makeup. The sample might be tiny, but it's enough for scientists to learn the temperature, pressure, and composition of the mantle in which the diamond formed.

Next time you see a diamond, you might wonder if it too has a tiny bit of ancient mantle from deep inside Earth.

Going Up?

Diamond crystals form deep within the mantle under intense pressures and temperatures. They come to Earth's surface in molten rock, or magma. The magma pulls diamonds from rock deep underground and rapidly carries them to the surface. The magma erupts onto Earth's surface in small, explosive volcanoes. Diamonds and other crystals and rocks from the mantle are in deep, carrot-shaped cones called kimberlite pipes that are part of these rare volcanoes.

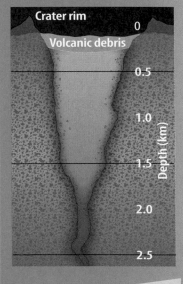

Crater rim

Volcanic debris

Depth (km)

0
0.5
1.0
1.5
2.0
2.5

It's Your Turn

RESEARCH Diamonds are the world's most popular gemstone. What other uses do diamonds have? Research the properties of diamonds and how they are used in industry. Report your findings to your class.

Earth's INTERIOR

 What are the interior layers of Earth?

 What evidence indicates that Earth has a solid inner core and a liquid outer core?

Vocabulary

crust p. 21

mantle p. 22

lithosphere p. 22

asthenosphere p. 22

core p. 24

magnetosphere p. 25

Florida NGSSS

LA.7.2.2.3 The student will organize information to show understanding (e.g., representing main ideas within text through charting, mapping, paraphrasing, summarizing, or comparing/contrasting);

SC.7.E.6.1 Describe the layers of the solid Earth, including the lithosphere, the hot convecting mantle, and the dense metallic liquid and solid cores.

SC.7.N.1.1 Define a problem from the seventh grade curriculum, use appropriate reference materials to support scientific understanding, plan and carry out scientific investigation of various types, such as systematic observations or experiments, identify variables, collect and organize data, interpret data in charts, tables, and graphics, analyze information, make predictions, and defend conclusions.

SC.7.N.1.3 Distinguish between an experiment (which must involve the identification and control of variables) and other forms of scientific investigation and explain that not all scientific knowledge is derived from experimentation.

SC.7.N.3.2 Identify the benefits and limitations of the use of scientific models.

Inquiry Launch Lab

SC.7.N.1.3
SC.7.N.3.2

10 minutes

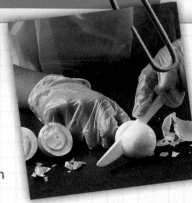

How can you model Earth's layers?

Earth is made of three main layers: the thin outer crust, the thick mantle, and the central core. You can use different objects to model these layers.

Procedure

1 Read and complete a lab safety form.

2 Place a **hard-cooked egg** on a **paper towel.** Use a **magnifying lens** to closely examine the surface of the egg. Is its shell smooth or rough? Record your observations.

3 Carefully peel away the shell from the egg.

4 Use the **plastic knife** to cut the egg in half. Observe the characteristics of the shell, the egg white, and the yolk.

5 Make a drawing of the egg's layers in the space provided. Which layers could represent layers of Earth? Label the layers as *crust, mantle,* or *core.*

Data and Observations

Think About This

1. What other objects could be used to model Earth's layers?

2. **Key Concept** Explain why a hard-cooked egg is a good model for Earth's layers.

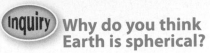
1. What do you think is inside Earth? Earth is thousands of kilometers thick. The deepest mines and wells in the world barely scratch Earth's surface. How do you think scientists learn about Earth's interior?

_____

_____

_____

_____

_____

_____

Clues to Earth's Interior

Were you ever given a gift and had to wait to open it? Maybe you tried to figure out what was inside by tapping on it or shaking it. Using methods such as these, you might have been able to determine the gift's contents. Scientists can't see what is inside Earth, either. But they can use indirect methods to discover what Earth's interior is like.

What's below Earth's surface?

Deep mines and wells give scientists hints about Earth's interior. The deepest mine ever constructed is a gold mine in South Africa. It is more than 3 km deep. People can go down the mine to explore the geosphere.

Drilled wells are even deeper. The deepest well is on the Kola Peninsula in Russia. It is more than 12 km deep. Drilling to such great depths is extremely difficult—it took more than 20 years to drill the Kola well. Even though people cannot go down in the well, they can send instruments down to make observations and bring samples to the surface. What have scientists learned about Earth's interior by studying mines and wells like the two mentioned above?

REVIEW VOCABULARY

observation
an act of recognizing and noting a fact or an occurrence

Temperature and Pressure Increase with Depth

One thing that workers notice in deep mines or wells is that it is hot inside Earth. In the South African gold mines, 3.5 km below Earth's surface, the temperature is about 53°C (127°F). The temperature at the bottom of the Kola well is 190°C (374°F). That's hot enough to bake cookies! No one has ever recorded the temperature of Earth's center, but it is estimated to be about 6,000°C. As shown in **Figure 7,** temperature within Earth increases with increasing depth.

Not only does temperature increase, but pressure also increases as depth increases inside Earth. This is due to the weight of the overlying rocks. The high pressure squeezes the rocks and makes them much denser than surface rocks.

> **Active Reading** 2. **Describe** Underline why pressure changes with depth within Earth.

High temperatures and pressures make it difficult to drill deep wells. The depth of the Kola well is less than 1 percent of the distance to Earth's center. Therefore, only a small part of the geosphere has been sampled. How can scientists learn about what is below the deepest wells?

Using Earthquake Waves

As you read earlier, scientists use indirect methods to study Earth's interior. They get most of their evidence by analyzing earthquake waves. Earthquakes release energy in the form of three types of waves. As these waves move through Earth, they are affected by the different materials they travel through. Some waves cannot travel through certain materials. Other waves change direction when they reach certain materials. By studying how the waves move, scientists are able to infer the density and composition of materials within Earth.

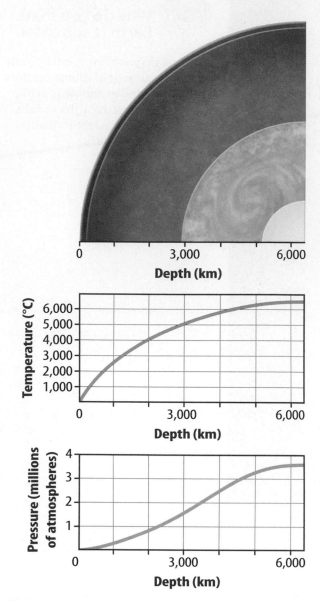

Figure 7 Temperature and pressure increase with depth in the geosphere.

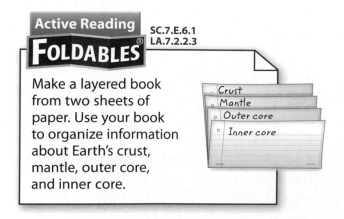

SC.7.E.6.1
LA.7.2.2.3

Active Reading

FOLDABLES

Make a layered book from two sheets of paper. Use your book to organize information about Earth's crust, mantle, outer core, and inner core.

Crust
Mantle
Outer core
Inner core

Earth's Layers

Differences in density resulted in materials within Earth forming layers. Each layer has a different composition, with the densest material in the center of Earth.

Crust

The brittle, rocky outer layer of Earth is called the **crust.** It is much thinner than the other layers, like the shell on a hard-cooked egg. It is the least dense layer of the geosphere. It is made mostly of elements of low mass, such as silicon and oxygen.

Crustal rocks are under oceans and on land. The crust under oceans is called oceanic crust. It is made of dense rocks that contain iron and magnesium. The crust on land is called continental crust. It is about four times thicker than oceanic crust. Continental crust is thickest under tall mountains. **Figure 8** shows a comparison of the two types of crust.

There is a distinct boundary between the crust and the layer beneath it. When earthquake waves cross this boundary, they speed up. This indicates that the lower layer is denser than the crust.

 3. Identify Label the different Earth crusts below.

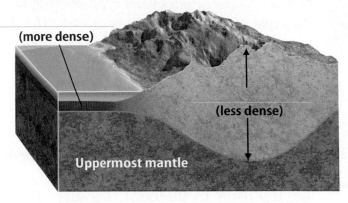

(more dense)

(less dense)

Uppermost mantle

Figure 8 Oceanic crust is thin and dense compared to continental crust.

Inquiry SC.7.N.1.3

Try It!

MiniLab *Which liquid is densest?* at connectED.mcgraw-hill.com

Apply It! After you complete the lab, answer these questions.

1. Why is the oceanic crust below the continental crust?

2. Which layer in the beaker represents the crust?

Plateaus

As you just read, plains are relatively flat and low. In contrast, plateaus are flat and high. **Plateaus** *are areas with low relief and high elevation.* Look again at **Figure 15** to see how a plateau differs from a plain.

Plateaus are much higher than the surrounding land and often have steep, rugged sides. They are less common than plains, but they are on every continent. Find some plateaus in different parts of the world in **Figure 16.**

Active Reading 6. **Describe** Highlight the description of a plateau.

Plateaus can form when forces within Earth uplift rock layers or cause collisions between sections of Earth's crust. For example, the highest plateau in the world is the Tibetan Plateau, called the "roof of the world." It is still being formed by collisions between India and Asia.

Plateaus also can be formed by volcanic activity. For example, the Columbia Plateau in the western United States is the result of the buildup of many successive lava flows.

Mountains

The tallest landforms of all are mountains. **Mountains** *are landforms with high relief and high elevation.* Look again at the world map in **Figure 16.** How many of Earth's well-know mountains can you find?

Mountains can form in several different ways. Some mountains form from the buildup of lava on the ocean floor. Eventually, the mountain grows tall enough to rise above the ocean's surface. The Hawaiian Islands are mountains that formed this way. Other mountains form when forces inside Earth fold, push, or uplift huge blocks of rocks. The Himalayas, the Rocky Mountains, and the Appalachian Mountains all formed from tremendous forces within Earth.

7. ⚫ **Visual Check Assess** Which of Earth's three major types of landforms—plains, plateaus, or mountains—covers most of Earth's land surface?

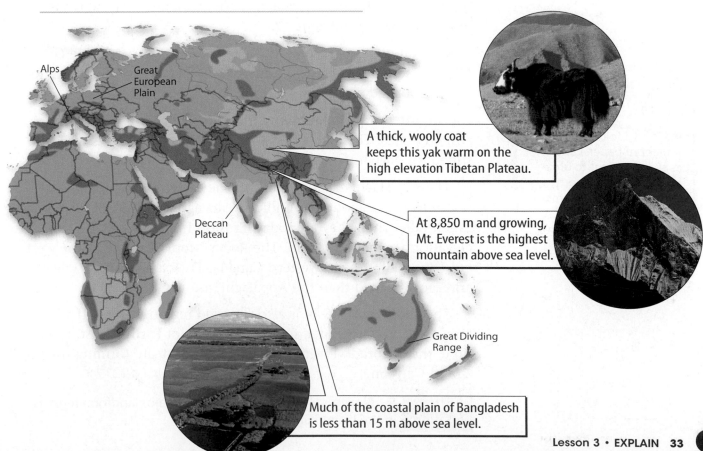

Alps

Great European Plain

Deccan Plateau

A thick, wooly coat keeps this yak warm on the high elevation Tibetan Plateau.

At 8,850 m and growing, Mt. Everest is the highest mountain above sea level.

Great Dividing Range

Much of the coastal plain of Bangladesh is less than 15 m above sea level.

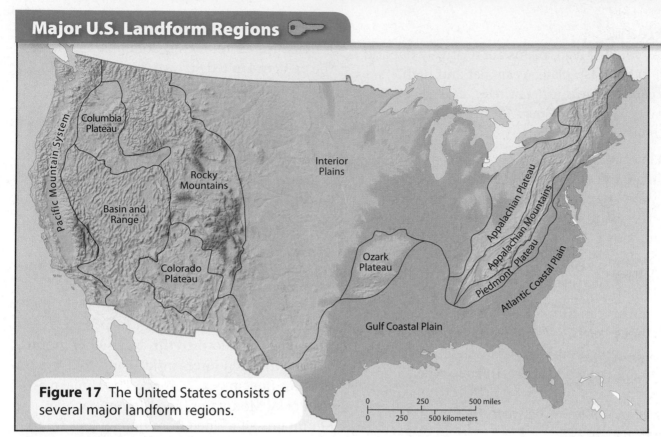

Figure 17 The United States consists of several major landform regions.

0   250   500 miles
0   250   500 kilometers

🔘 8. ✔ **Visual Check Survey** (Circle) the landform region you live in.

Major Landform Regions in the United States

From flat plains to towering mountains, the United States has a variety of landforms. The major landform regions in the United States are shown in **Figure 17.**

Coastal plains are along much of the East Coast and the Gulf Coast. These plains formed millions of years ago when sediments were deposited on the ocean floor.

The interior plains make up much of the central part of the United States. This flat, grassy area has thick soils and is well suited for growing crops and grazing animals.

The Appalachian Mountains, in the eastern United States, began forming about 480 million years ago (mya). They were once much taller than they are today. Erosion has reduced their average elevation to about 2,000 m. The Rocky Mountains are in the western United States and western Canada. They are younger, taller, and more rugged than the Appalachians.

The Colorado Plateau is also a rugged region. It formed when forces within Earth lifted up huge sections of Earth's crust. Over time, the Colorado River cut through the plateau, forming the Grand Canyon.

Active Reading

FOLDABLES® LA.7.2.2.3

Make a tri-fold book from a sheet of paper. Label it as shown. Use it to organize your notes about Earth's major landforms.

| Plains | Plateaus | Mountains |

Active Reading 9. **Name** <u>Underline</u> at least three major landform regions in the United States.

Landforms are topographic features formed by processes that shape Earth's surface.

Major landforms include flat plains, high plateaus, and rugged mountains.

Major landform regions in the United States include the Appalachian Mountains, the Great Plains, the Colorado Plateau, and the Rocky Mountains.

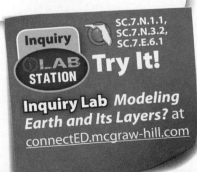

Inquiry **iLAB STATION** **Try It!**
SC.7.N.1.1,
SC.7.N.3.2,
SC.7.E.6.1

Inquiry Lab *Modeling Earth and Its Layers?* at connectED.mcgraw-hill.com

Use Vocabulary

1 **Plains** and mountains are examples of _____ formed by processes that shape Earth's surface. SC.7.E.6.2

2 A(n) _____ is a landform with high relief and high elevation.

3 **Distinguish** between a plain and a plateau.

_____

_____

Understand Key Concepts

4 A landform with low relief and high elevation is a

   (A) mountain.   (C) plateau.

   (B) plain.   (D) topography.

5 **Describe** any landforms that are near your school.

_____

_____

Interpret Graphics

6 **Summarize** Copy and fill in the graphic organizer below to identify the major types of landforms. LA.7.2.2.3

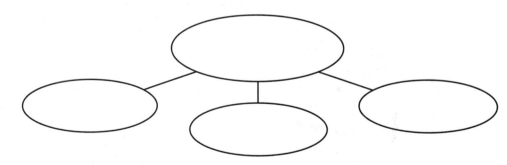

Critical Thinking

7 **Suggest** a way to model plains, plateaus, and mountains by using sheets of cardboard.

_____

_____

_____

8 **Evaluate** the drawbacks and benefits of living in the mountains.

_____

_____

Think About It! Over geologic time, Earth has developed three major layers, the crust, the mantle, and the core, which are continuously being altered.

 ## Key Concepts Summary

Vocabulary

LESSON 1 Spherical Earth

- Earth's major systems include the atmosphere, hydrosphere, biosphere, and **geosphere.**
- All four major Earth systems interact by exchanging matter and energy. A change in one Earth system affects all other Earth systems.
- **Gravity** caused particles to come together to form a spherical Earth.

sphere p. 11
geosphere p. 12
gravity p. 13
density p. 15

LESSON 2 Earth's Interior

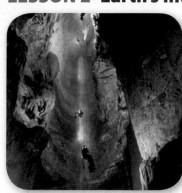

- Earth's interior layers include the **crust, mantle,** and **core.**
- By analyzing earthquake waves, scientists have determined that the outer core is liquid and the inner core is solid.

crust p. 21
mantle p. 22
lithosphere p. 22
asthenosphere p. 22
core p. 24
magnetosphere p. 25

LESSON 3 Earth's Surface

- Earth's major **landforms** include **plains, plateaus,** and **mountains.** Plains have low relief and low elevation. Plateaus have low relief and high elevation. Mountains have high relief and high elevation.
- Plains, plateaus, and mountains are all found in the United States.

landform p. 30
plain p. 32
plateau p. 33
mountain p. 33

FOLDABLES® Chapter Project

Assemble your lesson Foldables as shown to make a Chapter Project. Use the project to review what you have learned in this chapter.

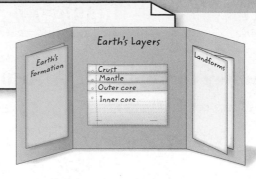

Use Vocabulary

1 Earth formed when _____ pulled together gas and dust that was spinning around the Sun.

2 The gravitational force is greater between similar-sized objects that have a higher _____.

3 The _____ is the largest of Earth's four systems.

4 Small amounts of melted material in the _____ produce flow in the mantle.

5 The least dense rocks on Earth are in the _____. **SC.7.E.6.1**

6 Liquid in the _____ produces Earth's magnetic field. **SC.7.E.6.1**

7 A topographic feature formed by processes that shape Earth's surface is a _____.

8 A(n) _____ has low relief and low elevation. **SC.7.E.6.2**

9 A landform that is high and flat is a(n) _____.

Link Vocabulary and Key Concepts

Use vocabulary terms from the previous page to complete the concept map.

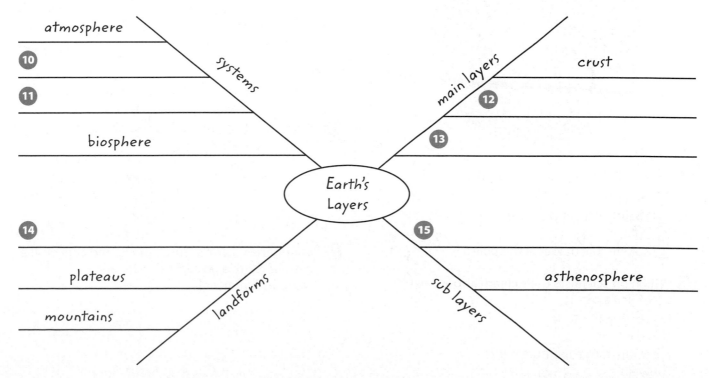

Fill in the correct answer choice.

🔑 Understand Key Concepts

1 What does the biosphere contain? **LA.7.2.2.3**
- Ⓐ air
- Ⓑ living things
- Ⓒ rocks
- Ⓓ water

2 What affects the strength of gravity between two objects? **LA.7.2.2.3**
- Ⓐ the density of the objects
- Ⓑ the mass of the objects
- Ⓒ the distance between the objects
- Ⓓ both the mass and the distance between the objects

3 The figure below shows Earth's layers. What does the red layer represent? **SC.7.E.6.1**
- Ⓐ asthenosphere
- Ⓑ crust
- Ⓒ lithosphere
- Ⓓ mantle

4 What is the shape of Earth? **SC.7.E.6.1**
- Ⓐ disklike
- Ⓑ slightly flattened sphere
- Ⓒ sphere
- Ⓓ sphere that bulges at the poles

5 Which do scientists use to learn about Earth's core? **SC.7.E.6.1**
- Ⓐ earthquake waves
- Ⓑ mines
- Ⓒ temperature measurements
- Ⓓ wells

Critical Thinking

6 **Explain** how gravity would affect you differently on a planet with less mass than Earth, such as Mercury. **LA.7.2.2.3**

7 **Compare** materials in the geosphere to materials in the atmosphere. **SC.7.E.6.1**

8 **Consider** How would Earth's layers be affected if all the materials that make up Earth had the same density? **SC.7.E.6.1**

9 **Relate** How do Earth's systems interact? **SC.7.E.6.1**

10 **Explain** why everything on or near Earth is pulled toward Earth's center. **LA.7.2.2.3**

11 **State** how the crust and the upper mantle are similar. **SC.7.E.6.1**

12 **Summarize** Earth's crust, mantle, and core on the basis of relative position, density, and composition. **SC.7.E.6.1**

13 **Explain** how a plateau differs from a plain.
SC.7.E.6.1

_____

_____

14 **Summarize** the characteristics of the landform regions labeled in the map below. LA.7.2.2.3

_____

_____

_____

Rocky Mountains

Ozark Plateau

Gulf Coastal Plain

15 **Evaluate** which type of landform is best suited for agriculture. LA.7.2.2.3

_____

Writing in Science

16 **Write** A song includes lyrics that usually rhyme. Write the lyrics to a song on a separate piece of paper about the elevation of landforms. LA.7.2.2.3

Big Idea Review

17 Identify and describe the different layers of Earth
SC.7.E.6.1

_____

_____

_____

18 How does Earth's core create Earth's magnetic field? SC.7.E.6.1

_____

_____

_____

Math Skills   MA.6.A.3.6

Solve One-Step Equations

19 A large weather balloon holds 3.00 m³ of air. The air in the balloon has a mass of 3.75 kg. What is the density of the air in the balloon?

Fill in the correct answer choice.

Multiple Choice

1 The asthenosphere is part of which of Earth's layers? **SC.7.E.6.1**

  Ⓐ crust

  Ⓑ mantle

  Ⓒ inner core

  Ⓓ outer core

2 Which force gave Earth its spherical shape? **SC.7.E.6.4**

  Ⓕ electricity

  Ⓖ friction

  Ⓗ gravity

  Ⓘ magnetism

Use the diagram below to answer question 3.

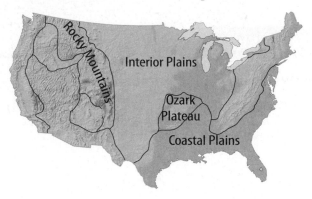

3 Which region on the map above was entirely eroded by water? **SC.7.E.6.2**

  Ⓐ Coastal Plains

  Ⓑ Interior Plains

  Ⓒ Ozark Plateau

  Ⓓ Rocky Mountains

4 Which describes Earth's asthenosphere? **SC.7.E.6.1**

  Ⓕ brittle

  Ⓖ fast-moving

  Ⓗ freeze-dried

  Ⓘ plastic

Use the diagram and the graphs below to answer question 5.

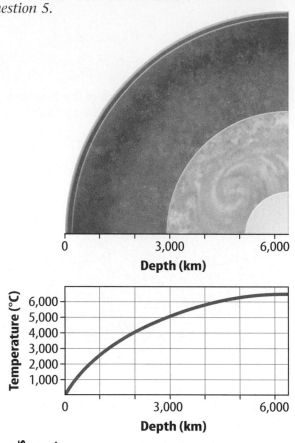

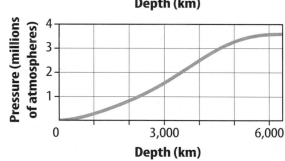

5 Which describes temperature and pressure at Earth's center? **SC.7.E.6.1**

  Ⓐ high pressure and high temperature

  Ⓑ high pressure and low temperature

  Ⓒ low pressure and high temperature

  Ⓓ low pressure and low temperature

6 How were the Rocky Mountains formed? **SC.7.E.6.2**

Ⓕ deposition

Ⓖ lava from the ocean floor

Ⓗ uplift, fold, or push forces inside Earth

Ⓘ wind erosion

7 Which is the correct order of Earth's layers from the surface to the center? **SC.7.E.6.1**

Ⓐ crust, core, mantle

Ⓑ crust, mantle, core

Ⓒ mantle, core, crust

Ⓓ mantle, crust, core

Use the diagram below to answer question 8.

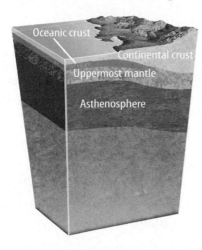

Oceanic crust
Continental crust
Uppermost mantle
Asthenosphere

8 Which of Earth's layers is NOT part of the lithosphere? **SC.7.E.6.1**

Ⓕ continental crust

Ⓖ oceanic crust

Ⓗ uppermost mantle

Ⓘ asthenosphere

9 Why do scientists theorize that Earth's core is made of metal? **SC.7.E.6.4**

Ⓐ Metals are the first elements created.

Ⓑ Metals are dense and would be pulled to the center as Earth formed.

Ⓒ Metals are magnetic and would be attracted to each other in the center as Earth formed.

Ⓓ Metals were pulled to the center of Earth as it formed because they are both dense and magnetic.

Use the diagrams below to answer questions 10-11.

10 What shaped the two landforms? **SC.7.E.6.2**

Ⓕ rock type

Ⓖ elevation

Ⓗ topography

Ⓘ moving water

11 How will the valley change over time? **SC.7.E.6.4**

Ⓐ It will get wider.

Ⓑ It will get deeper.

Ⓒ It will get wider and deeper.

Ⓓ It will fill up with water and rock.

NEED EXTRA HELP?

If You Missed Question...	1	2	3	4	5	6	7	8	9	10	11
Go to Lesson...	2	1	3	2	2	3	2	2	2	3	3

FLORIDA NGSSS

Benchmark Mini-Assessment   Chapter 1 • Lesson 1

mini BAT

Multiple Choice *Bubble the correct answer.*

Gravitational Force between Pairs of Objects

Pair	Mass of Each Object in Pair (kg)	Distance between Objects (m)
A	1,000 kg	30 m
B	1,000 kg	200 m
C	5,000 kg	30 m
D	5,000 kg	200 m

1. In the table above, which pair of objects has the weakest gravitational force between them? **SC.7.N.1.1**

(A) A

(B) B

(C) C

(D) D

2. Density is described by the following formula:

$$D = \frac{m}{V}$$

If a substance has a density of 2.5 g/mL and the volume of the substance is 8 mL, what is the mass of the substance? **MA.6.A.3.6**

(F) 0.3 g

(G) 3.2 g

(H) 8 g

(I) 20 g

3. Which of Earth's systems includes underground water resources such as aquifers and springs? **SC.7.E.6.1**

(A) atmosphere

(B) biosphere

(C) geosphere

(D) hydrosphere

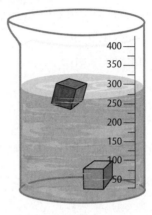

4. Felicia placed two cubes of the same size in a beaker of water as shown in the figure above. What can Felicia conclude about the cubes? **SC.7.N.1.1**

(F) The cubes have the same mass.

(G) One cube is denser than the other.

(H) The cube on the bottom of the beaker will eventually float.

(I) The force of gravity is greater on the cube that floats.

Copyright © Glencoe/McGraw-Hill, a division of The McGraw-Hill Companies, Inc.

Multiple Choice *Bubble the correct answer.*

Use the image and graphs below to answer questions 1 and 2.

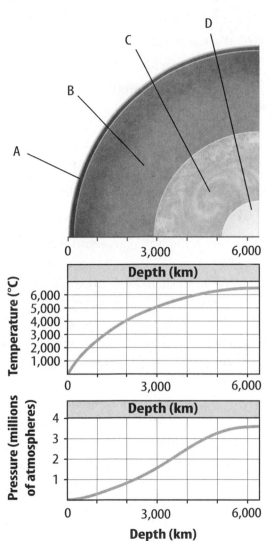

1. In which layer of Earth are temperature and pressure greatest? **SC.7.E.6.1**

(A) A

(B) B

(C) C

(D) D

2. How do temperature and pressure change as the depth of the geosphere increases? **SC.7.E.6.1**

(F) Temperature and pressure decrease.

(G) Temperature and pressure increase.

(H) Temperature decreases, and pressure increases.

(I) Temperature increases, and pressure decreases.

3. In which layer of Earth would you expect to find the least dense materials? **SC.7.E.6.1**

(A) crust

(B) mantle

(C) inner core

(D) outer core

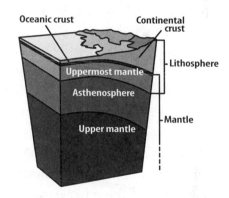

4. Which of Earth's upper layers, as shown in the image above, flows very slowly? **SC.7.E.6.1**

(F) lithosphere

(G) asthenosphere

(H) continental crust

(I) oceanic crust

Copyright © Glencoe/McGraw-Hill, a division of The McGraw-Hill Companies, Inc.

Benchmark Mini-Assessment   **Chapter 1 • Lesson 3**

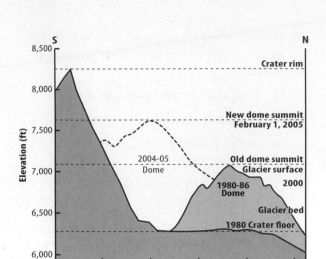

1. In 1980, Mount St. Helens, a volcano in Washington State, erupted. Based on the topographic profile above, how has the crater of Mount St. Helens changed since 1980? **SC.7.E.6.5**

Ⓐ The elevation and the relief of the crater have decreased over time.

Ⓑ The elevation and the relief of the crater have increased over time.

Ⓒ The elevation of the crater has decreased, but the relief is high.

Ⓓ The elevation of the crater has increased, but the relief remains low.

2. Which landform has low relief and elevation? **SC.7.E.6.1**

Ⓕ the Colorado Plateau

Ⓖ the Great Plains

Ⓗ the Appalachian Mountains

Ⓘ the Rocky Mountains

3. How do plains form? **SC.7.E.6.1**

Ⓐ from faults in rock layers

Ⓑ from the folding of rock layers

Ⓒ from sediment deposited by rivers

Ⓓ from the uplift of rock layers

4. How are relief and elevation related? **SC.7.E.6.1**

Ⓕ Elevation is height above sea level, and relief describes how the elevation of an area varies.

Ⓖ Elevation refers to the shape of a given area, and relief describes height above sea level.

Ⓗ A landform that has high elevation always has high relief.

Ⓘ The terms relief and elevation are interchangeable.

Copyright © Glencoe/McGraw-Hill, a division of The McGraw-Hill Companies, Inc.

Notes

Notes

Is it a mineral?

Earth contains many different kinds of minerals. Check off the things in the list below that you consider to be minerals.

_____ human-made diamond	_____ granite	_____ cement
_____ glacial ice	_____ sandstone	_____ water
_____ bone	_____ milk	_____ charcoal
_____ ruby	_____ coal	_____ brick
_____ clay	_____ quartz	_____ gold
_____ sugar crystal	_____ salt crystal	_____ soil

Explain your thinking. Describe your rule or reasoning for deciding whether something is a mineral.

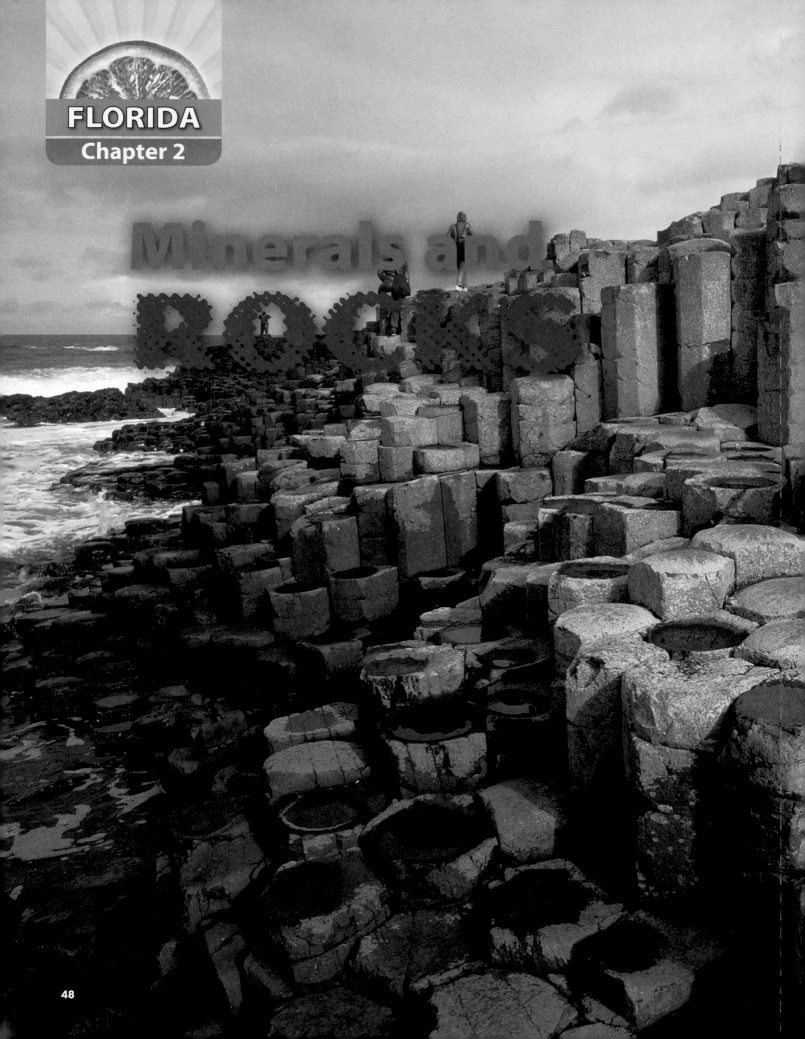

Minerals and ROCKS

1 The Practice of Science

6 Earth Structures

Think About It!

How are minerals and rocks formed, identified, classified, and used?

This is Giant's Causeway in Ireland. Columns of rocks, such as these, are present in several places on Earth. Some look like staircases, such as the rocks pictured here. Others look like a pile of telephone poles that have been knocked over.

1 Do you think they formed naturally?

_____

_____

_____

2 Do you think they are always the same shape?

_____

_____

_____

3 What do you think caused these rocks to form in this way?

_____

_____

_____

_____

Get Ready to Read

What do you think about minerals and rocks?

Before you read, decide if you agree or disagree with each of these statements. As you read this chapter, see if you change your mind about any of the statements.

	AGREE	DISAGREE
1 Minerals generally are identified by observing their color.	☐	☐
2 Minerals are made of crystals.	☐	☐
3 Once a rock forms, it lasts forever.	☐	☐
4 All rocks form when melted rock cools and changes into a solid.	☐	☐
5 All rock types are related through the rock cycle.	☐	☐
6 Rocks move at a slow and constant rate through the rock cycle.	☐	☐

There's More Online!
Video • Audio • Review • ⓘLab Station • WebQuest • Assessment • Concepts in Motion • Multilingual eGlossary

MINERALS

ESSENTIAL QUESTIONS

 How do minerals form?

 What properties can be used to identify minerals?

 What are some uses of minerals in everyday life?

Vocabulary

mineral p. 51

crystal structure p. 52

crystallization p. 53

streak p. 55

luster p. 55

cleavage p. 55

fracture p. 55

ore p. 57

 Florida NGSSS

LA.7.2.2.3 The student will organize information to show understanding (e.g., representing main ideas within text through charting, mapping, paraphrasing, summarizing, or comparing/contrasting);

LA.7.4.2.2 The student will record information (e.g., observations, notes, lists, charts, legends) related to a topic, including visual aids to organize and record information, as appropriate, and attribute sources of information;

MA.6.A.3.6 Construct and analyze tables, graphs, and equations to describe linear functions and other simple relations using both common language and algebraic notation.

SC.7.N.1.3 Distinguish between an experiment (which must involve the identification and control of variables) and other forms of scientific investigation and explain that not all scientific knowledge is derived from experimentation.

SC.7.N.1.3

Inquiry Launch Lab

20 minutes

Is everything crystal clear?

Do you have several shirts that are nearly the same color or style? If so, you probably know the subtle differences that make each unique. The same is true for the thousands of minerals on Earth. To most people, many of these minerals look exactly the same. In this lab, you'll examine four transparent minerals and demonstrate that not everything is crystal clear.

Procedure

1. Read and complete a lab safety form.
2. Draw a data table in which to record your observations after each step.
3. Use a **magnifying lens** to examine each **mineral.**
4. Place each mineral over this sentence, and observe the words.
5. Place the minerals in a **small bowl** of warm water for 2–3 minutes. Take the minerals out of the water and dry them with **paper towels.** Examine each mineral.
6. Carefully place one drop of **dilute hydrochloric acid** on each mineral. Record your observations. Use the paper towels to wipe the minerals dry.

Data and Observations

Think About This

1. How are the minerals the same? How are they different?

2. **Key Concept** How do you think each mineral might be used in everyday life?

1. Minerals are all around you every day. Sometimes you can even pick up a mineral right from the ground! But most minerals need to mined from beneath Earth's surface, which requires large equipment and areas of land. What is a mineral? Where can you find minerals? What can you use them for?

_____

_____

_____

_____

_____

What is a mineral?

Do you ever drink mineral water? Maybe you take vitamins and minerals to stay healthy. The word _mineral_ has many common meanings, but for geologists, scientists who study Earth and the materials of which it is made, this word has a very specific definition.

A **mineral** _is a naturally occurring, inorganic solid that has a crystal structure and a definite chemical composition._ In order for a substance to be classified as a mineral, it must have all five of the characteristics listed in the definition above. Samples of pyrite and coal are shown in **Figure 1**. Coal is made of ancient compressed plant material. Pyrite crystals are made of the elements iron and sulfur. One of these is a mineral, but the other is not. By considering each of the five characteristics of minerals, you can determine which sample is the mineral. The information on the next page will help you do this.

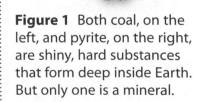

Figure 1 Both coal, on the left, and pyrite, on the right, are shiny, hard substances that form deep inside Earth. But only one is a mineral.

Active Reading **3. Infer** Which one do you think is a mineral?

_____

Active Reading **2. Point Out** (Circle) the five characteristics that define a mineral.

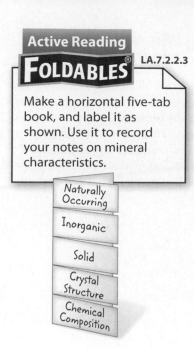

Active Reading

FOLDABLES® LA.7.2.2.3

Make a horizontal five-tab book, and label it as shown. Use it to record your notes on mineral characteristics.

Naturally Occurring

Inorganic

Solid

Crystal Structure

Chemical Composition

Math Skills MA.6.A.3.6

Use Ratios

A ratio compares numbers. For example, in the chemical formula for water, H_2O, the number *2* is called a subscript. The subscript tells you how many atoms of that element are in the formula. A symbol with no subscript means that element has one atom. So, the ratio of hydrogen (H) atoms to oxygen (O) atoms in H_2O is 2:1. This is read *two to one.*

Practice

4. Quartz has the formula SiO_2. What is the ratio of silicon (Si) atoms to oxygen (O) atoms in quartz?

Characteristics of Minerals

To be classified as a mineral, a substance must form naturally. Materials made by people are not considered minerals. Diamonds that form deep beneath Earth's surface are minerals, but diamonds that are made in a laboratory are not. As shown in **Figure 2,** these two types of diamonds can look very similar.

Materials that contain carbon and were once alive are organic. Minerals cannot be organic. This means that a mineral cannot have once been alive, and it cannot contain anything that was once alive, such as plant parts.

A mineral must be solid. Liquids and gases are not considered to be minerals. So, while solid ice is a mineral, water is not.

Figure 2 Natural and artificial diamonds look very much alike, but only the natural diamond on the right is a mineral.

A mineral must have a crystal structure. *The atoms in a crystal are arranged in an orderly, repeating pattern called a* **crystal structure.** This organized structure produces smooth faces and sharp edges on a crystal. The faces and edges of the pyrite crystals shown in **Figure 1** are produced by this internal atomic structure.

A mineral is made of specific amounts of elements. A chemical formula shows how much of each element is present in the mineral. For example, pyrite is made of the elements iron (Fe) and sulfur (S). There always must be one iron atom for every two sulfur atoms. Therefore, the chemical formula for pyrite is FeS_2.

Look again at **Figure 1.** Because the plants that turned into the coal were once alive, coal is not a mineral. The pyrite has all five characteristics of a mineral, so it is a mineral.

Mineral Formation

How do atoms form minerals? Atoms within a liquid join together to form a solid. **Crystallization** *is the process by which atoms form a solid with an orderly, repeating pattern.* Crystallization can happen in two main ways.

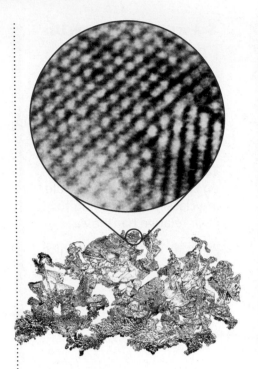

 5. Locate Underline how minerals form.

Crystallization from Magma When melted rock material—called magma—cools, some of the atoms join together and form solid crystals. As the liquid continues to cool, atoms are added to the surface of the crystals. The longer it takes the magma to cool, the more atoms are added to the crystal. Large crystals grow when the magma cools slowly. If the magma cools quickly, there is only enough time for small crystals to grow.

Crystallization from Water Many substances, such as sugar and salt, dissolve in water, especially if the water is warm. When water cools or evaporates, the particles of the dissolved substances come together again in the solution and crystallize. The gold shown in **Figure 3** formed this way. The orderly arrangement of atoms in this mineral is visible using a very powerful microscope.

Figure 3 🔑 The orderly atomic structure of gold crystals produces these neat rows of atoms.

 6. Compare and Contrast Complete the diagram below to compare crystallization *from magma* with crystallization *from water.*

Crystallization

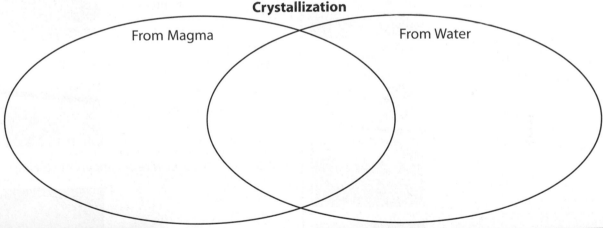

From Magma

From Water

Inquiry 🌴 SC.7.N.1.3

①LAB STATION **Try It!**

MiniLab *How does your garden grow?* at connectED.mcgraw-hill.com

Apply It! After you complete the lab, answer these questions.

1. Summarize How did the minerals form in this activity?

Foliated Metamorphic Rocks Recall that crystals form in a variety of shapes. Minerals with flat shapes, such as mica, produce a texture called foliation. **Foliation** [foh lee AY shun] *results when uneven pressures cause flat minerals to line up, giving the rock a layered appearance.* Eventually distinct bands of light and dark minerals form, as shown in the sample of gneiss [NISE] in **Figure 12.** Foliation is the most obvious characteristic of metamorphic rocks.

Nonfoliated Metamorphic Rocks Marble, another type of metamorphic rock, does not exhibit foliation. The grains in the marble pictured in **Figure 12** are not flattened like the grains in gneiss. The calcite crystals that make up marble became blocklike and square when exposed to high temperatures and pressure. Marble has a nonfoliated texture.

Rocks in Everyday Life

Rocks are abundant natural resources that are used in many ways based on their physical characteristics. Some igneous rocks are hard and durable, such as the granite used to construct the fountain shown in **Figure 13.** The igneous rock pumice is soft but contains small pieces of hard glass, which makes it useful for polishing and cleaning.

Natural layering makes sedimentary rock a high-quality building stone. Both sandstone and limestone are used in buildings. The building pictured in **Figure 14** is made of sandstone. Limestone also is used to make cement, which is then used in construction applications, including building highways.

Foliated metamorphic rocks split into flat pieces. Slate makes durable, fireproof roofing shingles, such as the ones shown in **Figure 15.** Other metamorphic rocks are used as art. Marble is soft enough to carve and often is used for making detailed sculptures.

Active Reading 7. **Determine** Underline some everyday uses for rocks.

Figure 13 Granite was used to build this fountain.

Figure 14 This building in Jordan was carved and constructed out of sandstone.

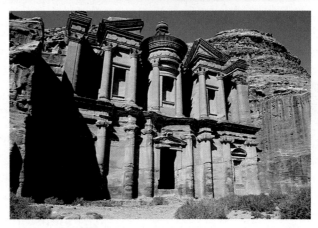

Figure 15 Slate sometimes is used on roofs like the ones on this house.

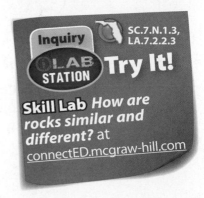

Inquiry SC.7.N.1.3, LA.7.2.2.3

LAB STATION **Try It!**

Skill Lab *How are rocks similar and different?* at connectED.mcgraw-hill.com

Interlocking crystals of different sizes are common in igneous rocks.

The individual grains that form sedimentary rocks can be mineral grains or fragments of other rocks.

Increases in temperature and pressure cause minerals to change in size and shape.

Use Vocabulary

1 **Distinguish** between lava and magma.

_____

_____

2 Loose grains of rock material are called _____.

3 **Use the term** *lithification* in a sentence.

_____

_____

Understand Key Concepts

4 **Explain** how sedimentary rocks form.

_____

_____

5 Which list shows the correct sequence? SC.7.E.6.2
   (A) shale, foliation, basalt
   (B) gneiss, lithification, shale
   (C) granite, metamorphism, gneiss
   (D) sandstone, lithification, sediment

6 Compare the formation of sandstone to the formation of gneiss.

_____

_____

Interpret Graphics

7 **Compile** List the two major textures of metamorphic rocks and why they form. LA.7.2.2.3

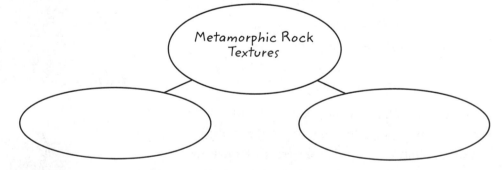

Metamorphic Rock Textures

Critical Thinking

8 **Classify** the following rock and justify your answer: light brown in color; small, sand-sized pieces of quartz cemented together in layers.

_____

_____

Details

Describe two characteristics geologists use to classify igneous rock.

Characteristic	Description
Texture	1.
Mineral composition	2.

Sequence the formation of sedimentary rock through the process of lithification. Use the term in parentheses in your explanation of each step.

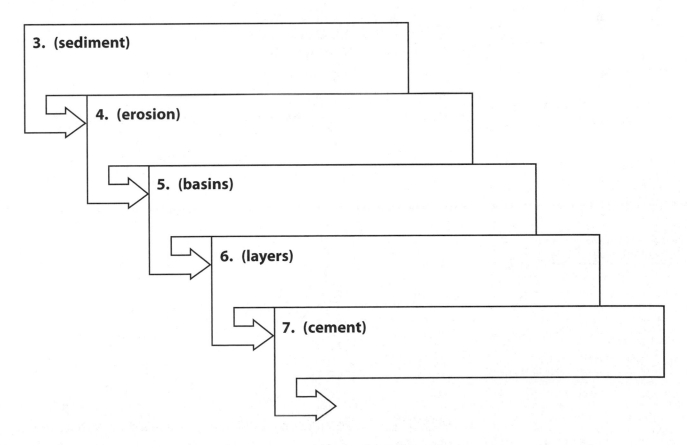

3. (sediment)

4. (erosion)

5. (basins)

6. (layers)

7. (cement)

8. Differentiate the grains of sedimentary rock from the grains of igneous rock.

_____

_____

_____

The Rock CYCLE

 How do surface processes contribute to the rock cycle?

How is the rock cycle related to plate tectonics?

Vocabulary

rock cycle p. 69

extrusive rock p. 70

intrusive rock p. 70

uplift p. 70

deposition p. 71

 Florida NGSSS

LA.7.2.2.3 The student will organize information to show understanding (e.g., representing main ideas within text through charting, mapping, paraphrasing, summarizing, or comparing/contrasting);

SC.7.E.6.2 Identify the patterns within the rock cycle and relate them to surface events (weathering and erosion) and sub-surface events (plate tectonics and mountain building).

SC.7.N.1.1 Define a problem from the seventh grade curriculum, use appropriate reference materials to support scientific understanding, plan and carry out scientific investigation of various types, such as systematic observations or experiments, identify variables, collect and organize data, interpret data in charts, tables, and graphics, analyze information, make predictions, and defend conclusions.

SC.7.N.1.3 Distinguish between an experiment (which must involve the identification and control of variables) and other forms of scientific investigation and explain that not all scientific knowledge is derived from experimentation.

Inquiry Launch Lab SC.7.N.1.1

20 minutes

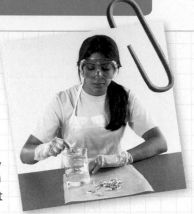

Do you "rock"?

Have you ever walked across a gravel road in your bare feet? If so, you know that rocks are hard. However, even though they are hard, rocks can change. How can you make a "rock" and model some of the changes that can turn it from one type into another?

Procedure

1. Read and complete a lab safety form.
2. Break **small candles** in half over a piece of **waxed paper.**
3. Drop the pieces of candle into very warm water.
4. After 10–20 seconds, use **forceps** to remove all the candle pieces and stack them back on the waxed paper.
5. Wrap the candles in the waxed paper and squeeze it tightly to press the warm "rock" pieces.

Think About This

1. What type of rock did you model? Explain.

2. What changed the "rocks" in step 2 to the "rock" in step 5? What type of rock formed?

3. **Key Concept** How might different processes contribute to the rock cycle?

Inquiry **Where do you think all the rock layers went?**

1. Monument Valley, Utah, is home to these towering rock formations. Over millions of years, processes on Earth's surface have worn away the surrounding rock layers, leaving the more resistant rocks behind. What do you think this landscape will look like in another few million years?

_____

_____

_____

_____

What is the rock cycle?

Do you have a recycling program at school? Or do you recycle at home? When materials such as paper or metal are recycled, they are used over again but not always for the same things. The metal from the beverage can you recycled yesterday might end up in a baseball bat.

Recycling also occurs naturally on Earth. The rock material that formed Earth 4.6 billion years ago is still here, but much of it has changed many times throughout Earth's history. _The series of processes that continually change one rock type into another is called the_ **rock cycle.**

As materials move through the rock cycle, they can take the form of igneous rocks, sedimentary rocks, or metamorphic rocks. At times, the material might not be rock at all. It might be sediment, magma, or lava, such as that pictured in **Figure 16.**

Active Reading **2. Compare** Underline how the rock cycle is similar to recycling.

Figure 16 Earth materials move through the rock cycle, changing both form and their location on Earth.

 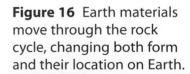

Figure 17 As rocks and rock material move slowly through the rock cycle, they are continually transformed from one rock type to another.

Active Reading **3. Describe** Record one path through the rock cycle that would result in the formation of a metamorphic rock.

Processes of the Rock Cycle

Mineral and rock formation are important processes in the rock cycle. The rock cycle is continuous, with no beginning or end. As shown in **Figure 17,** some processes take place on Earth's surface, and others take place deep beneath Earth's surface.

Cooling and Crystallization

Melted rock material is present both on and below Earth's surface. *When lava erupts and cools and crystallizes on Earth's surface, the igneous rock that forms is called* **extrusive rock.** *When magma cools and crystallizes inside Earth, the igneous rock that forms is called* **intrusive rock.**

Uplift

If intrusive rocks form deep within Earth, how are they ever exposed at the surface? **Uplift** *is the process that moves large amounts of rock up to Earth's surface and to higher elevations.* Uplift is driven by Earth's tectonic activity and often is associated with mountain building.

SCIENCE USE V. COMMON USE

intrusive

Science Use igneous rock that forms as a result of injecting magma into an existing rock body

Common Use the condition of being not welcome or invited

Weathering and Erosion

Uplift brings rocks to Earth's surface where they are exposed to the environment. Glaciers, wind, and rain, along with the activities of some organisms, start to break down exposed rocks. The same glaciers, wind, and rain also carry sediment to low-lying areas, called basins, by the process of erosion.

Deposition

Eventually, glaciers, wind, and water slow down enough that they can no longer transport the sediment. *The process of laying down sediment in a new location is called* **deposition.** Deposition forms layers of sediment. As time passes, more and more layers are deposited.

 4. **NGSSS Check** Explain Highlight how surface processes are involved in the rock cycle. **SC.7.E.6.2**

Compaction and Cementation

The weight of overlying layers of sediment pushes the grains of the bottom layers closer together. This process is called compaction. Sedimentary rocks have tiny spaces, called pores, between the grains. Pores sometimes contain water and dissolved minerals. When these minerals crystallize, they cement the grains together. **Figure 17** shows the path of sediment from weathering and erosion to compaction and cementation.

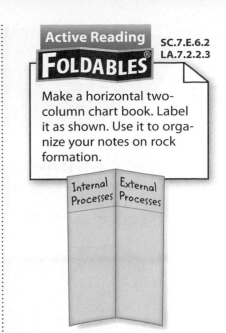

Active Reading SC.7.E.6.2 LA.7.2.2.3

FOLDABLES®

Make a horizontal two-column chart book. Label it as shown. Use it to organize your notes on rock formation.

Internal Processes | External Processes

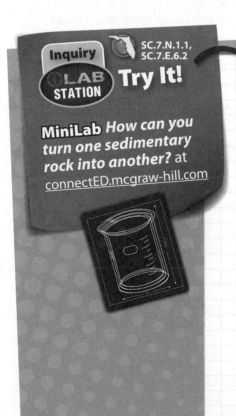

Inquiry SC.7.N.1.1, SC.7.E.6.2

LAB STATION Try It!

MiniLab *How can you turn one sedimentary rock into another?* at connectED.mcgraw-hill.com

Apply It! After you complete the lab, answer these questions.

1. **State** How did the lab materials change through weathering?

Through deposition?

Through compaction and cementation?

Shale

Slate

Phyllite

Schist

Gneiss

Increasing temperature and pressure

Figure 18 🔑 Rocks change form under high temperatures and pressure.

Active Reading

5. State How do the characteristics of the rocks change with increased pressure and temperature?

_____

_____

_____

Temperature and Pressure

Rocks subjected to high temperatures and pressure undergo metamorphism. This usually occurs far below Earth's surface. The progression of metamorphism can be observed in some rocks, as shown in **Figure 18.** As temperature and pressure increase, the sedimentary rock shale, shown at the top of **Figure 18,** changes to the metamorphic rock slate.

The rocks shown in **Figure 18,** slate, phyllite, schist, and gneiss, change from shale with increasing temperature and pressure. If the temperature is high enough, the rock melts and becomes magma. Igneous rocks form as the magma cools, and the material continues through the rock cycle.

Rocks and Plate Tectonics

The theory of plate tectonics states that Earth's surface is broken into rigid plates. The plates move as a result of Earth's internal thermal energy and convection in the mantle. The theory explains the movement of continents. It also explains earthquakes, volcanoes, and the formation of new crust. These events occur at plate boundaries, where tectonic plates interact.

Igneous rock forms where volcanoes occur and where plates move apart. Where plates collide, rocks are subjected to intense pressure and can undergo metamorphism. Colliding plates also can cause uplift or can push rock deep below Earth's surface, where it melts and forms magma. At Earth's surface, uplifted rocks are exposed and weathered. Weathered rock forms sediment, which eventually can form sedimentary rock.

Processes within Earth that move tectonic plates also drive part of the rock cycle. The rock cycle also includes surface processes. As long as these processes exist, the rock cycle will continue.

6. NGSSS Check **Infer** Underline how the rock cycle is related to plate tectonics. **SC.7.E.6.2**

Visual Summary

Weathering and erosion are important processes in the rock cycle.

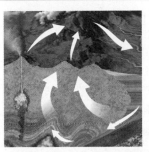

Uplift contributes to rock cycle processes on Earth's surface.

Plate tectonic activity contributes to rock cycle processes beneath Earth's surface.

Inquiry SC.7.N.1.3

Try It!

Skill Lab *Design a Forensic Investigation* at connectED.mcgraw-hill.com

Use Vocabulary

1 **Distinguish** between intrusive igneous rocks and extrusive igneous rocks.

_____

_____

2 **Define** *deposition* in your own words.

_____

2 **Use the term** *rock cycle* in a sentence.

_____

Understand Key Concepts

4 Which term refers to breaking rocks apart? SC.7.E.6.2

(A) cementation   (C) deposition

(B) crystallization   (D) weathering

5 **Classify** each of the following terms as Earth materials or rock cycle processes: magma, crystallization, sedimentary rocks, sediment, uplift, cementation. SC.7.E.6.2

_____

_____

_____

Interpret Graphics

6 **Sequence** Fill in the graphic organizer below and sequence the following terms: erosion, compaction and cementation, sedimentary rock, deposition, weathering. LA.7.2.2.3

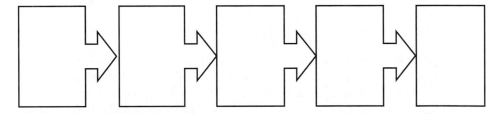

Critical Thinking

7 **Infer** where you would look along a tectonic plate to find the youngest rock.

_____

Think About It! Minerals and rocks form through natural constructive and destructive processes, have practical uses in everyday life, and are valued for their beauty. Minerals can be identified based on their physical properties. Rocks are classified based on their physical characteristics and how they formed.

Key Concepts Summary

Vocabulary

LESSON 1 Minerals

- **Minerals** form when solids crystallize from molten material or from solutions.

- Properties such as color, **streak,** hardness, and **cleavage** are used to identify minerals. Unique properties such as magnetism, reaction to acid, and fluorescence can also be used to identify certain minerals.

- Minerals are used to make everyday products such as toothpaste and makeup. Metals are used in cars and buildings. Gemstones are valued for their beauty.

mineral p. 51
crystal structure p. 52
crystallization p. 53
streak p. 55
luster p. 55
cleavage p. 55
fracture p. 55
ore p. 57

LESSON 2 Rocks

- Rocks are classified based on their **texture** and composition.

- Igneous rocks form when **magma** or **lava** solidifies. Sedimentary rocks form when **sediments** are **lithified.** Metamorphic rocks form when parent rocks are changed by thermal energy, pressure, or hot fluids.

- Rocks are used in construction, abrasives, and art.

rock p. 61
grain p. 61
magma p. 62
lava p. 62
texture p. 62
sediment p. 63
lithification p. 63
foliation p. 65

LESSON 3 The Rock Cycle

- Surface processes break down existing rocks into sediment. They transport this sediment to locations where it undergoes **deposition** and can be recycled to make more rocks.

- Thermal energy is released at plate boundaries. This thermal energy provides the energy needed for making igneous and metamorphic rocks. It also drives the forces that expose rocks to processes occurring on Earth's surface.

rock cycle p. 69
extrusive rock p. 70
intrusive rock p. 70
uplift p. 70
deposition p. 71

Active Reading

FOLDABLES® **Chapter Project**

Assemble your lesson Foldables as shown to make a Chapter Project. Use the project to review what you have learned in this chapter.

Internal Processes		
Rocks	Formation	Texture
Igneous		
Sedimentary		

Naturally Occurring
Inorganic
Solid
Crystal

Rocks   **Minerals**

Use Vocabulary

1 A mineral deposit that can be mined for a profit is a(n) _____ .

2 How does color differ from streak?

_____

_____

3 Loose rock and mineral fragments are called

_____ .

4 Define the word *rock* in your own words.

_____

_____

5 The process that brings rocks formed deep within Earth to the surface is called _____ .

6 Relate the words *deposition* and *lithification*.

_____

_____

Link Vocabulary and Key Concepts

Use vocabulary terms from the previous page to complete the concept map.

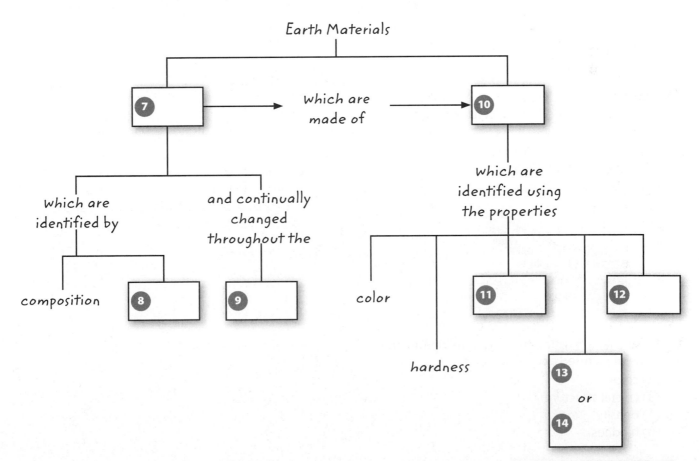

Earth Materials

7 → which are made of → **10**

7 which are identified by

and continually changed throughout the

composition   **8**

9

10 which are identified using the properties

color

hardness

11

12

13 or **14**

Fill in the correct answer choice.

🔑 Understand Key Concepts

1   Based on what you know about the Mohs hardness scale and mineral hardness, which mineral would make a good sandpaper? SC.7.E.6.2
   Ⓐ fluorite
   Ⓑ gypsum
   Ⓒ quartz
   Ⓓ talc

2   Which describes a way that minerals form? SC.7.E.6.2
   Ⓐ changing from gas to liquid
   Ⓑ changing from liquid to gas
   Ⓒ changing from liquid to solid
   Ⓓ changing from solid to gas

3   Which are mineral resources? SC.7.E.6.2
   Ⓐ gemstones and wood
   Ⓑ metals and cotton
   Ⓒ metals and gemstones
   Ⓓ wood and cotton

4   What can you learn from a mineral's chemical formula? SC.7.E.6.2
   Ⓐ composition
   Ⓑ crystal structure
   Ⓒ texture
   Ⓓ hardness

5   Which type of rock forms from magma and contains large interlocking crystals? SC.7.E.6.2
   Ⓐ extrusive igneous
   Ⓑ intrusive igneous
   Ⓒ foliated metamorphic
   Ⓓ nonfoliated metamorphic

6   What characteristic can be used to identify the mineral pictured above? SC.7.E.6.2
   Ⓐ color
   Ⓑ crystal structure
   Ⓒ density
   Ⓓ hardness

Critical Thinking

7   **Compare** the textures of the three main types of rocks. SC.7.E.6.2

_____

_____

_____

8   **Infer** why there is no longer a rock cycle on the Moon. SC.7.E.6.2

_____

_____

9   **Evaluate** the relative worth of minerals that are valued for their beauty and those that are used for practical purposes. SC.7.E.6.2

_____

_____

10   **Analyze** the relative usefulness of slate and granite for building a roof. SC.7.E.6.2

_____

_____

11   **Predict** what would happen to a sample of gneiss if it were heated enough to melt the mineral grains. SC.7.E.6.2

_____

12   **Construct** a flow chart showing the formation of quartzite starting with a mountain and ending with quartzite. (Quartzite is metamorphosed sandstone.) SC.7.E.6.2

13 **Critique** the rock cycle diagram below. Include one feature of the diagram that you find useful and one feature that could be improved. **SC.7.E.6.2**

_____

_____

_____

Igneous rock

Magma

Cooling

Melting

Weathering and erosion

High temperatures and pressure

Melting

Sediment

Weathering and erosion

Weathering and erosion

Compaction and cementation

Metamorphic rock

High temperatures and pressure

Sedimentary rock

14 How do geologists identify minerals? **LA.7.2.2.3**

_____

_____

_____

Writing in Science

15 **Write** On a separate piece of paper, write an acrostic poem based on the term *rock cycle*. Acrostic poems are written without rhyming. The letters of the given term form the first letter of each line of the poem, which, when read downward, spells out the term. **LA.7.2.2.3**

Big Idea Review

16 How is rock classification related to the rock cycle? **SC.7.E.6.2**

_____

_____

_____

_____

_____

17 What are some minerals and rocks that you use every day? **LA.7.2.2.3**

_____

_____

_____

_____

Math Skills   MA.6.A.3.6

Use Ratios

18 The ratio of iron (Fe) to chromium (Cr) to oxygen (O) in the mineral chromite is 1:2:4. What is the formula for chromite?

FLORIDA BIG IDEAS

1 **The Practice of Science**
3 **The Role of Theories, Laws, Hypotheses, and Models**
6 **Earth Structures**

How have natural events changed Earth over time?

Molten rock flowing everywhere, steaming volcanoes, asteroid and meteorite showers—ancient Earth was a very different place compared to today.

1 Can you think of a place where this environment exists?

_____

_____

_____

2 Do you think you could live there?

_____

_____

_____

3 What do you think has happened during Earth's long history to make it look as it does today?

_____

_____

_____

What do you think about the history of Earth?

Before you read, decide if you agree or disagree with each of these statements. As you read this chapter, see if you change your mind about any of the statements.

	AGREE	DISAGREE
1 All rocks contain fossils.	☐	☐
2 Humans produce all radioactive materials.	☐	☐
3 When Earth first formed, oceans were much larger than they are today.	☐	☐
4 Earth's early atmosphere was different from Earth's present-day atmosphere.	☐	☐
5 Fish were the first organisms in the oceans.	☐	☐
6 Asteroids no longer crash into Earth.	☐	☐

 **Connect**ED

There's More Online!
Video • Audio • Review • ⓘLab Station • WebQuest • Assessment • Concepts in Motion • Multilingual eGlossary

87

Geologic TIME

ESSENTIAL QUESTIONS

 What evidence supports the idea that Earth is very old?

 What evidence did scientists use to develop the geologic time scale?

 How does the geologic time scale compare to the human time scale?

Vocabulary

principle of superposition p. 90

fossil p. 90

radioactive decay p. 91

half-life p. 91

geologic time scale p. 92

 Florida NGSSS

LA.7.2.2.3 The student will organize information to show understanding (e.g., representing main ideas within text through charting, mapping, paraphrasing, summarizing, or comparing/contrasting);

MA.6.A.3.6 Construct and analyze tables, graphs, and equations to describe linear functions and other simple relations using both common language and algebraic notation.

SC.7.E.6.3 Identify current methods for measuring the age of Earth and its parts, including the law of superposition and radioactive dating.

SC.7.E.6.4 Explain and give examples of how physical evidence supports scientific theories that Earth has evolved over geologic time due to natural processes.

SC.7.N.1.2 Differentiate replication (by others) from repetition (multiple trials).

SC.7.N.1.3 Distinguish between an experiment (which must involve the identification and control of variables) and other forms of scientific investigation and explain that not all scientific knowledge is derived from experimentation.

SC.7.N.1.5 Describe the methods used in the pursuit of a scientific explanation as seen in different fields of science such as biology, geology, and physics.

SC.7.N.1.6 Explain that empirical evidence is the cumulative body of observations of a natural phenomenon on which scientific explanations are based.

Inquiry Launch Lab

SC.7.N.1.5, SC.7.N.1.6

15 minutes

What happened?

It might be hard to imagine that rocks change over time. How can you model rock layers to show how Earth processes change rock?

Procedure

1. Read and complete a lab safety form.

2. Use at least four balls of **clay** to make a model of a sequence of rock layers.

3. Use **colored pencils** to make a detailed drawing of your model below. Exchange models with another student.

4. Spend about 5 minutes changing your partner's rock sequence.

5. Get your original rock sequence back. What happened to the rocks?

Think About This

1. How did you change your partner's rock sequence?

2. Draw and explain what happened to your original rock sequence.

3. **Key Concept** How do you think rocks show that Earth has changed?

1. Geologists study the type of rock and special characteristics of the rock to learn about Earth's long history. What might be in the layers that can be used by scientists to learn about Earth's history? What makes each layer different from the next layer?

_____

_____

_____

_____

_____

_____

Evidence That Earth Has Changed

You have changed in many ways throughout your life. What kind of evidence could you use to investigate these changes? Maybe you have some old family photographs, some clothes that you wore years ago that no longer fit, or even a book or toy that you loved when you were younger. These items can be used to show you have changed. Similarly, scientists search for evidence that Earth has changed over time.

Scientists study Earth's past, and they develop ideas and theories about how it formed and how it has changed. Some-times when scientists make new discoveries, they must change their theories. As time passes, scientists either improve or revise scientific theories completely.

Active Reading 2. **Identify** <u>Underline</u> the sentences that explain why scientific theories are sometimes revised.

What kinds of evidence do scientists use to show that Earth has changed? Some evidence is under your feet. As shown in **Figure 1,** rock layers above and below Earth's surface hold the clues to Earth's past.

Figure 1 These rock layers have been folded and deformed. This is one type of evidence that tells of Earth's past.

Rock Layers and Fossils

Weathering breaks rocks exposed at Earth's surface into smaller pieces called sediment. Over time, gravity, water, and wind carry sediment downhill and deposit it in low areas called basins. Eventually, layers of sediment form. The increasing weight of the sediment slowly causes the layers to compress, forming layers of rock. The rock layers, such as those shown in the photograph at the beginning of this lesson, formed over millions of years. Therefore, scientists know that Earth must be very old.

 3. NGSSS Check Explain How are rocks evidence of Earth's age? SC.7.E.6.3, SC.7.E.6.4

_____

_____

_____

Because new sediment layers are always collecting on top of older layers, the oldest layer is usually on the bottom. Geologists use this observation to organize rock layers according to their ages. They use the **principle of superposition,** *which states that in rock layers that have not been folded or deformed, the oldest rock layers are on the bottom.* The principle of superposition cannot give the actual ages of rock layers in years, called the absolute age. Instead, it gives the relative ages of rock layers, which tells you whether the layers are younger or older than other rock layers.

Sometimes as sediment builds up, it buries organisms within the layers. Under certain conditions, the organisms become rock along with the layers, as shown in **Figure 2.** *The preserved remains or evidence of past living organisms are called* **fossils.** Many fossils represent species that no longer live on Earth.

Figure 2  This fossil crinoid was an animal that lived in a shallow ocean that once covered part of North America.

Inquiry

LAB STATION LA.7.2.2.3 **Try It!**

MiniLab
What was that? at connectED.mcgraw-hill.com

Apply It! After you complete the lab, answer these questions.

1. What was a common characteristic of marine organisms?

2. What was a common characteristic of land organisms?

3. How are fossils used by scientists?

Radioactivity

Tiny particles called atoms make up all matter on Earth. An element is a substance that contains only one type of atom. Most elements are stable, which means they remain unchanged under normal conditions. But some elements are unstable. Over time they decay, or break down, and form different elements. **Radioactive decay** *is the process by which one element naturally changes into another.* The decay occurs when the atom's nucleus ejects particles. The original element is called the parent element, and the new element that forms is called the daughter element.

Active Reading

4. Rephrase How do unstable elements change?

_____

_____

A radioactive element decays at a rate that is constant for that particular element. Scientists have calculated these rates. *The* **half-life** *of an element is the time required for half of the amount of a radioactive parent element to decay into a stable daughter element.* The graph in **Figure 3** shows how the percentage of parent atoms decreases as parent atoms decay and form daughter atoms.

Because radioactive elements decay at a constant rate, we can use them as clocks to measure time. First, scientists must know the element's half-life. Then, by comparing the amount of parent element to the amount of daughter element in a sample, scientists can calculate the age of the sample. Analyzing radioactive elements shows that some rocks are billions of years old.

  5. NGSSS Check Identify What evidence supports the idea that Earth is very old? **SC.7.E.6.3, SC.7.E.6.4**

_____

_____

_____

Math Skills MA.6.A.3.6

Use Percentages

You can use percentages to calculate how much of the original element is left after each half-life. In a sample of a radioactive element, 50 percent of the atoms decay after each half-life. If you start with 1,000 atoms of a radioactive element, how many atoms of that element will be left after one half-life?

1. Change the percentage to a decimal by moving the decimal point two places to the left.

   50% = 0.50

2. For each half-life, multiply the number of starting atoms by the decimal number.

   1,000 atoms × 0.50 = 500 atoms

Practice

6. If you start with 1,000 atoms, how many atoms would remain after 2 half-lives?

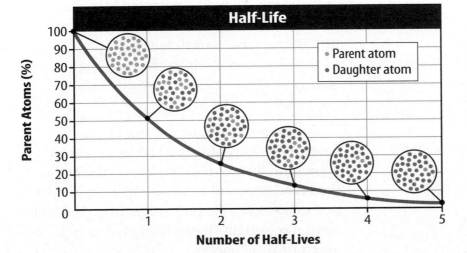

Half-Life

Parent Atoms (%)

- Parent atom
- Daughter atom

Number of Half-Lives

Figure 3 As time passes, more and more parent atoms decay and form daughter atoms. With each half-life, the percentage of parent atoms decreases by half.

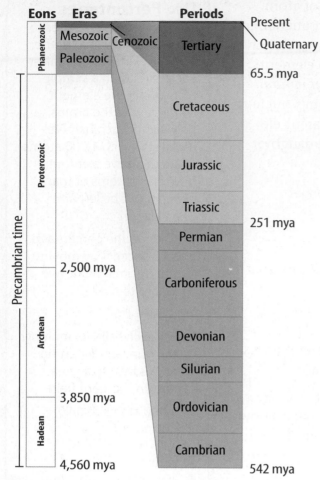

Figure 4 The divisions of geologic time are based on changes in rocks and fossils.

Active Reading

7. Classify Provide an example of an eon, an era in that eon, and a period within that era.

_____

_____

_____

WORD ORIGIN

Cambrian
from Latin *Cambria*, the Roman name for Wales

The Geologic Time Scale

During the 1800s, geologists organized the rocks and the rock layers they studied. They based their organization on superposition and the fossils present in the rock layers. They observed that older rock layers usually were positioned below younger rock layers. So, when they made charts showing the rock layers, they put the oldest rock layers at the bottom. They also assigned names to the rock layers. Each name represented a certain time in Earth's history. Rocks and rock layers that were all about the same age were assigned the same name. Eventually, they drew a standardized chart that represented all of the individual time periods in Earth's history. *The* **geologic time scale** *is a visual record of Earth's history, with the individual units based on changes in the rocks and fossils.* It is always drawn with the oldest rocks at the bottom and the youngest rocks at the top, as shown in **Figure 4**.

 8. NGSSS Check Recognize Highlight the evidence that scientists use to develop the geologic time scale. **SC.7.N.1.3**

Eons, Eras, Periods, and Epochs

You can divide the history of your life into years, months, weeks, and days. You might use special events, such as your first day of school, to mark divisions in your life. The geologic time scale shows the units used to describe Earth's history. Geologists divide Earth's history into eons, eras, periods, and epochs. They use the fossil record to mark geologic divisions. The divisions are not all the same length but mark places in the rock record where there are significant changes in the types of fossils present in the rocks.

For example, the beginning of the **Cambrian** period is marked by an abrupt appearance of complex life-forms. The end of the Permian period is marked by a significant and catastrophic die-off of organisms. You will learn more about these changes in Lessons 2 and 3 of this chapter.

Comparing Time Scales

Earth is 4.6 billion years old. Can you imagine that much time? You probably think only in terms of your own lifetime and maybe the lives of your parents and grandparents. Comparing the geologic time scale to something familiar, such as 1 year, can help.

Imagine all of Earth's history taking place within 1 year, starting January 1. The first tiny organisms float in the ocean on February 21. The earliest animals crawl onto land on November 20. The last of the dinosaurs become extinct on December 25. And human ancestors appear on the afternoon of December 31. Humans have experienced only a small part of Earth's history.

Active Reading

9. Summarize How does the geologic time scale compare to the human time scale?

Active Reading

FOLDABLES®   LA.7.2.2.3

Make a vertical four-tab Foldable. Label the sections as illustrated. Cut them to show the relationships among the units used to describe geologic time.

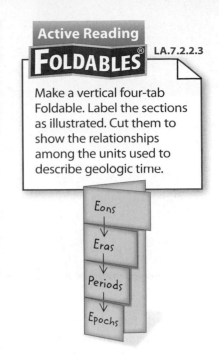

Eons

Eras

Periods

Epochs

Observing Slow and Rapid Changes

According to the geologic time scale, Earth has changed over billions of years. Most of these changes occurred slowly. For example, it takes millions of years for a mountain range to erode. However, some changes occur rapidly. The mountain in **Figure 5** became over 300 m tall in 1 year. It formed when eruptions from fissures, or cracks in the ground, formed a volcano in the middle of a cornfield. Likewise, an earthquake can change Earth's surface in just seconds.

Figure 5 Mount Paricutín formed when ash and lava erupted in this corn field in Michoacán, Mexico. Everything was buried in the town of Paricutín except the steeple of a church.

Visual Summary

Many ancient organisms that lived on Earth have been preserved in rocks.

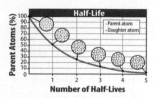

Radioactive elements contained within rocks can be used to calculate the age of rocks.

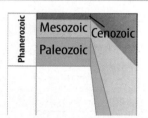

Eons, eras, periods and epochs are divisions of the geologic time scale.

Use Vocabulary

1 The preserved remains of organisms are called _____.

2 **Define** *geologic time scale* in your own words.

3 **Use the terms** *radioactive decay* and *half-life* in a sentence. SC.7.E.6.3

Understand Key Concepts 🔑

4 What is the longest unit of time on the geologic time scale?

(A) eon (C) era

(B) epoch (D) period

5 **Relate** How were fossils used in the development of the geologic time scale? SC.7.N.1.5

6 **Compare** the ages of rocks determined by the principle of superposition and ages determined by radioactive decay. SC.7.E.6.3

Interpret Graphics

7 **Sequence** Use the graphic organizer to correctly order the following three phrases: *new discoveries are made, people develop scientific theories, scientific theories are modified.* LA.7.2.2.3

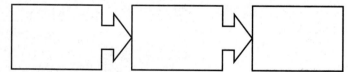

Critical Thinking

8 **Suggest** how the geologic time scale might be different if there were no radioactive elements. SC.7.E.6.3

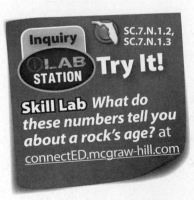

Inquiry SC.7.N.1.2, SC.7.N.1.3

①LAB STATION **Try It!**

Skill Lab *What do these numbers tell you about a rock's age?* at connectED.mcgraw-hill.com

Math Skills MA.6.A.3.6

9 If you begin with 140 g of an element, how much remains after 3 half-lives? [Hint: Use the same process for grams as for atoms.]

Sequence Put in order the concepts related to the study of Earth's past.

1. Rock forms in _____ over

_____ .

2. In undisturbed layers, the oldest layers are

_____ . This is called the

principle of _____ .

3. _____ embedded in rock layers

provide clues about _____

_____ .

Sort Order units of the geologic time scale. Write eras in three different colors and periods in the colors of the eras in which they occurred.

4. Eons	5. Eras	6. Periods

7. **Analyze** Compare the human time scale within geologic time.

_____

_____

_____

_____

ANCIENT EARTH

- How did gravity affect Earth's formation?
- How did the oceans and atmosphere form?
- What conditions made early Earth able to support life?
- How did environmental changes affect the evolution of life?

Vocabulary

Hadean eon p. 98

Archean eon p. 100

protocontinent p. 100

Proterozoic eon p. 102

Florida NGSSS

LA.7.2.2.3 The student will organize information to show understanding (e.g., representing main ideas within text through charting, mapping, paraphrasing, summarizing, or comparing/contrasting);

LA.7.4.2.2 The student will record information (e.g., observations, notes, lists, charts, legends) related to a topic, including visual aids to organize and record information, as appropriate, and attribute sources of information;

SC.7.E.6.4 Explain and give examples of how physical evidence supports scientific theories that Earth has evolved over geologic time due to natural processes.

SC.7.N.1.3 Distinguish between an experiment (which must involve the identification and control of variables) and other forms of scientific investigation and explain that not all scientific knowledge is derived from experimentation.

Inquiry Launch Lab   SC.7.N.1.3

20 minutes

How might Earth's crust have formed?

Earth today is much different from Earth billions of years ago. One major difference between early Earth and present-day Earth is temperature. Early Earth was extremely hot. But as the planet cooled, things changed.

Procedure

1. Read and complete a lab safety form.
2. Add 2–3 **wax sticks** to a 500-mL **beaker** half-filled with water.
3. Set the beaker on a **hot plate.** Turn on the hot plate.
4. Heat the water until all of the wax melts. Turn off the hot plate.
5. Put on **heat-resistant gloves** and remove the beaker from the hot plate.
6. As the water cools, observe the contents of the beaker. Record your observations below.

Data and Observations

Think About This

1. Describe what you observed as the wax melted.

2. What happened to the wax as the water cooled?

3. **Key Concept** How do you think Earth's continental crust might have formed?

1. Over the past 4.6 billion years, the surface of Earth has changed many times. The surface began as a sea of molten rock; it then changed to a surface covered by warm, shallow oceans; and finally it became a combination of deep oceans, shallow seas, and continents. Where might a scene like the one shown occur on Earth? Can organisms survive in this situation?

_____

_____

_____

_____

_____

_____

_____

Earth's Earliest History

How do you know about the beginning of your life? You probably don't remember it, but you can search for evidence to learn about it. Maybe there are photographs, video or tape recordings, or family stories that hold some clues. Earth, however, is so old that no one knows how it formed. So, scientists must search for evidence of its beginning.

Gravity and the Solar System

Before Earth or even the solar system existed, a cloud of gas, ice, and dust, called a nebula (NEB yuh luh), floated in space, as shown in **Figure 6.**

First, gravity pulled the particles together into a flattened disk shape that began to rotate. Then, the material in the center of the disk became dense, and the Sun formed. Finally, the pieces of material remaining in the disk attracted each other, and the planets formed. Some smaller bits of rock and ice remained as asteroids and comets.

Figure 6 Scientists hypothesize that the solar system formed when a nebula was pulled together by gravity.

Active Reading **2. Identify** Match the correct statement with the formation of the solar system.
  A. Sun formed
  B. Flattened disk
  C. Planets formed
  D. Nebula

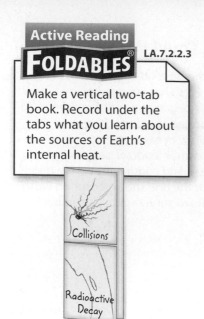

Make a vertical two-tab book. Record under the tabs what you learn about the sources of Earth's internal heat.

REVIEW VOCABULARY

thermal energy
energy that moves from one place to another because of differences in temperature

Figure 7 The seas of molten rock that covered Hadean Earth might have looked like this pool of lava.

Spherical Earth

Earth became larger as more particles came together. Collisions produce **thermal energy,** so colliding particles warmed the new planet. Because the hot rocks that made up ancient Earth were soft enough to flow, gravity pulled them into the shape of a sphere. Asteroids continued to crash into the surface, making Earth even hotter.

 3. NGSSS Check Find Highlight what effect gravity had on Earth's formation. **SC.7.E.6.4**

The Hadean Eon

Collisions with asteroids were not the only source of thermal energy. The young planet had large amounts of radioactive elements, and thermal energy is a by-product of radioactive decay. These two sources of thermal energy made early Earth much hotter than it is today. Because it was so hot, that time in history was named after the Greek god of the underworld, Hades (HAY deez). *The first 640 million years of Earth history are called the* **Hadean** *(hay DEE un)* **eon.**

Formation of Earth's Core

At first, Earth was a mixture of solid particles. When it became hot enough to melt metal, the denser metal began to flow because of the force of gravity. This is how Earth developed distinct layers of different materials. For example, when molten iron and small amounts of other metals flowed toward Earth's center, they formed the core.

Active Reading **4. Relate** How did gravity affect Earth after its formation?

_____

_____

_____

A hot Earth cools.

Throughout the Hadean eon, fewer and fewer asteroids struck Earth. And, as radioactive elements decayed and formed stable elements, less radioactive material was present. As a result, Earth began to cool. It continues to cool even today.

Seas of Molten Rock

How might Hadean Earth's surface have been different from Earth's surface today? Recall that sometimes volcanoes form when thermal energy escapes from inside Earth. Volcanoes produce lava, such as that shown in **Figure 7.** Because Earth was much hotter during the Hadean eon, scientists hypothesize that there must have been more lava produced. In fact, they think that a sea of molten rock covered Earth's surface.

Changes in the Seas

The pool of lava shown in **Figure 7** appears much like a small version of the huge molten seas of the Hadean eon. As Earth cooled, small islands of solid rock might have floated on the sea's surface. But motion in the molten seas or asteroid impacts would have destroyed them.

Earth continued to cool. Some of the material in the molten seas started to solidify, forming an ancient crust. This crust was not like Earth's crust today. Because of its composition, the crust would have melted very easily.

Active Reading **5. Identify** ⬭Circle the sentences that explain why Earth's earliest crust is no longer present today.

The Ancient Atmosphere

Lava was not the only thing that came from Hadean Earth's interior. Like present-day volcanic activity, Hadean eruptions produced gases. These volcanic gases formed Earth's earliest atmosphere. This atmosphere would have been poisonous for modern organisms. It contained water vapor, carbon dioxide, and poisonous gases, but no oxygen.

 6. NGSSS Check Describe How did Earth's early atmosphere form? SC.7.E.6.4

_____

_____

_____

_____

Inquiry LA.7.2.2.3

①LAB STATION Try It!

MiniLab *How might Earth's early atmosphere have formed?* at connectED.mcgraw-hill.com

Apply It! After you complete the lab, answer these questions.

1. What did the carbon dioxide represent?

2. How did temperature affect the release of the gas?

3. What impact did temperature have on the formation of Earth's early atmosphere?

Figure 8 During the Archean eon, the first continental crust formed protocontinents, which were smaller than present-day continents.

 7. Contrast How do the protocontinents compare with today's continents?

_____

_____

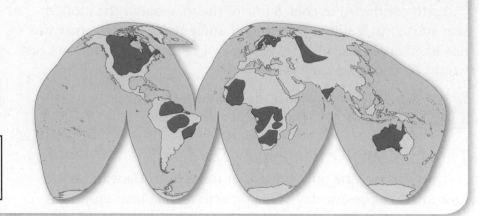

■ Protocontinents

■ Present-day Continents

The Archean Eon

The period of time that **occurred** *from about 4 to 2.5 million years ago is called the* **Archean** (ar KEE un) **eon.** Earth continued to cool after the Hadean eon. And during the Archean eon, Earth had its first solid surface. Portions of this first crust are still present on most of Earth's continents, as shown in **Figure 8.** Though Earth was cooler than during the Hadean eon, it still produced about twice as much internal thermal energy as present-day Earth. So far, the oldest rocks discovered on Earth formed during the Archean eon.

Extensive Volcanic Activity

You might have already learned that ocean crust forms when magma rises to the surface through cracks in the ocean floor. During the Archean eon, extensive volcanic activity formed Earth's first oceanic crust. Convection currents, formed by the rising and sinking of hot material below Earth's surface, moved the crust along Earth's surface in much the same way that convection moves tectonic plates today. But, with more thermal energy driving convection, the crust moved faster.

The Earliest Continents

Along with the first oceanic crust, the first continental crust formed during the Archean eon. As shown in **Figure 8,** the Archean continents were smaller than present-day continents, and there were more of them. *Scientists call the small, early continents* **protocontinents.** Throughout the Archean eon, convection caused collisions between these protocontinents. Sometimes they came together and formed larger landmasses.

ACADEMIC VOCABULARY

occur
(verb) to come into being—as an event, to come to pass

Active Reading

FOLDABLES® Chapter Project

Assemble your lesson Foldables as shown to make a Chapter Project. Use the project to review what you have learned in this chapter.

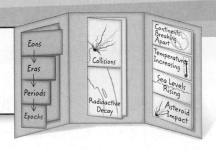

Use Vocabulary

1 **Define** the principle of superposition in your own words.

_____

_____

2 **Use the term** *fossil* in a sentence.

_____

3 **Distinguish** between the Hadean and the Proterozoic.

_____

_____

4 **Explain** radioactive decay in your own words.

_____

5 Humans are living during the _____ .

6 **Use the term** *extinct* in a sentence.

_____ .

Link Vocabulary and Key Concepts

Use vocabulary terms from the previous page to complete the concept map.

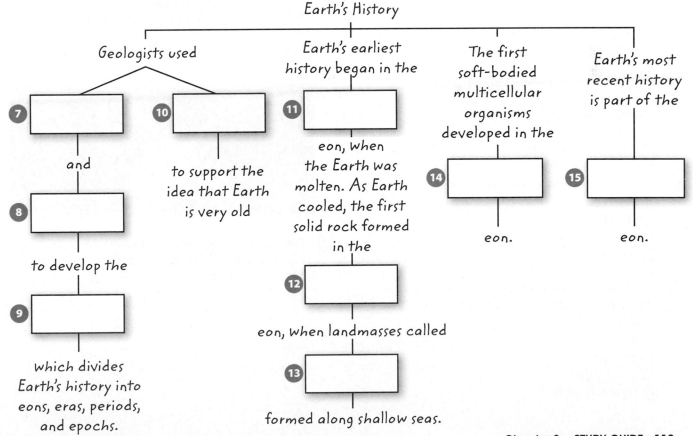

Fill in the correct answer choice.

Understand Key Concepts

1 What evidence could you use to determine which rock layer is oldest in the figure below? SC.7.E.6.3

- Ⓐ observation of layer order
- Ⓑ observation of layer colors
- Ⓒ measurement of layer thicknesses
- Ⓓ classification of particle size in layers

2 What is needed to make a fossil? SC.7.E.6.4
- Ⓐ rock and sediment
- Ⓑ sand and sediment
- Ⓒ an organism and water
- Ⓓ an organism and sediment

3 After 2 half-lives of a radioactive element, how much of the parent element is left? SC.7.E.6.3
- Ⓐ one-half
- Ⓑ one-third
- Ⓒ one-fourth
- Ⓓ one-eighth

4 Over time, which of these break down? SC.7.E.6.4
- Ⓐ all elements
- Ⓑ stable elements
- Ⓒ parent elements
- Ⓓ daughter elements

5 What molten metal formed most of Earth's core? SC.7.E.6.1
- Ⓐ iron
- Ⓑ nickel
- Ⓒ cobalt
- Ⓓ hematite

6 What product of volcanic eruptions made the atmosphere's composition more similar to that of today? SC.7.E.6.4
- Ⓐ oxygen
- Ⓑ nitrogen
- Ⓒ water vapor
- Ⓓ carbon dioxide

Critical Thinking

7 **Suggest** how scientists might revise their estimate for Earth's age if they discovered a 6-billion-year-old rock. SC.7.N.2.1

8 **Explain** how rock layers containing older fossils can occur above rock layers that contain younger fossils. SC.7.E.6.3

9 **Choose** what evidence you might use to determine the relative age of two rock layers in your neighborhood. SC.7.E.6.3

10 **Hypothesize** why the oldest rocks on Earth are from the Archean eon, not the Hadean eon. SC.7.E.6.4

11 **Describe** how the event above might impact Earth's surface, atmosphere, and organisms. SC.7.E.6.4

12 **Discuss** why there are no fossils of Hadean age. SC.7.E.6.4

_____

_____

13 **Suppose** that all the internal heat had completely escaped from inside Earth. How would this change Earth's surface? SC.7.E.6.1

_____

_____

_____

14 **Support** the decision to name the evolutionary changes occurring at the start of the Phanerozoic eon the Cambrian Explosion. LA.7.2.2.3

_____

_____

_____

Writing in Science

15 **Write** A haiku is a poem with three lines. The lines contain five, seven, and five syllables respectively. Write a haiku on a separate piece of paper about the changes Earth has gone through since it formed. LA.7.2.2.3

Big Idea Review

16 Will the natural events that change Earth's surface continue indefinitely? SC.7.E.6.5

_____

_____

_____

17 How have natural events changed over time? SC.7.E.6.4

_____

_____

_____

Math Skills   MA.6.A.3.6

Using Percentages

18 If you begin with 68 g of an isotope, how many grams of the original isotope will remain after 4 half-lives?

19 After 6 days, only 25 percent of a sample of a radioactive element remains. What is the half-life of the element?

20 Fifty atoms of a radioactive element remain after 5 half-lives. How many atoms were in the original sample?

Fill in the correct answer choice.

Multiple Choice

1 Which is NOT a cause of the Cambrian Explosion? **SC.7.E.6.4**

Ⓐ Continents were breaking apart.

Ⓑ Temperatures were increasing.

Ⓒ Sea levels were rising.

Ⓓ Ice caps were increasing.

Use the figure below to answer questions 2 and 3.

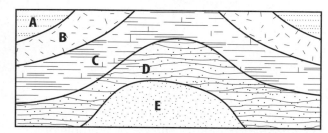

2 According to the principle of superposition, which rock layer in this cross section is youngest? **SC.7.E.6.3**

Ⓕ layer A

Ⓖ layer B

Ⓗ layer C

Ⓘ layer D

3 Which is probably true of fossils in layer B?
SC.7.E.6.3

Ⓐ They are younger than fossils in layer A.

Ⓑ They are younger than fossils in layer C.

Ⓒ They are older than fossils in layer D.

Ⓓ They are older than fossils in layer E.

4 Which force do scientists hypothesize formed our solar system? **SC.7.E.6.4**

Ⓕ collision

Ⓖ electromagnetism

Ⓗ gravity

Ⓘ magnetism

5 Which caused the formation of Earth's earliest atmosphere? **SC.7.E.6.4**

Ⓐ blue-green algae

Ⓑ colliding continents

Ⓒ multicellular organisms

Ⓓ volcanic activity

6 During which event does evidence show that life on Earth diversified greatly? **SC.7.E.6.4**

Ⓕ Cambrian Explosion

Ⓖ Hadean Earth

Ⓗ Oxygen Catastrophe

Ⓘ Snowball Earth

Use the graph below to answer question 7.

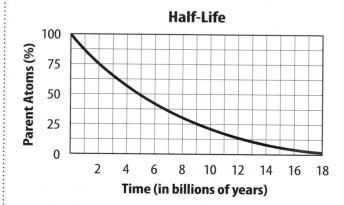

7 What is the half-life of the radioactive element shown in the graph? **SC.7.E.6.3**

Ⓐ 2 billion years

Ⓑ 5 billion years

Ⓒ 9 billion years

Ⓓ 12 billion years

8 Which conditions led to the evolution of new species during the Cambrian? **SC.7.E.6.4**

Ⓕ decreasing temperatures

Ⓖ unchanging positions of the continents

Ⓗ the presence of warm, shallow seas

Ⓘ multiple asteroid impacts

9 Which process or action is responsible for Earth's continual surface changes over the last 4.6 billion years? SC.7.E.6.5

(A) plate tectonics

(B) asteroid impacts

(C) volcanic activity

(D) mountain building

10 What do geologists use to divide geologic time into eons, eras, and periods? SC.7.E.6.3

(F) fossils

(G) radioactive dating

(H) changes in rocks and fossils

(I) radioactive dating and the principle of superposition

Use the table below to answer questions 11 and 12.

Some Radioactive Elements		
Parent Element	**Daughter Element**	**Half-life (in years)**
Carbon-14	Nitrogen-14	5,730
Uranium-235	Lead-207	704 million

11 Which element in the table above would be better for dating a sample from Precambrian time? SC.7.E.6.3

(A) carbon-14

(B) lead-207

(C) nitrogen-14

(D) uranium-235

12 A sample of wood contains 25 percent of the carbon-14 that a living plant contains. What is the approximate age of the wood? SC.7.E.6.3

(F) 2,865 years

(G) 5,730 years

(H) 11,460 years

(I) 17,190 years

13 A geologist is trying to determine the relative age of rock layers in a certain area, but he or she cannot rely on the principle of superposition. Which clue will help the geologist determine whether a rock layer has been overturned? SC.7.E.6.3

(A) Igneous rock usually forms below sedimentary rock.

(B) Rock layers erode at an average rate of 1.5 m per year.

(C) The largest sediment particles are usually deposited near the bottom of a rock layer.

(D) Metamorphic rocks are the lightest type of rock, so they will be deposited at the top of a rock layer.

14 Which term states that the oldest rock layer is found at the bottom in an undisclosed stack of rock layers? SC.7.E.6.3

(F) half-life

(G) geologic time scale

(H) superposition

(I) radioactive decay

Need Extra Help?

If You Missed Question...	1	2	3	4	5	6	7	8	9	10	11	12	13	14
Go to Lesson...	3	1	1	2	2	2	1	2	2	3	1	1	1	1

Multiple Choice *Bubble the correct answer.*

Use the figure below to answer questions 1 and 2.

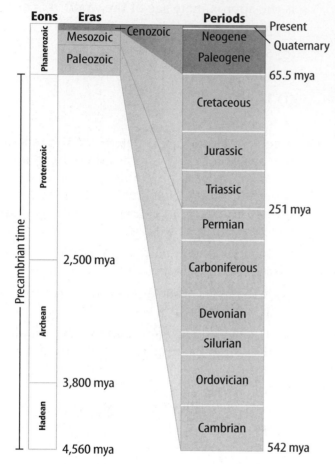

1. Researchers find a fossil that dates to 260 mya. According to the image above, this fossil is from the SC.7.E.6.4

   Ⓐ Cambrian period.

   Ⓑ Cretaceous period.

   Ⓒ Paleogene period.

   Ⓓ Permian period.

2. Which of these spanned the longest amount of time? SC.7.E.6.4

   Ⓕ Archean eon

   Ⓖ Paleozoic era

   Ⓗ Cretaceous period

   Ⓘ Quaternary period

3. Four fossils are found in different layers of an undisturbed deposit of sediment. Which fossil would be the youngest? SC.7.E.6.3

   Ⓐ the fossil of a crinoid in the bottom layer

   Ⓑ the fossil of a snail in the top layer

   Ⓒ the fossil of a fish in a layer just below the top layer

   Ⓓ the fossil of a sponge in a layer just above the bottom layer

4. What happens to parent atoms over time? SC.7.E.6.4

   Ⓕ They decay and form daughter atoms.

   Ⓖ They equal the number of daughter atoms.

   Ⓗ They multiply in number with the daughter atoms.

   Ⓘ They remain stable while forming daughter atoms.

Copyright © Glencoe/McGraw-Hill, a division of The McGraw-Hill Companies, Inc.

Multiple Choice *Bubble the correct answer.*

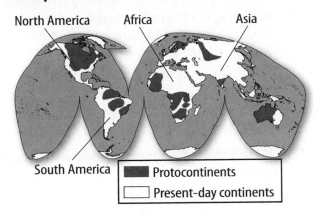

North America Africa Asia

South America

■ Protocontinents
□ Present-day continents

1. In the image above, which present-day continent contains the least amount of early continental crust from the Archean eon? **SC.7.E.6.4**

   (A) Africa
   (B) Asia
   (C) North America
   (D) South America

2. Which process was most harmful to the organisms that evolved to live on earliest Earth? **SC.7.E.6.4**

   (F) condensation
   (G) photosynthesis
   (H) asteroid impacts
   (I) volcanic eruptions

3. Why are fossils from the Proterozoic eon rare? **SC.7.E.6.4**

   (A) All Proterozoic organisms lived in the oceans.
   (B) All Proterozoic organisms were unicellular.
   (C) No sediment is left from the Proterozoic eon.
   (D) Proterozoic organisms had no hard parts, such as shells.

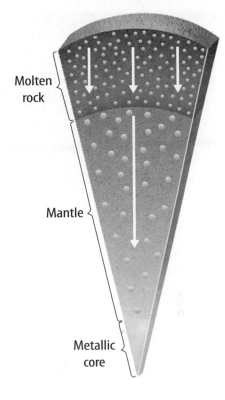

Molten rock

Mantle

Metallic core

4. The image above shows the formation of Earth's core. The arrows represent the movement of **SC.7.E.6.1**

   (F) cooling metal.
   (G) frozen water.
   (H) molten lava.
   (I) radioactive elements.

Copyright © Glencoe/McGraw-Hill, a division of The McGraw-Hill Companies, Inc.

Multiple Choice *Bubble the correct answer.*

1. The image above represents a hypothesized event at the end of the Mesozoic era. Which organisms likely benefited the most from this event? **SC.7.N.3.2**

   (A) dinosaurs

   (B) mammals

   (C) flying reptiles

   (D) land plants

2. What environmental conditions contributed to the development of trilobites and other multicellular organisms at the beginning of the Phanerozoic eon? **SC.7.E.6.4**

   (F) atmospheric dust

   (G) developing continents

   (H) deep, cold oceans

   (I) warm, shallow seas

3. Which species would most likely become extinct after a sudden period of global warming? **SC.7.E.6.4**

   (A) a species that is adapted to cool summers

   (B) a species that is adapted to hot grasslands

   (C) a species that is adapted to intense sunlight

   (D) a species that is adapted to warm-water lakes

Devonian   Pennsylvanian

4. The images above illustrate **SC.7.E.6.5**

   (F) the formation of Africa.

   (G) the formation of the Alps.

   (H) the formation of the supercontinent Pangaea.

   (I) the formation of the supercontinent Rodinia.

Copyright © Glencoe/McGraw-Hill, a division of The McGraw-Hill Companies, Inc.

Notes

Notes

Moving Plates

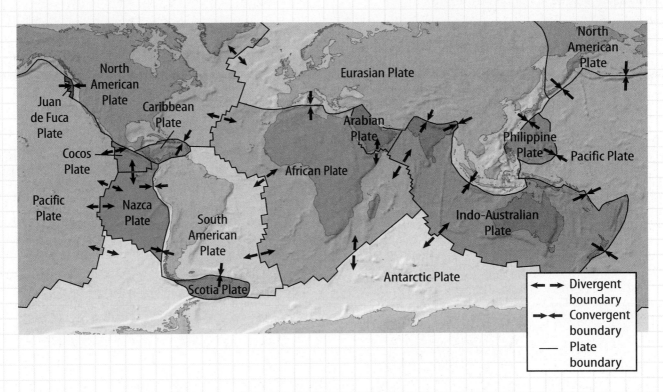

Divergent boundary
Convergent boundary
— **Plate boundary**

You may have heard about the theory of plate tectonics. This theory describes how Earth's surface is divided into large plates that move slowly in relation to each other. Put an *X* next to the things you think can result from this plate motion.

_____ volcanic eruptions	_____ hurricanes	_____ movement of continents
_____ earthquakes	_____ mountain ranges	_____ deep-ocean trenches
_____ tornadoes	_____ eclipses	_____ new ocean floor
_____ island chains	_____ coral reefs	_____ The Grand Canyon
_____ blizzards	_____ deserts	_____ undersea mountains

Explain why you selected the things marked with an *X*.

Plate
TECTONICS

FLORIDA BIG IDEAS

1 The Practice of Science

6 Earth Structures

| **Think About It!** | **How do you think volcanoes are associated with plate tectonics?** |

Iceland is home to many active volcanoes like this one. This eruption is called a fissure eruption. This occurs when lava erupts from a long crack, or fissure, in Earth's crust.

1 Why do you think the crust is breaking apart here?

_____

_____

2 What factors do you think determine where a volcano will form?

_____

_____

3 How do you think volcanoes are associated with plate tectonics?

_____

_____

_____

| **Get Ready to Read** | **What do you think about plate tectonics?** |

Before you read, decide if you agree or disagree with each of these statements. As you read this chapter, see if you change your mind about any of the statements.

AGREE DISAGREE

1 India has always been north of the equator. ☐ ☐

2 All the continents once formed one supercontinent. ☐ ☐

3 The seafloor is flat. ☐ ☐

4 Volcanic activity occurs only on the seafloor. ☐ ☐

5 Continents drift across a molten mantle. ☐ ☐

6 Mountain ranges can form when continents collide. ☐ ☐

The Continental Drift
HYPOTHESIS

 What evidence supports continental drift?

 Why did scientists question the continental drift hypothesis?

Vocabulary

Pangaea p. 127

continental drift p. 127

Florida NGSSS

LA.7.2.2.3 The student will organize information to show understanding (e.g., representing main ideas within text through charting, mapping, paraphrasing, summarizing, or comparing/contrasting);

SC.7.E.6.4 Explain and give examples of how physical evidence supports scientific theories that Earth has evolved over geologic time due to natural processes.

SC.7.N.1.1 Define a problem from the seventh grade curriculum, use appropriate reference materials to support scientific understanding, plan and carry out scientific investigation of various types, such as systematic observations or experiments, identify variables, collect and organize data, interpret data in charts, tables, and graphics, analyze information, make predictions, and defend conclusions.

SC.7.N.1.6 Explain that empirical evidence is the cumulative body of observations of a natural phenomenon on which scientific explanations are based.

SC.7.N.1.7 Explain that scientific knowledge is the result of a great deal of debate and confirmation within the science community.

 Launch Lab SC.7.N.1.1, SC.7.N.1.6

20 minutes

Can you put together a peel puzzle?

Early map makers observed that the coastlines of Africa and South America appeared as if they could fit together like pieces of a puzzle. Scientists eventually discovered that these continents were once part of a large landmass. Can you use an orange peel to illustrate how continents may have fit together?

Procedure 🥽 🧤 🧼

1. Read and complete a lab safety form.
2. Carefully peel an **orange,** keeping the orange-peel pieces as large as possible.
3. Set the orange aside.
4. Refit the orange-peel pieces back together in the shape of a sphere.
5. After successfully reconstructing the orange peel, disassemble your pieces.
6. Trade the entire orange peel with a classmate and try to reconstruct his or her orange peel.

Think About This

1. Which orange peel was easier for you to reconstruct? Why?

2. Look at a world map. Do the coastlines of any other continents appear to fit together?

3. 🔑 **Key Concept** What additional evidence would you need to prove that all the continents might have once fit together?

Inquiry How did this happen?

1. In Iceland, elongated cracks called rift zones are easy to find. Why do rift zones occur here? Iceland is above an area of the seafloor where Earth's crust is breaking apart. Earth's crust is constantly on the move. Scientists realized this long ago, but they could not prove how or why this happened. Where do you think rift zones normally occur? Why do you think they are usually in these locations?

_____

_____

_____

_____

_____

_____

Pangaea

Did you know that Earth's surface is on the move? Can you feel it? Each year, North America moves a few centimeters farther away from Europe and closer to Asia. That is several centimeters, or about the thickness of this book. Even though you don't necessarily feel this motion, Earth's surface moves slowly every day.

Nearly 100 years ago Alfred Wegener (VAY guh nuhr), a German scientist, began an important investigation that continues today. Wegener wanted to know whether Earth's continents were fixed in their positions. He proposed that *all the continents were once part of a supercontinent called* **Pangaea** (pan JEE uh). Over time Pangaea began breaking apart, and the continents slowly moved to their present positions. Wegener proposed the hypothesis of **continental drift**, *which suggests that continents are in constant motion on the surface of Earth.*

Alfred Wegener observed the similarities of continental coastlines now separated by ocean basins. Look at the outlines of Africa and South America in **Figure 1.** Notice how they could fit together like pieces of a puzzle. Hundreds of years ago mapmakers noticed this jigsaw-puzzle pattern as they made the first maps of the continents.

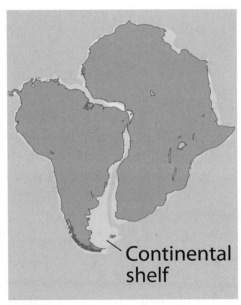

Continental shelf

Figure 1 The eastern coast of South America mirrors the shape of the west coast of Africa.

Active Reading **2. Apply** Complete the paragraph below.

A German scientists named _____ proposed that all of the continents were once part of a supercontinent called _____. He recognized that the eastern coast of _____ matched the shape of the western coast of _____. His hypothesis that continents are in constant motion on the surface of the Earth is called _____.

Summarize Scientists use the scientific method to prove or modify their theories. Fill in the chart below by summarizing each step in the process scientists used to prove the continental drift theory.

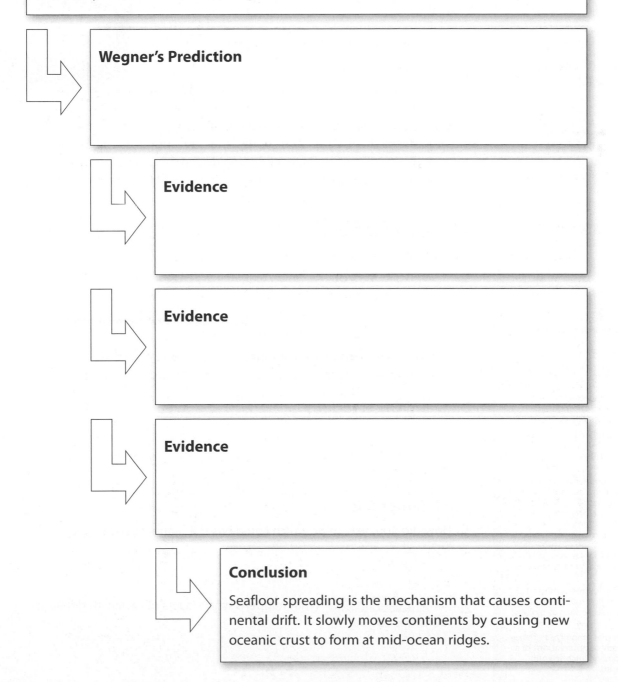

Observation

Magma rises through cracks in the seafloor, called mid-ocean ridges, and cools as soon as it hits the cold ocean water. As new seafloor crust forms, it pushes the older oceanic crust away from the mid-ocean ridge.

Wegner's Prediction

Evidence

Evidence

Evidence

Conclusion

Seafloor spreading is the mechanism that causes continental drift. It slowly moves continents by causing new oceanic crust to form at mid-ocean ridges.

The Theory of PLATE TECTONICS

ESSENTIAL QUESTIONS

 What is the theory of plate tectonics?

 What are the three types of plate boundaries?

 Why do tectonic plates move?

Vocabulary

plate tectonics p. 143

lithosphere p. 144

divergent plate boundary p. 145

transform plate boundary p. 145

convergent plate boundary p. 145

subduction p. 145

convection p. 148

ridge push p. 149

slab pull p. 149

 Florida NGSSS

LA.7.2.2.3 The student will organize information to show understanding (e.g., representing main ideas within text through charting, mapping, paraphrasing, summarizing, or comparing/contrasting);

MA.6.A.3.6 Construct and analyze tables, graphs, and equations to describe linear functions and other simple relations using both common language and algebraic notation.

SC.7.E.6.4 Explain and give examples of how physical evidence supports scientific theories that Earth has evolved over geologic time due to natural processes.

SC.7.E.6.5 Explore the scientific theory of plate tectonics by describing how the movement of Earth's crustal plates causes both slow and rapid changes in Earth's surface, including volcanic eruptions, earthquakes, and mountain building.

SC.7.E.6.7 Recognize that heat flow and movement of material within Earth causes earthquakes and volcanic eruptions, and creates mountains and ocean basins.

SC.7.N.1.1 Define a problem from the seventh grade curriculum, use appropriate reference materials to support scientific understanding, plan and carry out scientific investigation of various types, such as systematic observations or experiments, identify variables, collect and organize data, interpret data in charts, tables, and graphics, analyze information, make predictions, and defend conclusions.

SC.7.N.1.5 Describe the methods used in the pursuit of a scientific explanation as seen in different fields of science such as biology, geology, and physics.

SC.7.N.1.6 Explain that empirical evidence is the cumulative body of observations of a natural phenomenon on which scientific explanations are based.

 SC.7.E.6.5

Inquiry Launch Lab

15 minutes

Can you determine density by observing buoyancy?

Density is the measure of an object's mass relative to its volume. Buoyancy is the upward force a liquid places on objects that are immersed in it. If you immerse objects with equal densities into liquids that have different densities, the buoyant forces will be different. An object will sink or float depending on the density of the liquid compared to the object. Earth's layers differ in density. These layers float or sink depending on density and buoyant force.

Procedure

1. Read and complete a lab safety form.

2. Obtain four **test tubes.** Place them in a **test-tube rack.** Add **water** to one test tube until it is ¾ full.

3. Repeat with the other test tubes using **vegetable oil** and **glucose syrup.** One test tube should remain empty.

4. Drop **beads of equal density into** each test tube. Observe what the object does when immersed in each liquid. Record your observations.

Data and Observations

Think About This

1. How did you determine which liquid has the highest density?

2. **Key Concept** What happens when layers of rock with different densities collide?

The Plate Tectonics Theory

When you blow into a balloon, the balloon expands and its surface area increases. Similarly, if oceanic crust continues to form at mid-ocean ridges and is never destroyed, Earth's surface area should increase. However, this is not the case. The older crust must be destroyed somewhere—but where?

By the late 1960s a more complete theory, called plate tectonics, was proposed. The theory of **plate tectonics** states that *Earth's surface is made of rigid slabs of rock, or plates, that move with respect to each other.* This new theory suggested that Earth's surface is divided into large plates of rigid rock. Each plate moves over Earth's hot and semi-plastic mantle.

 2. NGSSS Check **Define** What is plate tectonics? **SC.7.E.6.5**

_____

_____

Geologists use the word *tectonic* to describe the forces that shape Earth's surface and the rock structures that form as a result. Plate tectonics provides an explanation for the occurrence of earthquakes and volcanic eruptions. When plates separate on the seafloor, earthquakes result and a mid-ocean ridge forms. When plates come together, one plate can dive under the other, causing earthquakes and creating a chain of volcanoes. When plates slide past each other, earthquakes can result.

Inquiry **How did these islands form?**

1. The photograph shows a chain of active volcanoes. These volcanoes make up the Aleutian Islands of Alaska. Just south of these volcanic islands is a 6 km-deep ocean trench. Why did these volcanic mountains form in a line? Can you predict where volcanoes are? Are they related to plate tectonics?

_____

_____

_____

_____

_____

_____

_____

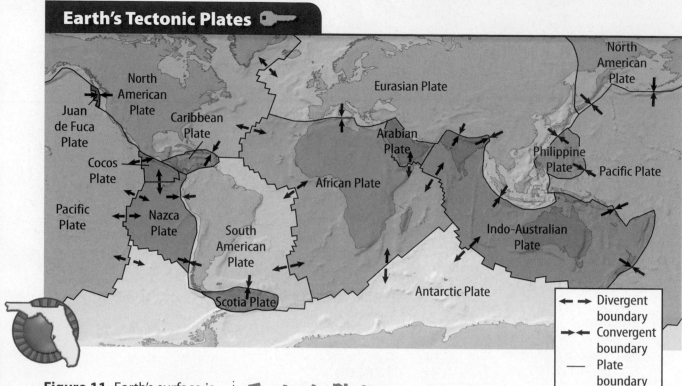

Earth's Tectonic Plates 🔑

North American Plate

Juan de Fuca Plate

Cocos Plate

Pacific Plate

Nazca Plate

Caribbean Plate

Eurasian Plate

Arabian Plate

African Plate

South American Plate

Scotia Plate

Philippine Plate

Pacific Plate

Indo-Australian Plate

Antarctic Plate

North American Plate

◄—— ——► Divergent boundary

——►◄—— Convergent boundary

—— Plate boundary

Figure 11 Earth's surface is broken into large plates that fit together like pieces of a giant jigsaw puzzle. The arrows show the general direction of movement of each plate.

Active Reading **3. Locate** (Circle) the name of the tectonic plate that contains the state of Florida.

SCIENCE USE V. COMMON USE

plastic
Science Use capable of being molded or changing shape without breaking

Common Use any of numerous organic, synthetic, or processed materials made into objects

Tectonic Plates

You read on the previous page that the theory of plate tectonics states that Earth's surface is divided into rigid plates that move relative to one another. These plates are "floating" on top of a hot and semi-plastic mantle. The map in **Figure 11** illustrates Earth's major plates and the boundaries that define them. The Pacific Plate is the largest plate. The Juan de Fuca Plate is one of the smallest plates. It is between the North American and Pacific Plates. Notice the boundaries that run through the oceans. Many of these boundaries mark the positions of the mid-ocean ridges.

Earth's outermost layers are cold and rigid compared to the layers within Earth's interior. *The cold and rigid outermost rock layer is called the* **lithosphere.** It is made up of the crust and the solid, uppermost mantle. The lithosphere is thin below mid-ocean ridges and thick below continents. Earth's tectonic plates are large pieces of lithosphere. These lithospheric plates fit together like the pieces of a giant jigsaw puzzle.

The layer of Earth below the lithosphere is called the asthenosphere (as THEE nuh sfihr). This layer is so hot that it behaves like a **plastic** material. This enables Earth's plates to move because the hotter, plastic mantle material beneath them can flow. The interactions between lithosphere and asthenosphere help to explain plate tectonics.

Active Reading **4. Recognize** What are Earth's outermost layers called?

Plate Boundaries

Place two books side by side and imagine each book represents a tectonic plate. A plate boundary exists where the books meet. How many different ways can you move the books with respect to each other? You can pull the books apart, you can push the books together, and you can slide the books past one another. Earth's tectonic plates move in much the same way.

Divergent Plate Boundaries

Mid-ocean ridges are located along divergent plate boundaries. *A **divergent plate boundary** forms where two plates separate.* When the seafloor spreads at a mid-ocean ridge, lava erupts, cools, and forms new oceanic crust. Divergent plate boundaries can also exist in the middle of a continent. They pull continents apart and form rift valleys. The East African Rift is an example of a continental rift.

Transform Plate Boundaries

The famous San Andreas Fault in California is an example of a transform plate boundary. *A **transform plate boundary** forms where two plates slide past each other.* As they move past each other, the plates can get stuck and stop moving. Stress builds up where the plates are "stuck." Eventually, the stress is too great and the rocks break, suddenly moving apart. This results in a rapid release of energy as earthquakes.

Convergent Plate Boundaries

***Convergent plate boundaries** form where two plates collide. The denser plate sinks below the more buoyant plate in a process called* **subduction.** The area where a denser plate descends into Earth along a convergent plate boundary is called a **subduction** zone.

When an oceanic plate and a continental plate collide, the denser oceanic plate subducts under the edge of the continent. This creates a deep ocean trench. A line of volcanoes forms above the subducting plate on the edge of the continent. This process can also occur when two oceanic plates collide. The older and denser oceanic plate will subduct beneath the younger oceanic plate. This creates a deep ocean trench and a line of volcanoes called an island arc.

When two continental plates collide, neither plate is subducted, and mountains such as the Himalayas in southern Asia form from uplifted rock. **Table 1** on the next page summarizes the interactions of Earth's tectonic plates.

5. NGSSS Check Identify <u>Underline</u> the three types of plate boundaries? **SC.7.E.6.5**

Active Reading

FOLDABLES® LA.7.2.2.3

Make a layered book using two sheets of notebook paper. Use it to organize information about the different types of plate boundaries and the features that form there.

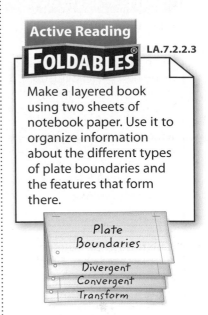

WORD ORIGIN

subduction
from Latin *subductus*, means "to lead under, removal"

Table 1 The direction of motion of Earth's plates creates a variety of features at the boundaries between the plates.

Active Reading **6. Illustrate** Draw arrows on each image in the center column to illustrate the motion of the plates. Use clues from the text in the first column to guide you.

Table 1 Interactions of Earth's Tectonic Plates

Plate Boundary	Relative Motion	Example
Divergent plate boundary When two plates separate and create new oceanic crust, a divergent plate boundary forms. This process occurs where the seafloor spreads along a mid-ocean ridge, as shown to the right. This process can also occur in the middle of continents and is referred to as a continental rifting.	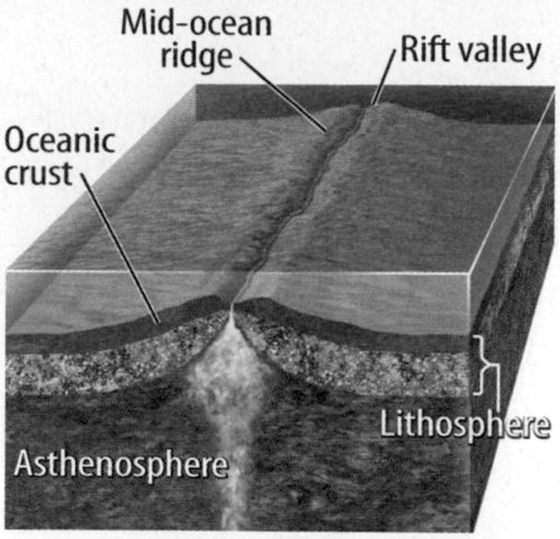	
Transform plate boundary Two plates slide horizontally past one another along a transform plate boundary. Earthquakes are common along this type of plate boundary. The San Andreas Fault, shown to the right, is part of the transform plate boundary that extends along the coast of California.	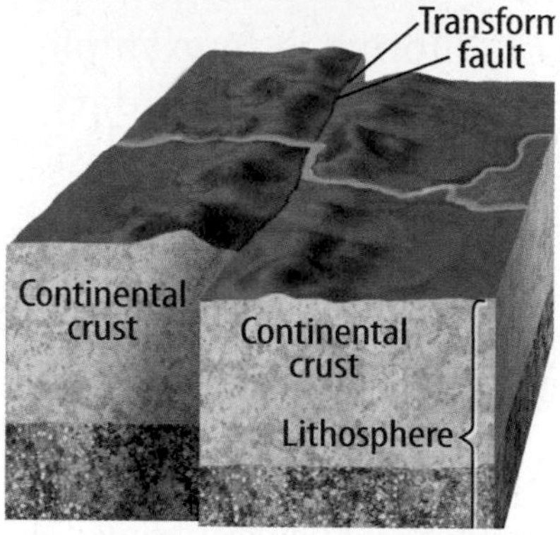	
Convergent plate boundary (ocean-to-continent) When an oceanic and a continental plate collide, they form a convergent plate boundary. The denser plate will subduct. A volcanic mountain, such as Mount Rainier in the Cascade Mountains, forms along the edge of the continent. This process can also occur where two oceanic plates collide, and the denser plate is subducted.	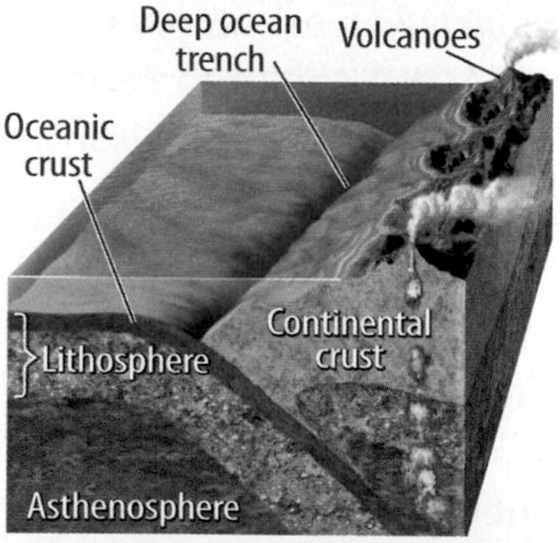	
Convergent plate boundary (continent-to-continent) Convergent plate boundaries can also occur where two continental plates collide. Because both plates are equally dense, neither plate will subduct. Both plates uplift and deform. This creates huge mountains like the Himalayas, shown to the right.	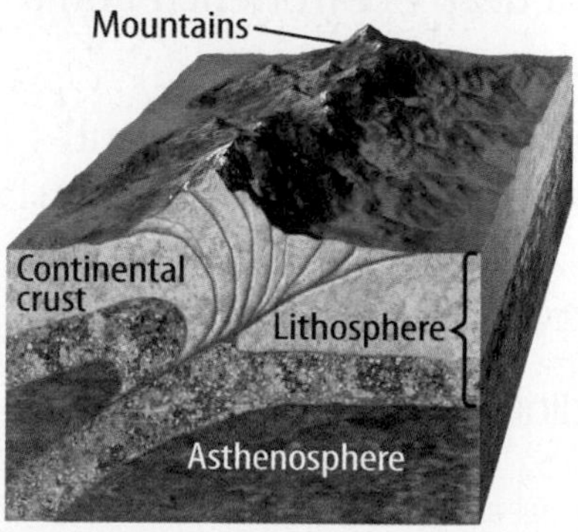	

Evidence for Plate Tectonics

When Wegener proposed the continental drift hypothesis, the technology used to measure how fast the continents move today wasn't yet available. Recall that continents move apart or come together at speeds of a few centimeters per year. This is about the length of a small paperclip.

Today, scientists can measure how fast continents move. A network of satellites orbiting Earth monitors plate motion. By keeping track of the distance between these satellites and Earth, it is possible to locate and determine how fast a tectonic plate moves. This network of satellites is called the Global Positioning System (GPS).

The theory of plate tectonics also provides an explanation for why earthquakes and volcanoes occur in certain places. Because plates are rigid, tectonic activity occurs where plates meet. When plates separate, collide, or slide past each other along a plate boundary, stress builds. A rapid release of energy can result in earthquakes. Volcanoes form where plates separate along a mid-ocean ridge or a continental rift or collide along a subduction zone. Mountains can form where two continents collide. **Figure 12** illustrates the relationship between plate boundaries and the occurrence of earthquakes and volcanoes. Refer back to the lesson opener photo. Find these islands on the map. Are they located near a plate boundary?

 7. NGSSS Check Identify Underline how earthquakes and volcanoes are related to the theory of plate tectonics? **SC.7.E.6.5 SC.7.E.6.7**

Figure 12 Notice that most earthquakes and volcanoes occur near plate boundaries.

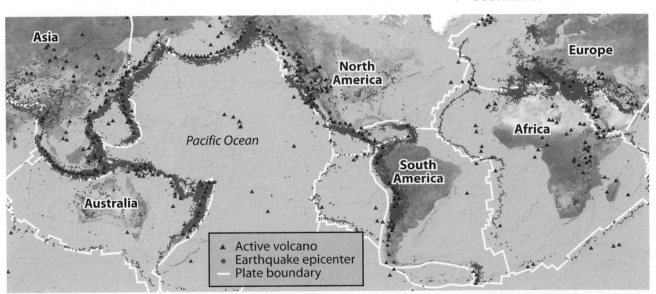

Asia

North America

Europe

Africa

Pacific Ocean

South America

Australia

▲ Active volcano
● Earthquake epicenter
— Plate boundary

✔ **8. Visual Check Locate** (Circle) any areas where earthquakes and volcanoes occur away from plate boundaries.

Figure 13 When water is heated, it expands. Less dense heated water rises because the colder water sinks, forming convection currents.

Plate Motion

The main objection to Wegener's continental drift hypothesis was that he could not explain why or how continents move. Scientists now understand that continents move because the asthenosphere moves underneath the lithosphere.

Convection Currents

You are probably already familiar with the process of **convection,** *the circulation of material caused by differences in temperature and density.* For example, the upstairs floors of homes and buildings are often warmer. This is because hot air rises while dense, cold air sinks. Look at **Figure 13** to see convection in action.

 9. Identify <u>Underline</u> the cause of convection?

Plate tectonic activity is related to convection in the mantle, as shown in **Figure 14.** Radioactive elements, such as uranium, thorium, and potassium, heat Earth's interior. When materials such as solid rock are heated, they expand and become less dense. Hot mantle material rises upward and comes in contact with Earth's crust. Thermal energy is transferred from hot mantle material to the colder surface above. As the mantle cools, it becomes denser and then sinks, forming a convection current. These currents in the asthenosphere act like a conveyor belt moving the lithosphere above.

 10. NGSSS Check Summarize Why do tectonic plates move?
SC.7.E.6.7

_____

_____

_____

_____

Inquiry

iLAB STATION

SC.7.N.1.1, SC.7.E.6.5

Try It!

MiniLab *How do changes in density cause motion?* at <u>connectED.mcgraw-hill.com</u>

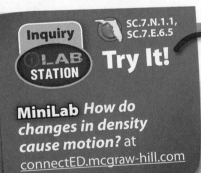

Apply It! After you complete the lab, answer these questions.

1. **Predict** Would this experiment work with a larger, heavier prune instead of a raisin? Why or why not?

2. **Infer** What would happen if the magma in Earth's interior became more dense when it was heated instead of less dense?

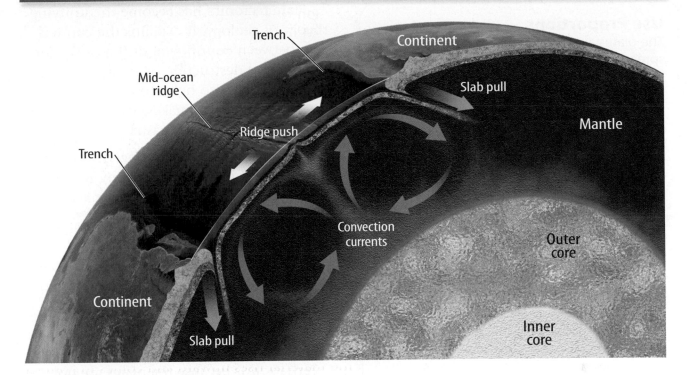

Trench
Mid-ocean ridge
Ridge push
Trench
Continent
Slab pull
Continent
Slab pull
Convection currents
Mantle
Outer core
Inner core

Forces Causing Plate Motion

How can something as massive as the Pacific Plate move? **Figure 14** shows the three forces that interact to cause plate motion. Scientists still debate over which of these forces has the greatest effect on plate motion.

Basal Drag Convection currents in the mantle produce a force that causes motion called basal drag. Notice in **Figure 14** how convection currents in the asthenosphere circulate and drag the lithosphere similar to the way a conveyor belt moves items along at a supermarket checkout.

Ridge Push Recall that mid-ocean ridges have greater elevation than the surrounding seafloor. Because mid-ocean ridges are higher, gravity pulls the surrounding rocks down and away from the ridge. *Rising mantle material at mid-ocean ridges creates the potential for plates to move away from the ridge with a force called* **ridge push.** Ridge push moves lithosphere in opposite directions away from the mid-ocean ridge.

Slab Pull As you read earlier in this lesson, when tectonic plates collide, the denser plate will sink into the mantle along a subduction zone. This plate is called a slab. Because the slab is old and cold, it is denser than the surrounding mantle and will sink. *As a slab sinks, it pulls on the rest of the plate with a force called* **slab pull.** Scientists are still uncertain about which force has the greatest influence on plate motion.

Figure 14 Convection occurs in the mantle underneath Earth's tectonic plates. Three forces act on plates to make them move: basal drag from convection currents, ridge push at mid-ocean ridges, and slab pull from subducting plates.

✔ 11. Visual Check
Describe What is happening to a plate that is undergoing slab pull?

Use Proportions

The plates along the Mid-Atlantic Ridge spread at an average rate of 2.5 cm/y. How long will it take the plates to spread 1 m? Use proportions to find the answer.

1 **Convert the distance to the same unit.**

$$1\ m = 100\ cm$$

2 **Set up a proportion:**

$$\frac{2.5\ cm}{1\ y} = \frac{100\ cm}{x\ y}$$

3 **Cross multiply and solve for x as follows:**

$$2.5\ cm \times x\ y = 100\ cm \times 1\ y$$

4 **Divide both sides by 2.5 cm.**

$$x = \frac{100\ cm\ y}{2.5\ cm}$$

$$x = 40\ y$$

Practice

12. The Eurasian plate travels the slowest, at about 0.7 cm/y. How long would it take the plate to travel 3 m? (1 m = 100 cm)

Figure 15 Seismic waves were used to produce this tomography scan. These colors show a subducting plate. The blue colors represent rigid materials with faster seismic wave velocities.

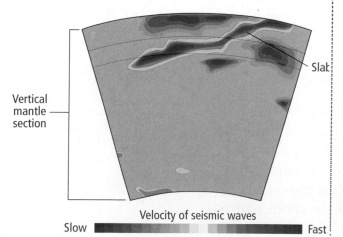

Vertical mantle section

Slab

Velocity of seismic waves

Slow Fast

A Theory in Progress

Plate tectonics has become the unifying theory of geology. It explains the connection between continental drift and the formation and destruction of crust along plate boundaries. It also helps to explain the occurrence of earthquakes, volcanoes, mountains, and ocean basins.

The investigation that Wegener began nearly a century ago is still being revised. Several unanswered questions remain.

• Is Earth the only planet in the solar system that has plate tectonic activity? Different hypotheses have been proposed to answer this. Extrasolar planets outside our solar system are also being studied.

• Why do some earthquakes and volcanoes occur far away from plate boundaries? Perhaps it is because the plates are not perfectly rigid. Different thicknesses and weaknesses exist within the plates. Also, the mantle is much more active than scientists originally understood.

• What forces dominate plate motion? Currently accepted models suggest that convection currents occur in the mantle. However, presently there is no way to measure or observe them.

• What will scientists investigate next? **Figure 15** shows an image produced by a new technique called anisotropy that creates a 3-D image of seismic wave velocities in a subduction zone. This developing technology might help scientists better understand the processes that occur within the mantle and along plate boundaries.

Active Reading   **13. Infer** Why does the theory of plate tectonics continue to change?

_____

_____

_____

_____

Visual Summary

Tectonic plates are made of cold and rigid slabs of rock.

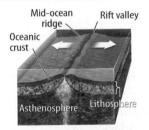

Mantle convection— the circulation of mantle material due to density differences—drives plate motion.

The three types of plate boundaries are divergent, convergent, and transform boundaries.

SC.7.N.1.1, SC.7.E.1.5-6, SC.7.E.6.4

Inquiry LAB STATION **Try It!**

Skill Lab *Movement of Plate Boundaries* at connectED.mcgraw-hill.com

Use Vocabulary

1 The theory that proposes that Earth's surface is broken into moving, rigid plates is called _____. **SC.7.E.6.5**

Understand Key Concepts

2 **Compare and contrast** the geological activity that occurs along the three types of plate boundaries. **SC.7.E.6.5**

_____

_____

_____

3 **Explain** why mantle convection occurs. **SC.7.E.6.7**

_____

_____

4 Tectonic plates move because of **SC.7.E.6.7**
- (A) convection currents.
- (B) Earth's increasing size.
- (C) magnetic reversals.
- (D) volcanic activity.

Interpret Graphics

5 **Determine Cause and Effect** Fill in the graphic organizer below to list the cause and effects of convection currents. **SC.7.E.6.7**

Critical Thinking

6 **Explain** why earthquakes occur at greater depths along convergent plate boundaries. **SC.7.E.6.5**

_____

_____

_____

Math Skills **MA.6.A.3.6**

7 Two plates in the South Pacific separate at an average rate of 15 cm/y. How far will they have separated after 5,000 years?

Think About It! The scientific theory of plate tectonics states that Earth's lithosphere is broken up into rigid plates that move over Earth's surface causing slow and rapid changes.

🔑 Key Concepts Summary

LESSON 1 The Continental Drift Hypothesis

- The puzzle piece fit of continents, fossil evidence, climate, rocks, and mountain ranges supports the hypothesis of **continental drift.**
- Scientists were skeptical of continental drift because Wegener could not explain the mechanism for movement.

Pangaea p. 127
continental drift p. 127

LESSON 2 Development of a Theory

- **Seafloor spreading** provides a mechanism for continental drift.
- Seafloor spreading occurs at **mid-ocean ridges.**
- Evidence of **magnetic reversal** in rock, thermal energy trends, and the discovery of seafloor spreading all contributed to the development of the theory of plate tectonics.

mid-ocean ridge p. 135
seafloor spreading p. 136
normal polarity p. 138
magnetic reversal p. 138
reversed polarity p. 138

LESSON 3 The Theory of Plate Tectonics

- Types of plate boundaries, the location of earthquakes, volcanoes, and mountain ranges, and satellite measurement of plate motion support the theory of **plate tectonics.**
- Mantle **convection, ridge push,** and **slab pull** are the forces that cause plate motion. Radioactivity in the mantle and thermal energy from the core produce the energy for convection.

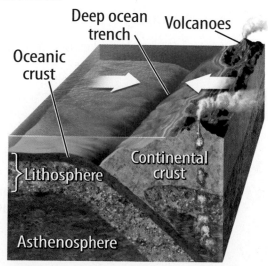

Deep ocean trench

Volcanoes

Oceanic crust

Continental crust

Lithosphere

Asthenosphere

plate tectonics p. 143
lithosphere p. 144
divergent plate boundary p. 145
transform plate boundary p. 145
convergent plate boundary p. 145
subduction p. 145
convection p. 148
ridge push p. 149
slab pull p. 149

Active Reading

FOLDABLES® Chapter Project

Assemble your lesson Foldables as shown to make a Chapter Project. Use the project to review what you have learned in this chapter.

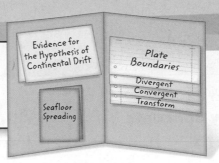

Evidence for the Hypothesis of Continental Drift

Seafloor Spreading

Plate Boundaries
- Divergent
- Convergent
- Transform

Use Vocabulary

1 The process in which hot mantle rises and cold mantle sinks is called _____.

2 What is the plate tectonics theory?

_____

_____

_____

3 What was Pangaea?

_____

_____

4 Identify the three types of plate boundaries and the relative motion associated with each type.

_____

_____

_____

_____

_____

5 Magnetic reversals occur when_____.

6 Explain seafloor spreading in your own words.

_____

_____

_____

Link Vocabulary and Key Concepts

Use vocabulary terms from the previous page to complete the concept map.

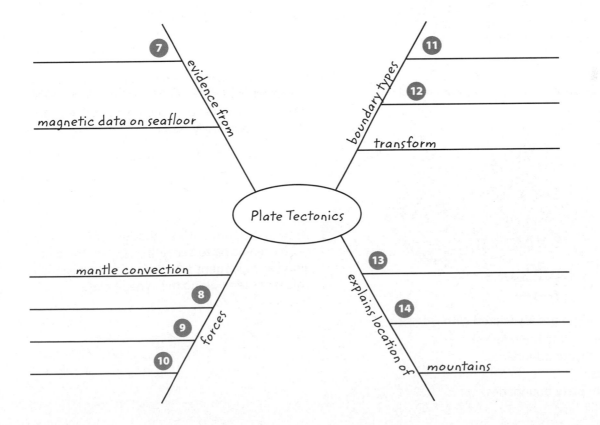

evidence from

7 _____

magnetic data on seafloor

boundary types

11 _____

12 _____

transform

Plate Tectonics

forces

mantle convection

8 _____

9 _____

10 _____

explains location of

13 _____

14 _____

mountains

Fill in the correct answer choice.

🔑 Understand Key Concepts

1 Alfred Wegener proposed the _____ hypothesis. SC.7.E.6.5
- (A) continental drift
- (B) plate tectonics
- (C) ridge push
- (D) seafloor spreading

2 Ocean crust is SC.7.E.6.5
- (A) made from submerged continents.
- (B) magnetically produced crust.
- (C) produced at the mid-ocean ridge.
- (D) produced at all plate boundaries.

3 What technologies did scientists NOT use to develop the theory of seafloor spreading? SC.7.N.1.6
- (A) echo-sounding measurements
- (B) GPS (global positioning system)
- (C) magnetometer measurements
- (D) seafloor thickness measurements

4 The picture below shows Pangaea's position on Earth approximately 280 million years ago. Where did geologists discover glacial features associated with a cooler climate? SC.7.E.6.4
- (A) Antarctica
- (B) Asia
- (C) North America
- (D) South America

Pangaea

5 Mid-ocean ridges are associated with SC.7.E.6.5
- (A) convergent plate boundaries.
- (B) divergent plate boundaries.
- (C) hot spots.
- (D) transform plate boundaries.

Critical Thinking

6 **Evaluate** The oldest seafloor in the Atlantic Ocean is located closest to the edge of continents, as shown in the image below. Explain how this age can be used to figure out when North America first began to separate from Europe. SC.7.E.6.5

_____

_____

_____

7 **Examine** the evidence used to develop the theory of plate tectonics. How has new technology strengthened the theory? SC.7.N.1.5

_____

_____

_____

8 **Explain** Sediments deposited by glaciers in Africa are surprising because Africa is now warm. How does the hypothesis of continental drift explain these deposits? SC.7.E.6.4

_____

_____

_____

9 **Draw** a diagram to show subduction of an oceanic plate beneath a continental plate along a convergent plate boundary. Explain why volcanoes form along this type of plate boundary. SC.7.E.6.5

10 **Infer** Warm peanut butter is easier to spread than cold peanut butter. How does knowing this help you understand why the mantle is able to deform in a plastic manner? **SC.7.E.6.7**

Writing in Science

11 **Predict** If continents continue to move in the same direction over the next 200 million years, how might the appearance of landmasses change? On another sheet of paper write a paragraph to explain the possible positions of landmasses in the future. Based on your understanding of the plate tectonic theory, is it possible that new supercontinents will form in the future? **SC.7.E.6.7**

Big Idea Review

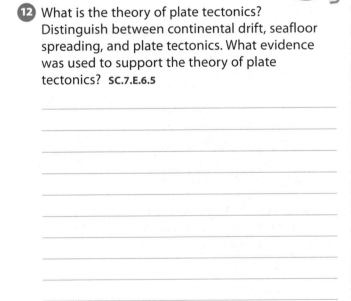

12 What is the theory of plate tectonics? Distinguish between continental drift, seafloor spreading, and plate tectonics. What evidence was used to support the theory of plate tectonics? **SC.7.E.6.5**

13 Use the image below to interpret how the theory of plate tectonics helps to explain formation of huge mountains like the Himalaya. **SC.7.E.6.5**

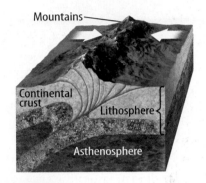

Mountains

Continental crust

Lithosphere

Asthenosphere

Math Skills MA.6.A.3.6

Use Proportions

14 Mountains on a convergent plate boundary may grow at a rate of 3 mm/y. How long would it take a mountain to grow to a height of 3,000 m? (1 m = 1,000 mm)

15 The North American Plate and the Pacific Plate have been sliding horizontally past each other along the San Andreas fault zone for about 10 million years. The plates move at an average rate of about 5 cm/y.
a. How far have the plates traveled, assuming a constant rate, during this time?

b. How far has the plate traveled in kilometers? (1 km = 100,000 cm)

Fill in the correct answer choice.

Multiple Choice

Use the diagram below to answer questions 1 and 2.

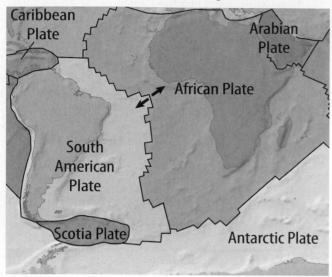

Caribbean Plate

Arabian Plate

African Plate

South American Plate

Scotia Plate

Antarctic Plate

1 In the diagram above, what does the irregular line between tectonic plates represent? **SC.7.E.6.5**

  (A) abyssal plain

  (B) island chain

  (C) mid-ocean ridge

  (D) polar axis

2 What do the arrows indicate? **SC.7.E.6.6**

  (F) magnetic polarity

  (G) ocean flow

  (H) plate movement

  (I) volcanic eruption

3 What evidence helped to support the theory of seafloor spreading? **SC.7.E.6.5**

  (A) magnetic equality

  (B) magnetic interference

  (C) magnetic north

  (D) magnetic polarity

4 Which plate tectonic process creates a deep ocean trench? **SC.7.E.6.5**

  (F) conduction

  (G) deduction

  (H) induction

  (I) subduction

5 What causes plate motion? **SC.7.E.6.7**

  (A) convection in Earth's mantle

  (B) currents in Earth's oceans

  (C) reversal of Earth's polarity

  (D) rotation on Earth's axis

6 Ocean basins will expand as new oceanic crust forms and moves away from a mid-ocean ridge during **SC.7.E.6.7**

  (F) continental drift.

  (G) magnetic reversal.

  (H) normal polarity.

  (I) seafloor spreading.

Use the diagram below to answer question 7.

Equator

Europe

North America

South America

Africa

Asia

India

Australia

Antarctica

Ice mass

7 What is the name of Alfred Wegener's ancient supercontinent pictured in the diagram above? **SC.7.E.6.5**

  (A) Caledonia

  (B) continental drift

  (C) *Glossopteris*

  (D) Pangaea

Use the diagram below to answer question 8.

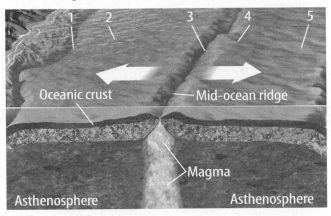

8 The numbers in the diagram represent sea-floor rock. Which represent the oldest rock? **SC.7.E.6.5**

(F) 1 and 5

(G) 2 and 4

(H) 3 and 4

(I) 4 and 5

9 Which part of the ocean basin contains the thickest sediment layer? **SC.7.E.6.2**

(A) abyssal plain

(B) deposition band

(C) mid-ocean ridge

(D) tectonic zone

10 What type of rock forms when lava cools and crystallizes on the seafloor? **SC.7.E.6.2**

(F) a fossil

(G) a glacier

(H) basalt

(I) magma

11 What is the dominant type of rock formed at mid-ocean ridges? **SC.7.E.6.2**

(A) granite

(B) basalt

(C) sediment

(D) sedimentary rock

Use the diagram below to answer questions 12 and 13.

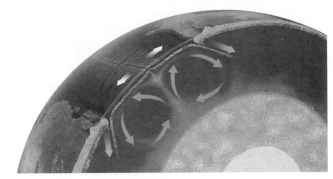

12 According to the theory of plate tectonics, what is the source of the force that causes the plates to move? **SC.7.E.6.7**

(F) convection currents

(G) magnetic reversal

(H) normal polarity

(I) ridge push

13 What is the energy source that powers plate movement? **SC.7.E.6.7**

(A) slab pull

(B) less dense continents

(C) magma and lava

(D) Earth's hot interior

NEED EXTRA HELP?

If You Missed Question...	1	2	3	4	5	6	7	8	9	10	11	12	13
Go to Lesson...	3	3	2	3	3	2	1	2	2	2	3	3	1

Benchmark Mini-Assessment   **Chapter 4 • Lesson 1**   mini BAT

Multiple Choice *Bubble the correct answer.*

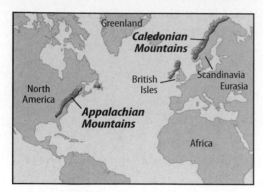

1. In the image above, what suggests that North America and Europe once were connected? **SC.7.E.6.4**

   (A) The continents have the same fossils.

   (B) The continents share the same year-round climate.

   (C) The Appalachian and Caledonian mountain ranges connect.

   (D) The rocks from the British Isles also exist in North America.

2. One continental portion of Pangaea became part of Asia. Which subcontinent joined Asia during the breakup of Pangaea? **SC.7.E.6.4**

   (F) Africa

   (G) Australia

   (H) Europe

   (I) India

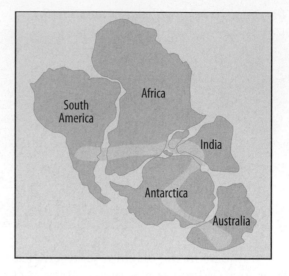

3. What is happening in the image above? **SC.7.E.6.5**

   (A) climate changes

   (B) continental drift

   (C) fossil formation

   (D) glacial movement

4. According to Wegener's theory, in which direction did the North American continental mass move? **SC.7.E.6.4**

   (F) northeast

   (G) northwest

   (H) due north

   (I) due west

Copyright © Glencoe/McGraw-Hill, a division of The McGraw-Hill Companies, Inc.

Benchmark Mini-Assessment   Chapter 4 • Lesson 2

Multiple Choice *Bubble the correct answer.*

Use the illustration below to answer questions 1–3.

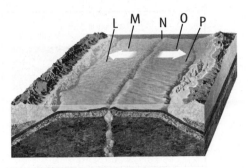

1. At which point in the ocean-floor diagram above will the oldest rock be found? **SC.7.E.6.3**

   (A) L

   (B) M

   (C) N

   (D) O

2. At what point will sediment have made the seafloor smoothest? **SC.7.E.6.7**

   (F) L

   (G) N

   (H) O

   (I) P

3. At which point is the lithosphere the thinnest? **SC.7.E.6.4**

   (A) M

   (B) N

   (C) O

   (D) P

4. What evidence was found to support the theory of seafloor spreading? **SC.7.E.6.5**

   (F) Earth's magnetic field has reversed direction many times.

   (G) Earth's magnetic field has stayed the same for hundreds of years.

   (H) More heat is released from beneath the abyssal plains than near mid-ocean ridges.

   (I) The iron-rich minerals found on the seafloor have proved not to be magnetic.

Copyright © Glencoe/McGraw-Hill, a division of The McGraw-Hill Companies, Inc.

Multiple Choice *Bubble the correct answer.*

Use the illustration below to answer questions 1 and 2.

1. At the boundaries of which plates would you expect to find volcanoes? **SC.7.E.6.5**

- (A) African and Arabian Plates
- (B) Eurasian and Indo-Australian Plates
- (C) North American and Caribbean Plates
- (D) Pacific and Nazca Plates

2. At the boundaries of which plates would you expect to find a rift? **SC.7.E.6.5**

- (F) African and South American Plates
- (G) Eurasian and African Plates
- (H) Eurasian and Indo-Australian Plates
- (I) Nazca and South American Plates

3. Which image shows a transform plate boundary? **SC.7.E.6.5**

(A)
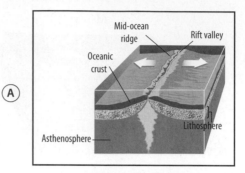

(B)
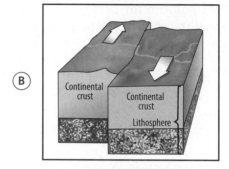

(C)
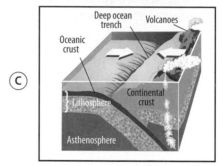

(D)
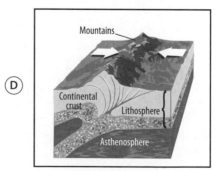

Copyright © Glencoe/McGraw-Hill, a division of The McGraw-Hill Companies, Inc.

Notes

Notes

Mountain Size

Three mountain climbers looked at Mount Everest, one of the tallest mountains in the world, and wondered how old it was. This is what they said:

Leo: I think it is one of the oldest mountains on Earth.

Ina: I think it is one of the younger mountains on Earth.

Pete: I think all the mountains on Earth are the same age..

Who do you most agree with? Explain why you agree.

Earth
DYNAMICS

Think About It!

How is Earth's surface shaped by plate motion?

You might think that seashells are found only near oceans. But some of the rocks in Mount Everest contain seashells from the ocean floor!

1 How do you think seashells got to the top of Mount Everest?

_____

_____

_____

2 What do you think are the different ways tectonic plates move to make mountains such as these—or deep-sea trenches, valleys, and plateaus?

_____

_____

3 How do you think Earth's surface is shaped by plate motion?

_____

_____

_____

Get Ready to Read

What do you think about the dynamic Earth?

	AGREE	DISAGREE
1 Forces created by plate motion are small and do not deform or break rocks.	☐	☐
2 Plate motion causes only horizontal motion of continents.	☐	☐
3 New landforms are created only at plate boundaries.	☐	☐
4 The tallest and deepest landforms are created at plate boundaries.	☐	☐
5 Metamorphic rocks formed deep below Earth's surface sometimes can be located near the tops of mountains.	☐	☐
6 Mountain ranges can form over long periods of time through repeated collisions between plates.	☐	☐
7 The centers of continents are flat and old.	☐	☐
8 Continents are continually shrinking because of erosion.	☐	☐

 **Connect ED**

There's More Online!
Video • Audio • Review • ⓘLab Station • WebQuest • Assessment • Concepts in Motion • Multilingual eGlossary   **165**

Forces That Shape EARTH

ESSENTIAL QUESTIONS

 How do continents move?

 What forces can change rocks?

How does plate motion affect the rock cycle?

Vocabulary

isostasy p. 168

subsidence p. 169

uplift p. 169

compression p. 169

tension p. 169

shear p. 169

strain p. 170

Florida NGSSS

LA.7.2.2.3 The student will organize information to show understanding (e.g., representing main ideas within text through charting, mapping, paraphrasing, summarizing, or comparing/contrasting);

SC.7.E.6.2 Identify the patterns within the rock cycle and relate them to surface events (weathering and erosion) and sub-surface events (plate tectonics and mountain building).

SC.7.N.1.1 Define a problem from the seventh grade curriculum, use appropriate reference materials to support scientific understanding, plan and carry out scientific investigation of various types, such as systematic observations or experiments, identify variables, collect and organize data, interpret data in charts, tables, and graphics, analyze information, make predictions, and defend conclusions.

SC.7.N.1.3 Distinguish between an experiment (which must involve the identification and control of variables) and other forms of scientific investigation and explain that not all scientific knowledge is derived from experimentation.

 Launch Lab
10 minutes

SC.7.N.1.1

Do rocks bend?

As Earth's continents move, rocks get smashed between them and bend or break. Land can take on different shapes, depending on the temperature and composition of the rocks and the size and direction of the force.

Procedure

1. Read and complete a lab safety form.
2. Spread out a **paper towel** on your work area, and place an unwrapped **candy bar** on the paper towel.
3. Gently pull on the edges of your candy bar. Observe any changes to the candy bar. Draw your observations below.
4. Reassemble your candy bar and gently squeeze the two ends of your candy bar together. Draw your observations.

Data and Observations

Think About This

1. How are the results of pulling and pushing different?

2. What would be different if the candy bar were warm? What if it were cold?

3. **Key Concept** What kinds of forces do you think can change rocks?

Inquiry **Can rocks talk?**

1. This campsite located in Thingvellir, Iceland, is in a grassy valley next to a rocky cliff. How might these geologic features occur so close to each other? What might be responsible for their formation?

_____

_____

_____

_____

_____

_____

Plate Motion

How far is your school from the nearest large mountain? If you live in the west or northeastern part of the United States, you are probably close to mountains. In contrast, most of the central region of the United States has no mountains. Why are these regions so different?

The Rocky Mountains in the west are high and have sharp peaks, but the Appalachian Mountains in the east are lower and gently rounded, as shown in **Figure 1.**

Mountains do not last forever. Weathering and erosion gradually wear them down. The Appalachian Mountains are shorter and smoother than the Rocky Mountains because they are older. They formed hundreds of millions of years ago. The Rockies formed just 50 to 100 million years ago.

Mountain ranges are produced by plate tectonics. The theory of plate tectonics states that Earth's surface is broken into rigid plates that move horizontally on Earth's more fluid upper mantle. Mountains and valleys form where plates collide, move away from each other, or slide past each other.

Active Reading 2. **Contrast** How are the Rocky Mountains different from the Appalachian Mountains?

_____

_____

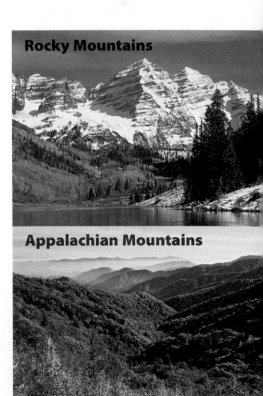

Rocky Mountains

Appalachian Mountains

Figure 1 The Rocky Mountains are located on the west coast of the United States, while the Appalachian Mountains can be found on the east coast.

Hot Spots!

Volcanoes on a Plate

Not all volcanoes form at plate boundaries. Some, called hot spot volcanoes, pop up in the middle of a tectonic plate. A hot spot volcano forms over a rising column of magma called a mantle plume. The origin of mantle plumes is still uncertain, but evidence shows they probably rise up from the boundary between Earth's mantle and core.

As a tectonic plate passes over a mantle plume, a volcano forms above the plume. The tectonic plate continues to move, and a chain of volcanoes forms. If the volcanoes are in the ocean and if they get large enough, they become islands, such as the Hawaiian Islands. Here is how this happens:

4 The oldest islands are farthest from the plume.

3 As the Pacific Plate moves, the islands formed by the hot spot are carried with it and away from the magma plume.

Direction of Pacific Plate motion

Hawaiian Ridge

2 The seamount continues to grow until it rises above the water and becomes an island.

Hawaii

1 Magma, which is less dense than the surrounding rock, rises to the seafloor and forms a seamount.

Niihau

Kauai
3.8 to 5.6 million years old

Oahu
2.2 to 3.3 million years old

Maui
less than 1.0 million years old

Molokai

Lanai

Kahoolawe

Direction of plate motion

Hawaii
started forming
0.8 million years ago.

It's Your Turn

RESEARCH Not all hot spots arise in oceans. Much of Yellowstone National Park lies inside the caldera of a gigantic volcano that sits on a hot spot. Is Yellowstone's hot spot still active?

Mountain BUILDING

 How do mountains change over time?

 How do different types of mountains form?

Vocabulary

folded mountain p. 186

fault-block mountain p. 186

uplifted mountain p. 187

 Launch Lab SC.7.N.1.1

10 minutes

What happens when Earth's tectonic plates diverge?

When tectonic plates diverge, the crust gets thinner. Sometimes large blocks of crust subside and form valleys. Blocks next to them move up and become fault-block mountains. This is how the Basin and Range Province in the western United States formed.

Procedure

1. Stand 5–6 **hardbound books** on a desk with the bindings vertical.

2. Using a **ruler,** measure the width and the height of the books, as shown. Record the results in a table below.

3. Holding the books together, tilt them sideways at the same time to about a 30° angle. Measure the width and the height of the books. Record the results in your table.

4. Tilt the books to about a 60° angle. Measure the width and the height of the books. Record the results in your table.

5. Draw a diagram of the tilted books below and label mountains, valleys, and faults.

Data and Observations

Think About This

1. How does the thickness of the crust relate to the height of a mountain?

2. **Key Concept** How do you think fault-block mountains form?

Florida NGSSS

LA.7.2.2.3 The student will organize information to show understanding (e.g., representing main ideas within text through charting, mapping, paraphrasing, summarizing, or comparing/contrasting);

MA.6.A.3.6 Construct and analyze tables, graphs, and equations to describe linear functions and other simple relations using both common language and algebraic notation.

SC.7.E.6.5 Explore the scientific theory of plate tectonics by describing how the movement of Earth's crustal plates causes both slow and rapid changes in Earth's surface, including volcanic eruptions, earthquakes, and mountain building.

SC.7.N.1.1 Define a problem from the seventh grade curriculum, use appropriate reference materials to support scientific understanding, plan and carry out scientific investigation of various types, such as systematic observations or experiments, identify variables, collect and organize data, interpret data in charts, tables, and graphics, analyze information, make predictions, and defend conclusions.

The Mountain-Building Cycle

Mountain ranges are built slowly, and they change slowly. Because they are the result of many different plate collisions over many millions of years, they are made of many different types of rocks. The processes of weathering and erosion can remove part or all of a mountain.

 Active Reading **2. Name** What process can remove part of a mountain?

_____

_____

Converging Plates

Recall that when plates collide at a plate boundary, a combination of folds, faults, and uplift creates mountains. Eventually, after millions of years, the forces that originally caused the plates to move together can become inactive. As shown in **Figure 15,** a single new continent is created from two old ones, and the plate boundary becomes inactive. With no compression at a convergent plate boundary, the mountains stop increasing in size.

(Inquiry) Is this a safe place to live?

1. Not all mountains are the same. Once you know how a mountain formed, you can predict what is likely to happen in the future. What possible dangers might these local residents face?

_____

_____

_____

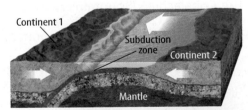

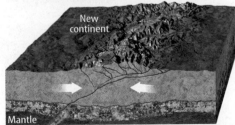

Figure 15 The forces that originally caused plates to move together eventually become inactive. A single continent is created from the two old ones, and the plate boundary becomes inactive.

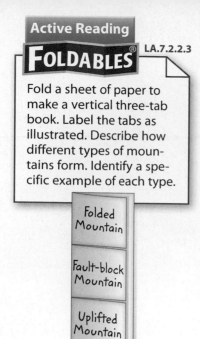

Fold a sheet of paper to make a vertical three-tab book. Label the tabs as illustrated. Describe how different types of mountains form. Identify a specific example of each type.

Folded Mountain

Fault-block Mountain

Uplifted Mountain

WORD ORIGIN

Appalachian
from the Apalachee *abalahci*, means "other side of the river"

Collisions and Rifting

Continents are continuously changing because Earth's tectonic plates are always moving. When continents split at a divergent plate boundary, they often break close to the place where they first collided. First a large split, or rift, forms. The rift grows, and seawater flows into it, forming an ocean.

Eventually plate motion changes again, and the continents collide. New mountain ranges form on top of or next to older mountain ranges. The cycle of repeated collisions and rifting can create old and complicated mountain ranges, such as the **Appalachian** Mountains.

Active Reading 3. **Identify** Where do plates tend to break apart?

_____

Figure 16 illustrates the history of the plate collisions and rifting that produced the mountain range as it is today. Rocks that make up mountain ranges such as the Appalachian Mountains record the history of plate motion and collisions that formed the mountains.

Weathering

The Appalachian Mountains are an old mountain range that stretches along most of the eastern United States. They are not as high and rugged as the Rocky Mountains in the west because they are much older. They are no longer growing. Weathering has rounded the peaks and lowered the elevations.

Formation of the Appalachians 🗝

Figure 16 The Appalachian Mountains formed over several hundred million years.

4. ✓ **Visual Check Identify** (Circle) the mountain range between Valley and Ridge and Piedmont.

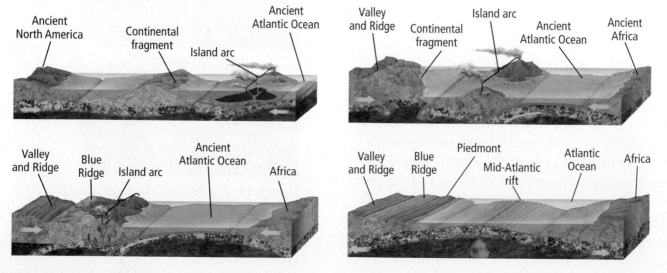

Erosion and Uplift

Over time, natural processes wear down mountains, smooth their peaks, and reduce their height. But some mountain ranges are hundreds of millions of years old. How do they last so long? Recall how isostasy works. As a mountain erodes, the crust under it must rise to restore the balance between what is left of the mountain and how it floats on the mantle. Therefore, rocks deep under continents rise slowly toward Earth's surface. In old mountain ranges, metamorphic rocks that formed deep below the surface are exposed on the top of mountains, such as the rocks in **Figure 17.**

 5. NGSSS Check Describe How can mountains change over time? **SC.7.E.6.2**

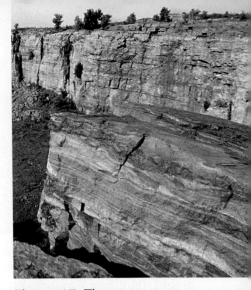

Figure 17 These metamorphic rocks formed deep below Earth's surface. After the material above them eroded, the rock rose due to isostasy. Now they are on Earth's surface.

Types of Mountains

You learned in the first lesson that stresses caused by plate movement can pull or compress crust. This is one way plate motion is involved in creating many types of mountains. But the effect of plate movement is also responsible for changing the positions of rocks and the rocks themselves within a mountain range.

Folded Mountains

Rocks that are deeper in the crust are warmer than rocks closer to Earth's surface. Deeper rocks are also under much more pressure. When rocks are hot enough or under enough pressure, folds form instead of faults, as shown in **Figure 18. Folded mountains** *are made of layers of rocks that are folded.* Folded mountains form as continental plates collide, folding and uplifting layers of rock. When erosion removes the upper part of the crust, folds are exposed on the surface.

The arrangement of the folds is not accidental. You can demonstrate this by taking a piece of paper and gently pushing the ends toward one another to form a fold. The fold is a long ridge that is **perpendicular** to the direction in which you pushed. Folded mountains are similar. The folds are perpendicular to the direction of the compression that created them.

ACADEMIC VOCABULARY

perpendicular
(*adjective*) being at right angles to a line or plane

Figure 18 Compression stresses folded these rocks.

Active Reading **6. Indicate** Draw arrows on the figure to show which direction the compression stress came from.

Fault-Block Mountains

Sometimes tension stresses within a continent create mountains. As tension pulls crust apart, faults form. At the faults, some blocks of crust fall and others rise, as shown in **Figure 19**. **Fault-block mountains** *are parallel ridges that form where blocks of crust move up or down along faults.*

The Basin and Range Province in the western United States consists of dozens of parallel fault-block mountains that are oriented north to south. The tension that created the mountains pulled in the east-west directions. One of these mountains is shown at the beginning of this lesson. Notice how a high, craggy ridge is right next to a valley. Somewhere between the two, there is a fault where huge movement once occurred.

Figure 19 In the middle of a continent, tension can pull crust apart. Where the crust breaks, fault-block mountains and valleys can form as huge blocks of Earth rise or fall.

Active Reading **7. Illustrate** Draw arrows on the lines above to indicate which way the tension is pulling.

8. NGSSS Check Specify How do folded and fault-block mountains form? **SC.7.E.6.5**

_____

_____

_____

Inquiry **LAB STATION** **Try It!** SC.7.N.1.1, SC.7.E.6.5

MiniLab *How do folded mountains form?* at connectED.mcgraw-hill.com

Apply It! After you complete the lab, answer these questions.

1. Name the type of stress you applied in the lab.

2. Contrast the plate tectonics involved with folded mountains and with fault-block mountains.

Uplifted Mountains

The granite on top of the Sierra Nevada's Mount Whitney was once 10 km below Earth's surface. Now it is on top of a 4,400-m-tall mountain! How did this happen? Mount Whitney is an example of an uplifted mountain. *When large regions rise vertically with very little deformation,* **uplifted mountains** *form.* The rocks in the Sierra Nevada are made of granite, which is an igneous rock originally formed several kilometers below Earth's surface. Uplift and erosion have exposed it.

Scientists do not fully understand how uplifted mountains form. One hypothesis proposes that cold mantle under the crust detaches from the crust and sinks deeper into the mantle, as shown in **Figure 20.** The sinking mantle pulls the crust and creates compression closer to the surface. As the crust thickens, the upper part of the crust rises to maintain isostasy. Sometimes it rises high enough to create huge mountain ranges. Geologists are designing experiments to test this hypothesis.

Volcanic Mountains

You might not think of volcanoes as mountains, but scientists consider volcanoes to be special types of mountains. In fact, some of the largest mountains on Earth are made by volcanic eruptions. As molten rock and ash erupt onto Earth's surface, they harden. Over time, many eruptions can build huge volcanic mountains such as the ones that make up the Hawaiian Islands.

Not all volcanic mountains erupt all the time. Some volcanic mountains are dormant, which means they might erupt again someday. Some volcanic mountains will never erupt again.

 9. NGSSS Check Describe How do uplifted and volcanic mountains form? **SC.7.E.6.5**

_____

_____

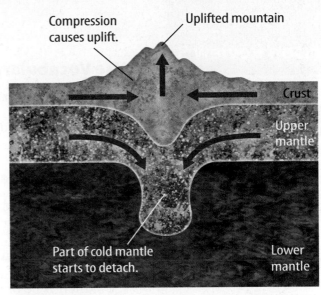

Compression causes uplift.

Uplifted mountain

Crust

Upper mantle

Part of cold mantle starts to detach.

Lower mantle

Figure 20 ⊙━ One possible explanation for how uplifted mountains form is that sinking mantle creates compression of the crust. The crust rises to regain isostasy, forming mountains.

Math Skills   MA.6.A.3.6

Use Proportions

An equation showing two equal ratios is a proportion. Some mountains in the Himalayas are rising 0.001 m/y. How long would it take the mountains to reach a height of 7,000 m?

1. Set up a proportion.

$$\frac{0.001 \text{ m}}{1y} = \frac{7,000 \text{ m}}{xy}$$

2. Cross multiply.

$$0.001x = 7,000$$

3. Divide both sides by 0.001.

$$\frac{0.001x}{0.001} = 7,000$$

4. Solve for x.

$$x = 7,000,000 \text{ y}$$

10. Practice

If the uplift rate of Mount Everest is 0.0006 m/y, how long did it take Mount Everest to reach a height of 8,848 m?

Visual Summary

Mountain ranges can be the result of repeated continental collision and rifting.

Tension stresses create mountain ranges that are a series of faults, ridges, and valleys.

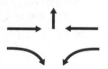

Uplifted mountains form as a result of compression near Earth's surface.

Inquiry SC.7.N.1.1, SC.7.E.6.5

iLAB STATION **Try It!**

Skill Lab *What tectonic processes are most responsible for shaping North America?* at connectED.mcgraw-hill.com

Use Vocabulary

1. Folded mountains are made by _____ stress. SC.7.E.6.5

2. Rocks formed deep inside Earth can be found at the surface due to

_____.

Understand Key Concepts 🔑

3. **Contrast** folded and fault-block mountains.

_____

_____

4. Which type of mountains form with little deformation? SC.7.E.6.5

   (A) fault-block mountains   (C) uplifted mountains

   (B) folded mountains   (D) volcanic mountains

Interpret Graphics

5. **Summarize** the plate tectonic events that built the Appalachian Mountains, using the graphic organizer below. SC.7.E.6.5

Critical Thinking

6. **Critique** the generalization that mountains only form at convergent boundaries. Explain the other processes that produce mountains.
   SC.7.E.6.5

_____

_____

_____

Math Skills   MA.6.A.3.6

7. Volcanoes in Hawaii began forming on the seafloor, about 5,000 m below the surface. If a volcano reaches the surface in 300,000 years, what was its rate of vertical growth per year?

Connect Use the graphic organizer below to include information you have learned about the different types of mountains.

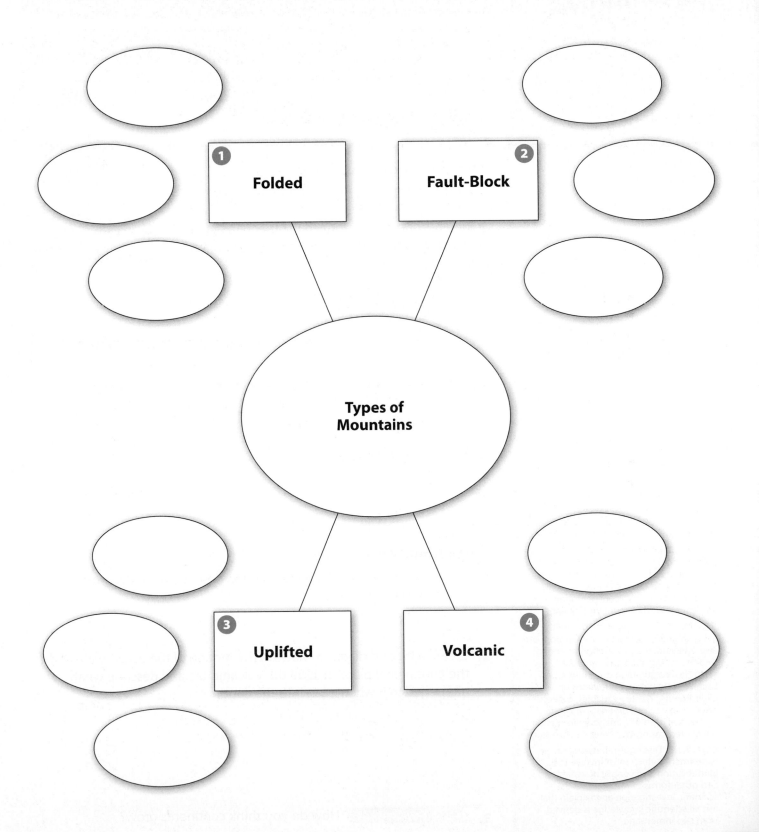

1 **Folded**

2 **Fault-Block**

Types of Mountains

3 **Uplifted**

4 **Volcanic**

Use the figure below to answer question 9.

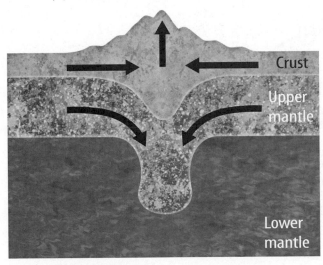

9 Which kind of mountain is shown in the figure? **SC.7.E.6.5**

Ⓐ fault-block

Ⓑ folded

Ⓒ uplifted

Ⓓ volcanic

10 What process most likely causes a continental rift? **SC.7.E.6.7**

Ⓕ compressional stress

Ⓖ movement of materials within Earth

Ⓗ isostasy and uplift of materials within Earth

Ⓘ subsidence and later uplift of surface materials

11 Which landforms are most likely to have coal, oil, and gas deposits? **SC.7.E.6.2**

Ⓐ basins

Ⓑ mountains

Ⓒ plains

Ⓓ plateaus

Use the figure below to answer questions 12 and 13.

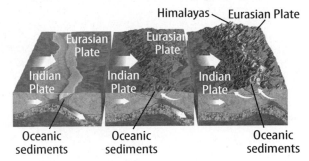

12 Identify the force involved in the formation of the Himalayas. **SC.7.E.6.5**

Ⓕ shear

Ⓖ tension

Ⓗ isostasy

Ⓘ compression

13 What would happen if the plate motion were reversed? **SC.7.E.6.5**

Ⓐ A volcanic arc would form.

Ⓑ New, younger mountains would form.

Ⓒ A continental rift would begin to form.

Ⓓ A deep oceanic trench would form offshore.

14 What explains the formation of the Hawaiian Islands? **SC.7.E.6.7**

Ⓕ an island arc at a plate boundary

Ⓖ an ocean plate moving over a hot spot

Ⓗ a mid-ocean ridge extending above the ocean surface.

Ⓘ a continental plate and an ocean plate meeting at mid ocean

NEED EXTRA HELP?

If You Missed Question...	1	2	3	4	5	6	7	8	9	10	11	12	13	14
Go to Lesson...	2	2	1	3	3	4	2	3	4	2	2	4	4	4

Multiple Choice *Bubble the correct answer.*

Use the image below to answer questions 1 through 4.

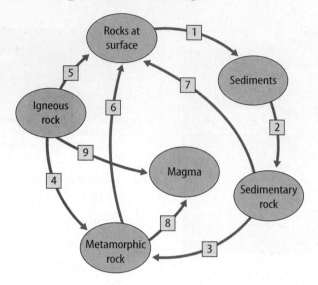

1. In the figure above, which numbers indicate processes that involve upward vertical motion of Earth's crust? **SC.7.E.6.2**

   (A) 1, 2, 3

   (B) 2, 3, 4

   (C) 5, 6, 7

   (D) 7, 8, 9

2. Which number in the figure above indicates a process that results in the breakdown of rock? **SC.7.E.6.2**

   (F) 1

   (G) 5

   (H) 7

   (I) 9

3. What do the processes labeled 3 and 4 have in common? **SC.7.E.6.2**

   (A) They involve compression, which is a type of stress on rock.

   (B) They involve failure, which causes the formation of faults.

   (C) They involve subduction, which takes rock deep inside Earth.

   (D) They involve uplift, which exposes rocks to erosion.

4. What process is indicated by labels 8 and 9? **SC.7.E.6.2**

   (F) burial

   (G) compression

   (H) subduction

   (I) uplift

Copyright © Glencoe/McGraw-Hill, a division of The McGraw-Hill Companies, Inc.

Benchmark Mini-Assessment   Chapter 5 • Lesson 2

Multiple Choice *Bubble the correct answer.*

Landform	Type of late boundary	Description
A	Convergent	The crust of two continental plates moves upward.
B	Convergent	The crust of an oceanic plate is pulled and pushed under another oceanic plate.
C	Convergent	The crust of an oceanic plate is pulled and pushed under the crust of a continental plate.
D	Divergent	Between oceanic plates, a crack forms and hotter mantle rock material is forced upward, pushing up cooler rock.

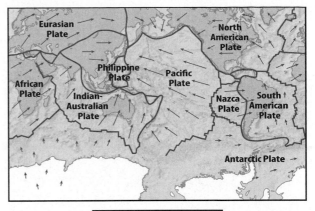

— Plate boundaries
← Plate movement

1. In the table above, which landforms are produced by compression? **SC.7.E.6.5**

(A) A

(B) D

(C) A, B, C

(D) B, C, D

2. When divergent boundaries occur within a continent, they can form a **SC.7.E.6.5**

(F) continental rift.

(G) mid-ocean ridge.

(H) mountain range.

(I) volcanic arc.

3. In the image above, what landform most likely forms where the Indian-Australian plate meets the Eurasian plate? **SC.7.E.6.5**

(A) mountains

(B) trench

(C) continental rift

(D) volcanic arc

4. Where do transform faults form? **SC.7.E.6.5**

(F) They form where continental plates collide.

(G) They form where tectonic plates collide under water.

(H) They form where one tectonic plate is pulled and pushed under another.

(I) They form where tectonic plates slide horizontally past each other.

Copyright © Glencoe/McGraw-Hill, a division of The McGraw-Hill Companies, Inc.

Benchmark Mini-Assessment Chapter 5 • Lesson 3

mini BAT

Multiple Choice *Bubble the correct answer.*

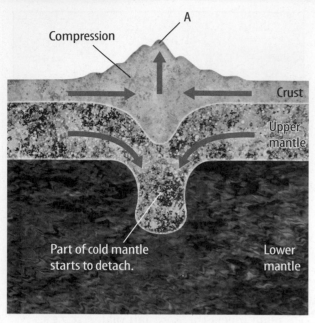

Compression

A

Crust

Upper mantle

Part of cold mantle starts to detach.

Lower mantle

1. In the image above, what type of mountain is represented by A? **SC.7.E.6.5**

   (A) fault-block mountain

   (B) folded mountain

   (C) uplifted mountain

   (D) volcanic mountain

2. What happens to mountains when the forces that caused the plates to collide become inactive? **SC.7.E.6.5**

   (F) They begin to divide.

   (G) They continue to grow.

   (H) They spread out.

   (I) They stop growing.

3. Corrine drew the above picture of the formation of the fault-block mountains. What error did Corrine make? **SC.7.E.6.5**

   (A) The arrows for the movement of the crust are facing the wrong direction.

   (B) The blocks move upward, not downward as shown in the figure.

   (C) The ridges shown in the figure should be perpendicular not parallel.

   (D) There are too many ridges and valleys shown in the figure.

4. Several layers of rock were pulled and pushed into the mantle. The rocks did not form faults due to high levels of heat and pressure. If the layer of rock above these layers eroded away, what type of mountain would remain? **SC.7.E.6.7**

   (F) fault-block mountain

   (G) folded mountain

   (H) uplifted mountain

   (I) volcanic mountain

Copyright © Glencoe/McGraw-Hill, a division of The McGraw-Hill Companies, Inc.

Multiple Choice *Bubble the correct answer.*

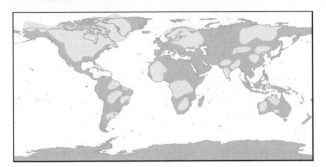

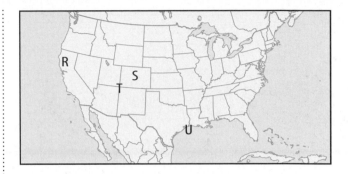

1. Which statement best describes the features of the continents as they are shown in the map above? **SC.7.E.6.5**

   (A) The lighter areas are older and flatter than the darker areas are.

   (B) The lighter areas undergo less erosion than the darker areas do.

   (C) The lighter areas are almost entirely volcanic while the darker areas are sedimentary rock.

   (D) The lighter areas are regions where continents grow faster while the darker areas exhibit little growth.

2. A scientist is studying a low, flat area filled with sediment. She is most likely examining a **SC.7.E.6.5**

   (F) basin.

   (G) craton.

   (H) plateau.

   (I) terrane.

3. The map above shows the continental United States. At which point on the map would you expect to find crust from other continental plates? **SC.7.E.6.5**

   (A) R

   (B) S

   (C) T

   (D) U

4. Which of the following processes might result in the formation of plateaus? **SC.7.E.6.5**

   (F) deposition

   (G) erosion

   (H) subsidence

   (I) uplift

Copyright © Glencoe/McGraw-Hill, a division of The McGraw-Hill Companies, Inc.

Notes

Where do volcanoes form?

Five friends were talking about volcanoes. They each had different ideas about where volcanoes form. This is what they said:

Izzy: I think most volcanoes form over hot spots in the ocean.

Eben: I think most volcanoes form in places where it is warm.

Anne: I think most volcanoes form in areas where earthquakes can occur.

Matt: I think most volcanoes form in areas where there are no earthquakes.

Cole: I think volcanoes can form anywhere on the earth. There are no particular areas where volcanoes are more likely to form.

Circle the friend with whom you most agree. Explain why you agree with that friend.

Earthquakes and Volcanoes

What causes earthquakes and volcanic eruptions?

Think About It!

The eruption of Eyjafjallajökull in Iceland in April of 2010 caused most of the air space in western and northern Europe to be closed for a five day period. If the ejected superheated particles of ash and dust were taken into airplane engines there could have been catastrophic consequences. Could something like this happen in Florida?

1 Why did Eyjafjallajökull erupt explosively?

_____
_____
_____

2 Can scientists predict earthquakes and volcanic eruptions?

_____
_____
_____

3 What causes earthquakes and volcanic activity?

_____
_____
_____
_____

Why do you think volcanoes erupt?

Get Ready to Read

Do you agree or disagree with each of these statements? As you read this chapter, see if you change your mind about any of the statements.

	AGREE	DISAGREE
1 Earth's crust is broken into rigid slabs of rock that move, causing earthquakes and volcanic eruptions.	☐	☐
2 Earthquakes create energy waves that travel through Earth.	☐	☐
3 All earthquakes occur on plate boundaries.	☐	☐
4 Volcanoes can erupt anywhere on Earth.	☐	☐
5 Volcanic eruptions are rare.	☐	☐
6 Volcanic eruptions only affect people and places close to the volcano.	☐	☐

There's More Online!
Video • Audio • Review • ⓘLab Station • WebQuest • Assessment • Concepts in Motion • Multilingual eGlossary

EARTHQUAKES

ESSENTIAL QUESTIONS

🔑 What is an earthquake?

🔑 Where do earthquakes occur?

🔑 How do daily activities impact the environment?

Vocabulary

earthquake p. 211

fault p. 213

seismic wave p. 214

focus p. 214

epicenter p. 214

primary wave p. 215

secondary wave p. 215

surface wave p. 215

seismologist p. 216

seismometer p. 217

seismogram p. 217

Florida NGSSS

LA.7.2.2.3 The student will organize information to show understanding (e.g., representing main ideas within text through charting, mapping, paraphrasing, summarizing, or comparing/contrasting);
MA.6.A.3.6 Construct and analyze tables, graphs, and equations to describe linear functions and other simple relations using both common language and algebraic notation.
SC.7.E.6.5 Explore the scientific theory of plate tectonics by describing how the movement of Earth's crustal plates causes both slow and rapid changes in Earth's surface, including volcanic eruptions, earthquakes, and mountain building.
SC.7.E.6.7 Recognize that heat flow and movement of material within Earth causes earthquakes and volcanic eruptions, and creates mountains and ocean basins.
SC.7.N.1.1 Define a problem from the seventh grade curriculum, use appropriate reference materials to support scientific understanding, plan and carry out scientific investigation of various types, such as systematic observations or experiments, identify variables, collect and organize data, interpret data in charts, tables, and graphics, analyze information, make predictions, and defend conclusions.
SC.7.N.1.3 Distinguish between an experiment (which must involve the identification and control of variables) and other forms of scientific investigation and explain that not all scientific knowledge is derived from experimentation.
SC.7.N.1.5 Describe the methods used in the pursuit of a scientific explanation as seen in different fields of science such as biology, geology, and physics.
SC.7.N.1.6 Explain that empirical evidence is the cumulative body of observations of a natural phenomenon on which scientific explanations are based.

SC.7.N.1.1

(Inquiry) Launch Lab

15 minutes

What causes earthquakes?

Earthquakes occur every day. On average, approximately 35 earthquakes happen on Earth every day. These earthquakes vary in severity. What causes the intense shaking of an earthquake? In this activity, you will simulate the energy released during an earthquake and observe the shaking that results.

Procedure 🥽 👐

1 Read and complete a lab safety form.

2 Tie two **large, thick rubber bands** together.

3 Loop one rubber band lengthwise around a **textbook.**

4 Use **tape** to secure a sheet of **medium-grained sandpaper** to the tabletop.

5 Tape a second sheet of sandpaper to the cover of the textbook.

6 Place the book on the table so that the sheets of sandpaper touch.

7 Slowly pull on the end of the rubber band until the book moves.

8 Observe and record what happens in the Data and Observations section below.

Data and Observations

Think About This

1. How does this experiment model the buildup of stress along a fault?

2. 🔑 **Key Concept** Why does the rapid movement of rocks along a fault result in an earthquake?

inquiry **Why did this building collapse?**

1. This building collapsed during the Loma Prieta earthquake that shook the San Francisco Bay area of California in 1989. The magnitude 7.1 earthquake produced severe shaking and damage. Freeways and buildings collapsed and a number of injuries and fatalities occurred. Why are earthquakes common in California?

_____

_____

_____

_____

_____

_____

What are earthquakes?

Have you ever tried to bend a stick until it breaks? When the stick snaps, it vibrates, releasing energy. Earthquakes happen in a similar way. **Earthquakes** *are the vibrations in the ground that result from movement along breaks in Earth's lithosphere*. These breaks are called faults.

2. NGSSS Check **Define** What is an earthquake? SC.7.E.6.7

_____

_____

Why do rocks move along a fault? The forces that move tectonic plates also push and pull on rocks along the fault. If these forces become large enough, the blocks of rock on either side of the fault can move horizontally or vertically past each other. The greater the force applied to a fault, the greater the chance of a large and destructive earthquake. **Figure 1** shows earthquake damage from the Northridge earthquake in 1994.

Figure 1 In 1994, the Northridge earthquake along the San Andreas Fault in California caused $20 billion in damage.

Inquiry **What makes an eruption explosive?**

1. Notice the red, hot "fire fountain" erupting from Kilauea volcano in Hawaii. Kilauea is the most active volcano in the world. Now recall the ash eruption pictured in the chapter opener. What do you think makes volcanoes erupt so differently?

What is a volcano?

Perhaps you have heard of some famous volcanoes such as Mount St. Helens, Kilauea, or Mount Pinatubo. All of these volcanoes have erupted within the last 30 years. *A* **volcano** *is a vent in Earth's crust through which melted—or molten—rock flows. Molten rock below Earth's surface is called* **magma.** Volcanoes are in many places worldwide. Some places have more volcanoes than others. In this lesson, you will learn about how volcanoes form, where they form, and about their structure and eruption style.

Active Reading

2. Define What is magma?

How do volcanoes form?

Volcanic eruptions constantly shape Earth's surface. They can form large mountains, create new crust, and leave a path of destruction behind. Scientists have learned that the movement of Earth's tectonic plates causes the formation of volcanoes and the eruptions that result.

Oceanic crust
Continental crust

Lithosphere

Magma

Asthenosphere

◀ **Figure 8**
During subduction, magma forms when one plate sinks beneath another plate.

Active Reading

3. Identify What type of boundary is shown in Figure 8?

Figure 9 When plates spread apart, it forces magma to the surface and creates new crust. The pillow lava shown in the photograph formed at the mid-ocean ridge. ▼

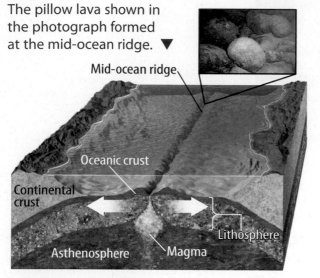

Mid-ocean ridge

Oceanic crust

Continental crust

Lithosphere

Asthenosphere

Magma

Kauai
Niihau
Oahu
Molokai
Maui
Lanai
Kahoolawe
Hawaii

Hot spot

◀ **Figure 10**
The farther each of the Hawaiian Islands is from the hot spot, the older the island is.

Active Reading

4. Name What type of boundary is shown in Figure 10?

Convergent Boundaries

Volcanoes can form along convergent plate boundaries. Recall that when two plates collide, the denser plate sinks, or subducts, into the mantle, as shown in **Figure 8.** The thermal energy below the surface and fluids driven off the subducting plate melt the mantle and form magma. Magma is less dense than the surrounding mantle and rises through cracks in the crust. This forms a volcano. *Molten rock that erupts onto Earth's surface is called* **lava.**

Divergent Boundaries

Lava erupts along divergent plate boundaries, too. Recall that two plates spread apart along a divergent plate boundary. As the plates separate, magma rises through the vent or opening in Earth's crust that forms between them. This process commonly occurs at a mid-ocean ridge and forms new oceanic crust, as shown in **Figure 9.** More than 60 percent of all volcanic activity on Earth occurs along mid-ocean ridges.

Hot Spots

Not all volcanoes form on or near plate boundaries. Volcanoes in the Hawaiian Island-Emperor Seamount chain are far from plate boundaries. *Volcanoes that are not associated with plate boundaries are called* **hot spots.** Geologists hypothesize that hot spots originate above a rising convection current from deep within Earth's mantle. They use the word *plume* to describe these rising currents of hot mantle material.

Figure 10 illustrates how a new volcano forms as a tectonic plate moves over a plume. When the plate moves away from the plume, the volcano becomes dormant, or inactive. Over time, a chain of volcanoes forms as the plate moves. The oldest volcano will be farthest away from the hot spot. The youngest volcano will be directly above the hot spot.

 5. NGSSS Check Explain How do volcanoes form? SC.7.E.6.2, SC.7.E.6.5

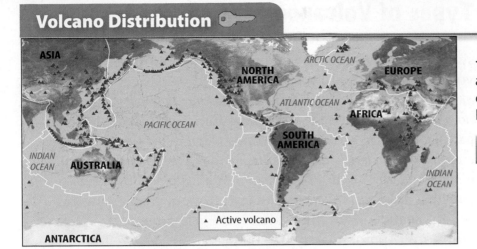

◀ **Figure 11** Most of the world's active volcanoes are located along convergent and divergent plate boundaries and hot spots.

Active Reading **6. Illustrate** Circle the Ring of Fire in Figure 11.

Where do volcanoes form?

The world's active volcanoes are shown in **Figure 11.** The volcanoes all erupted within the last 100,000 years. Notice that most volcanoes are close to plate boundaries.

Ring of Fire

The Ring of Fire represents an area of earthquake and volcanic activity that surrounds the Pacific Ocean. When you compare the locations of active volcanoes and plate boundaries in **Figure 11,** you can see that volcanoes are mostly along convergent plate boundaries where plates collide. They also are located along divergent plate boundaries where plates separate. Volcanoes also can occur over hot spots, like Hawaii, the Galapagos Islands, and Yellowstone National Park in Wyoming.

Volcanoes in the United States

There are 60 potentially active volcanoes in the United States. Most of these volcanoes are part of the Ring of Fire. Alaska, Hawaii, Washington, Oregon, and northern California all have active volcanoes, such as Mount Redoubt in Alaska. A few of these volcanoes have produced violent eruptions, like the explosive eruption of Mount St. Helens in 1980.

The United States Geological Survey (USGS) has established three volcano observatories to monitor the potential for future volcanic eruptions in the United States. Because large populations of people live near volcanoes such as Mount Rainier in Washington, shown in **Figure 12,** the USGS has developed a hazard assessment program. Scientists monitor earthquake activity, changes in the shape of the volcano, gas emissions, and the past eruptive history of a volcano to evaluate the possibility of future eruptions.

Active Reading **7. Explain** Where is the Ring of Fire?

_____

_____

_____

_____

Figure 12 Mount Rainier is an active volcano in the Cascade Mountains of the Pacific Northwest. Many people live in Seattle, which is in close proximity to the volcano. ▼

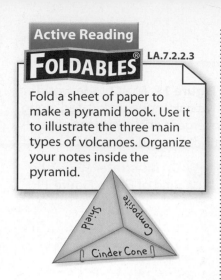

Active Reading

FOLDABLES® LA.7.2.2.3

Fold a sheet of paper to make a pyramid book. Use it to illustrate the three main types of volcanoes. Organize your notes inside the pyramid.

Types of Volcanoes

Volcanoes are classified based on their shape and size, as shown in **Table 4.** Magma composition and eruptive style of the volcano contribute to the shape. **Shield volcanoes** *are common along divergent plate boundaries and oceanic hot spots. Shield volcanoes are large with gentle slopes of basaltic lavas.* **Composite volcanoes** *are large, steep-sided volcanoes that result from explosive eruptions of andesitic and rhyolitic lava and ash along convergent plate boundaries.* **Cinder cones** *are small, steep-sided volcanoes that erupt gas-rich, basaltic lavas.* Some volcanoes are classified as supervolcanoes—volcanoes that have very large and explosive eruptions. Approximately 630,000 years ago, the Yellowstone Caldera in Wyoming ejected more than 1,000 km³ of rhyolitic ash and rock in one eruption. This eruption produced nearly 2,500 times the volume of material erupted from Mount St. Helens in 1980.

 8. NGSSS Check Summarize What determines the shape of a volcano? SC.7.E.6.7

_____

_____

Table 4 Geologists classify volcanoes based on their size, shape, and eruptive style.

Active Reading **9. Classify** Label the volcano types below.

Table 4 Volcanic Features

Large, shield-shaped volcano with gentle slopes made from basaltic lavas

Large, steep-sided volcano made from a mixture of andesitic and rhyolitic lava and ash

Small, steep-sided volcano; made from moderately explosive eruptions of basaltic lavas

Large volcanic depression formed when a volcano's summit collapses or is blown away by explosive activity

Volcanic Eruptions

When magma surfaces, it might erupt as a lava flow, such as the lava shown in **Figure 13,** erupting from Kilauea volcano in Hawaii. Other times magma might erupt explosively, sending **volcanic ash**—*tiny particles of pulverized volcanic rock and glass*—high into the atmosphere. **Figure 13** also shows Mount St. Helens in Washington, erupting violently in 1980. Why do some volcanoes erupt violently while others erupt quietly?

Eruption Style

Magma chemistry determines a volcano's eruptive style. The explosive behavior of a volcano is affected by the amount of dissolved gases, specifically the amount of water vapor, a magma contains. It also is affected by the silica, SiO_2, content of magma.

Magma Chemistry Magmas that form in different volcanic environments have unique chemical compositions. Silica is the main chemical compound in all magmas. Differences in the amount of silica affect magma thickness and its **viscosity**—*a liquid's resistance to flow.*

Magma that has a low silica content also has a low viscosity and flows easily like warm maple syrup. When the magma erupts, it flows as fluid lava that cools, crystallizes, and forms the volcanic rock basalt. This type of lava commonly erupts along mid-ocean ridges and at oceanic hot spots, such as Hawaii.

Magma that has a high silica content has a high viscosity and flows like sticky toothpaste. This type of magma forms when rocks rich in silica melt or when magma from the mantle mixes with continental crust. The volcanic rocks andesite and rhyolite form when intermediate and high silica magmas erupt from subduction zone volcanoes and continental hot spots.

 10. NGSSS Check Specify What factors affect eruption style? SC.7.E.6.7

Quiet Eruption

Violent Eruption

Figure 13 Lavas that are low in silica and the amount of dissolved gases erupt quietly. Explosive eruptions result from lava and ash that are high in silica and dissolved gases.

11. ✓ Visual Check Describe Where would each type of eruption pictured above occur?

Figure 14 Notice the holes in this pumice rock.

 12. Infer What caused the holes?

_____

_____

_____

Dissolved Gases The presence of **dissolved** gases in magma contributes to how explosive a volcano can be. This is similar to what happens when you shake a can of soda and then open it. The bubbles come from the carbon dioxide that is dissolved in the soda. The pressure inside the can decreases rapidly when you open it. Trapped bubbles increase in size rapidly and escape as the soda erupts from the can.

All magmas contain dissolved gases. These gases include water vapor and small amounts of carbon dioxide and sulfur dioxide. As magma moves toward the surface, the pressure from the weight of the rock above decreases. As pressure decreases, the ability of gases to stay dissolved in the magma also decreases. Eventually, gases can no longer remain dissolved in the magma and bubbles begin to form. As the magma continues to rise to the surface, the bubbles increase in size and the gas begins to escape. Because gases cannot easily escape from high-viscosity lavas, this combination often results in explosive eruptions. When gases escape above ground, the lava, ash, or volcanic glass that cools and crystallizes has holes. These holes, shown in **Figure 14,** are a common feature in the volcanic rock pumice.

Inquiry

SC.7.N.1.3, SC.7.N.1.5

LAB STATION **Try It!**

MiniLab *Can you model the movement of magma?* at connectED.mcgraw-hill.com

Apply It! After you complete the lab, answer these questions.

1. Recall The explosive behavior of a volcano is affected by _____ and _____ in magma content.

2. Define What is viscosity?

3. Apply What gases are included in magma?

Effects of Volcanic Eruptions

On average, about 60 different volcanoes erupt each year. The effects of lava flows, ash fall, pyroclastic flows, and mudflows can affect all life on Earth. Volcanoes enrich rock and soil with valuable nutrients and help to regulate climate. Unfortunately, they also can be destructive and sometimes even deadly.

Lava Flows Because lava flows are relatively slow moving, they are rarely deadly. But lava flows can be damaging. Mount Etna in Sicily, Italy, is Europe's most active volcano. **Figure 15** shows a fountain of fluid, hot lava erupting from one of the volcano's many vents. In May 2008, the volcano began spewing lava and ash in an eruption lasting over six months. Although lavas tend to be slow moving, they threaten communities nearby. People who live on Mount Etna's slopes are used to evacuations due to frequent eruptions.

Ash Fall During an explosive eruption, volcanoes can erupt large volumes of volcanic ash. Ash columns can reach heights of more than 40 km. Recall that ash is a mixture of particles of pulverized rock and glass. Ash can disrupt air traffic and cause engines to stop mid-flight as shards of rock and ash fuse onto hot engine blades. Ash also can affect air quality and can cause serious breathing problems. Large quantities of ash erupted into the atmosphere also can affect climate by blocking out sunlight and cooling Earth's atmosphere.

Mudflows The thermal energy a volcano produces during an eruption can melt snow and ice on the summit. This meltwater can then mix with mud and ash on the mountain to form mudflows. Mudflows also are called lahars. Mount Redoubt in Alaska erupted on March 23, 2009. Snow and meltwater mixed to form the mudflows shown in **Figure 16**.

▲ **Figure 15** Mount Etna is one of the world's most active volcanoes. People that live near the volcano are accustomed to frequent eruptions of both lava and ash.

Active Reading **13. Recall** (Circle) three things ash fall can affect.

◀ **Figure 16** Many of the steep-sided composite volcanoes are covered with seasonal snow. When a volcano becomes active, the snow can melt and mix with mud and ash to form a mudflow like the one shown here in the Cook Inlet, Alaska.

▲ **Figure 17** A pyroclastic flow travels down the side of Mount Mayon in the Philippines. Pyroclastic flows are made of hot (*pyro*) volcanic particles (*clast*).

Active Reading **14. Locate** Place an arrow showing where the pyroclastic flow is on the volcano in Figure 17.

Figure 18 In 1991, Mount Pinatubo erupted. The greatest concentration of sulfur dioxide gas from the eruption is shown below in blue. ▼

15. ✓ **Visual Check**

Tally How many degrees did the temperature decrease between 1991 and 1992?

Pyroclastic Flow Explosive volcanoes can produce fast-moving avalanches of hot gas, ash, and rock called pyroclastic (pi roh KLAS tihk) flows. Pyroclastic flows travel at speeds of more than 100 km/hr and with temperatures greater than 1000°C. In 1980, Mount St. Helens produced a pyroclastic flow that killed 58 people and destroyed 1 billion km^3 of forest. Mount Mayon in the Phillipines erupts frequently, producing pyroclastic flows like the one shown in **Figure 17**.

Predicting Volcanic Eruptions

Unlike earthquakes, volcanic eruptions can be predicted. Moving magma can cause ground deformation, a change in shape of the volcano, and a series of earthquakes called an earthquake swarm. Volcanic gas emissions can increase. Ground and surface water near the volcano can become more acidic. Geologists study these events, in addition to satellite and aerial photographs, to assess volcanic hazards.

Volcanic Eruptions and Climate Change

Volcanic eruptions affect climate when volcanic ash in the atmosphere blocks sunlight. High-altitude wind can move ash around the world. In addition, sulfur dioxide gases released from a volcano form sulfuric acid droplets in the upper atmosphere. These droplets reflect sunlight into space, resulting in lower temperatures as less sunlight reaches Earth's surface. **Figure 18** shows the result of sulfur dioxide gas in the atmosphere from the 1991 eruption of Mt. Pinatubo.

Active Reading **16. Determine** How do volcanic eruptions affect climate?

_____

_____

_____

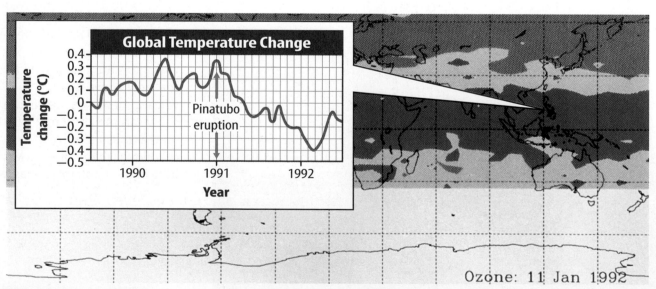

Lesson Review 2

Visual Summary

Volcanoes form when magma rises through cracks in the crust and erupts from vents on Earth's surface.

Magma with low amounts of silica and low viscosity erupts to form shield volcanoes.

Magma with high amounts of silica and high viscosity erupts explosively to form composite cones.

SC.6.N.1.4, SC.8.N.3.1, LA.6.2.2.3

Inquiry **⊙LAB STATION** **Try It!**

Skill Lab *The Dangers of Mount Rainier* at connectED.mcgraw-hill.com

Use Vocabulary

1. **Compare and contrast** *lava* and *magma*. SC.7.E.6.2

_____

_____

2. **Explain** the term *viscosity*.

_____

3. Pulverized rock and ash that erupts from explosive volcanoes is called _____. SC.7.E.6.2

Understand Key Concepts 🔑

4. **Identify** places where volcanoes form. SC.7.E.6.5

_____

_____

5. **Compare** the three main types of volcanoes. SC.7.E.6.7

_____

_____

6. What type of lava erupts from shield volcanoes?
   - (A) andesitic
   - (B) basaltic
   - (C) granitic
   - (D) rhyolitic

Interpret Graphics

7. Complete the graphic organizer to illustrate the four types of eruptive products that can result from a volcanic eruption. LA.7.2.2.3

Eruptive Products

Critical Thinking

8. **Compare** the shapes of composite volcanoes and shield volcanoes. Why are their shapes and eruptive styles so different?

_____

_____

9. **Explain** how explosive volcanic eruptions can cause climate change. What might happen if Yellowstone Caldera erupted today?

_____

_____

Think About It! Internal energy and movement of material within Earth causes both earthquakes along plate boundaries where plates slide past each other, collide, or separate and volcanoes at subduction zones, mid-ocean ridges, and hot spots.

🔑 Key Concepts Summary

Vocabulary

LESSON 1 Earthquakes

- Earthquakes commonly occur on or near tectonic plate boundaries.
- Earthquakes are used to study the composition and structure of Earth's interior and to identify the location of active faults.
- Earthquakes are monitored using **seismometers** and described using the Richter magnitude scale, the moment magnitude scale, and the Modified Mercalli scale.

earthquake p. 211
fault p. 213
seismic wave p. 214
focus p. 214
epicenter p. 214
primary wave p. 215
secondary wave p. 215
surface wave p. 215
seismologist p. 216
seismometer p. 217
seismogram p. 217

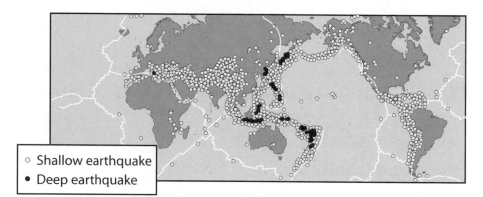

○ Shallow earthquake
● Deep earthquake

LESSON 2 Volcanoes

- Molten **magma** is forced upward through cracks in the crust, erupting from volcanoes.
- The eruption style, size, and shape of a volcano depends on the composition of the magma, including the amount of dissolved gas.
- Volcanoes are classified as **cinder cones, shield volcanoes,** and **composite cones.**

volcano p. 225
magma p. 225
lava p. 226
hot spot p. 226
shield volcano p. 228
composite volcano p. 228
cinder cone p. 228
volcanic ash p. 229
viscosity p. 229

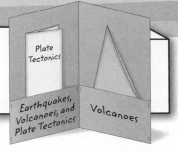

Active Reading

FOLDABLES® **Chapter Project**

Assemble your lesson Foldables as shown to make a Chapter Project. Use the project to review what you have learned in this chapter.

Plate Tectonics

Earthquakes, Volcanoes, and Plate Tectonics

Volcanoes

Use Vocabulary

1 A volcano with gently sloping sides is a(n)

_____ .

2 Write a sentence using the terms *seismic waves*, *P-waves*, and *S-waves*.

_____

_____

3 Magma that erupts quietly is

_____ .

Magma most likely to erupt explosively is

_____ .

4 Volcanic activity that does not occur near a plate boundary happens at a(n)

_____ .

5 Molten rock inside Earth is called

_____ .

6 What is used to record ground motion during an earthquake?

_____

_____

7 What marks the exact location where an earthquake occurs?

_____

What is the place on Earth's surface directly above it?

_____

8 A type of seismic wave that has movement similar to an ocean wave is a(n)

_____ .

9 A mixture of pulverized ash, rock, and gas ejected during explosive eruptions is called a(n)

_____ .

Link Vocabulary and Key Concepts

Use vocabulary terms from the previous page to complete the concept map.

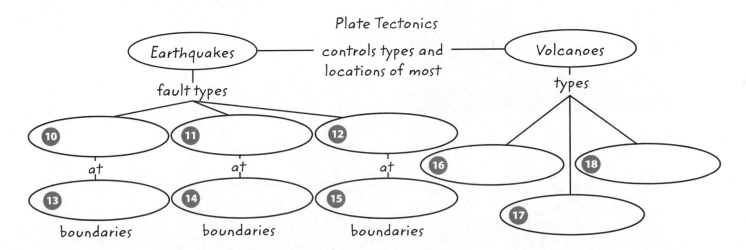

FLORIDA BIG IDEAS

1 The Practice of Science
6 Earth Structures

How do human activities impact the environment?

More than 6 billion people live on Earth. Every day, people all over the world travel, eat, use water, and participate in recreational activities.

1 What resources do you think people need and use?

_____

_____

2 What do you think might happen if any resources run out?

_____

_____

_____

3 How do you think human activities impact the environment?

_____

_____

_____

What do you think about the environment?

Do you agree or disagree with each of these statements? As you read this chapter, see if you change your mind about any of the statements.

	AGREE	DISAGREE
1 Earth can support an unlimited number of people.	☐	☐
2 Humans can have positive and negative impacts on the environment.	☐	☐
3 Deforestation does not affect soil quality.	☐	☐
4 Most trash is recycled.	☐	☐
5 Sources of water pollution are always easy to identify.	☐	☐
6 The proper method of disposal for motor oil is down the drain.	☐	☐
7 The greenhouse effect is harmful to life on Earth.	☐	☐
8 Air pollution can affect human health.	☐	☐

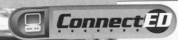
Connect ED

There's More Online!
Video • Audio • Review • ①Lab Station • WebQuest • Assessment • Concepts in Motion • Multilingual eGlossary   **245** 🔊

People and the ENVIRONMENT

What is the relationship between resource availability and human population growth?

How do daily activities impact the environment?

Vocabulary

population p. 247

carrying capacity p. 248

 Florida NGSSS

SC.7.E.6.6 Identify the impact that humans have had on Earth, such as deforestation, urbanization, desertification, erosion, air and water quality, changing the flow of water.

SC.7.N.1.1 Define a problem from the seventh grade curriculum, use appropriate reference materials to support scientific understanding, plan and carry out scientific investigation of various types, such as systematic observations or experiments, identify variables, collect and organize data, interpret data in charts, tables, and graphics, analyze information, make predictions, and defend conclusions.

SC.7.N.1.3 Distinguish between an experiment (which must involve the identification and control of variables) and other forms of scientific investigation and explain that not all scientific knowledge is derived from experimentation.

LA.7.2.2.3 The student will organize information to show understanding (e.g., representing main ideas within text through charting, mapping, paraphrasing, summarizing, or comparing/contrasting).

Inquiry Launch Lab

SC.7.N.1.1

20 minutes

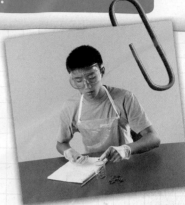

What happens as populations increase in size?

In the year 200, the human population was about 250 million people. By the year 2000, it had increased to more than 6 billion. By 2050, it is projected to be more than 9 billion. However, the amount of space available on Earth will remain the same.

Procedure

1. Read and complete a lab safety form.
2. Place 10 **dried beans** in a **100-mL beaker.**
3. At the start signal, double the number of beans in the beaker. There should now be 20 beans.
4. Make a table to record your data. Indicate the number of beans added and the total number of beans in the beaker after each addition.
5. Double the number of beans each time the start signal sounds. Continue until the stop signal sounds.

Data and Observations

Think About This

1. Can you add any more beans to the beaker? Why or why not?

2. How many times did you have to double the beans to fill the beaker?

3. **Key Concept** How might the growth of a population affect the availability of resources, such as space?

inquiry **What's the impact?**

1. This satellite image shows light coming from the United States at night. You can see where large cities are located. What do you think the dark areas represent? When you turn on the lights at night, where does the energy to power the lights come from? How might this daily activity impact the environment?

_____
_____
_____
_____
_____
_____
_____

Population and Carrying Capacity

Have you ever seen a sign such as the one shown in **Figure 1?** The sign shows the population of a city. In this case, population means how many people live in the city. Scientists use the term *population,* too, but in a slightly different way. For scientists, *a* **population** *is all the members of a species living in a given area.* You are part of a population of humans. The other species in your area, such as birds or trees, each make up a separate population.

The Human Population

When the first American towns were settled, most had low populations. Today, some of those towns are large cities, crowded with people. For example, the population of the city of Miami, Florida increased from about 1,700 people in 1900 to over 430,000 in 2009. In a similar way, Earth was once home to relatively few humans. Today, about 6.7 billion people live on Earth. The greatest increase in human population occurred during the last few centuries.

 2. Investigate How much has the population of your town increased over the last 10 years?

_____
_____
_____

Figure 1 This sign shows the population of a city.

Active Reading **3. Compare** How is this use of the word *population* similar to the scientific use?

_____
_____
_____

Figure 2 🔑 Human population "exploded" in the last few hundred years.

4. Compare How does the rate of human population growth from the years 200 to 1800 compare to the rate of growth from 1800 to 2000?

_____

_____

_____

_____

WORD ORIGIN

population
from Latin *populus*, means "people"

Active Reading

FOLDABLES LA.7.2.2.3

Use a sheet of paper to make a small vertical shutterfold. Draw the arrows on each tab and label as illustrated. Use the Foldable to discuss how human population growth is related to resources.

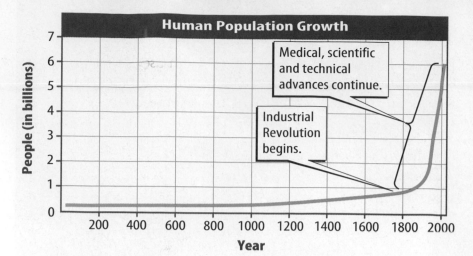

Human Population Growth

Medical, scientific and technical advances continue.

Industrial Revolution begins.

People (in billions) — 0, 1, 2, 3, 4, 5, 6, 7

Year — 200, 400, 600, 800, 1000, 1200, 1400, 1600, 1800, 2000

Population Trends

Have you ever heard the phrase *population* explosion? Population explosion describes the sudden rise in human population that has happened in recent history. The graph in **Figure 2** shows how the human population has changed. The population increased at a fairly steady rate for most of human history. In the 1800s, the population began to rise sharply.

What caused this sharp increase? Improved health care, clean water, and other technological advancements mean that more people are living longer and reproducing. In the hour or so it might take you to read this chapter, about 15,000 babies will be born worldwide.

Active Reading **5. Identify** (Circle) the factors that contributed to the increase in human population.

Population Limits

Every human being needs certain things, such as food, clean water, and shelter, to survive. People also need clothes, transportation, and other items. All the items used by people come from resources found on Earth. Does Earth have enough resources to support an unlimited number of humans?

Earth has limited resources. It cannot support a population of any species in a given environment beyond its carrying capacity. **Carrying capacity** *is the largest number of individuals of a given species that Earth's resources can support and maintain for a long period of time.* If the human population continues to grow beyond Earth's carrying capacity, eventually Earth will not have enough resources to support humans.

 6. NGSSS Check Assess What is the relationship between the availability of resources and human population growth? SC.7.E.6.6

_____

_____

Inquiry How do people use land?

1. Study this photo of an area in the Florida Everglades. List three ways people use land. What other possible land uses are not shown in the photo? What impact do humans have on land resources?

Active Reading **2. Identify** Underline ways deforestation impacts the land.

Using Land Resources

What do the metal in staples and the paper in your notebook have in common? Both come from resources found in or on land. People use land for timber production, agriculture, and mining. All of these activities impact the environment.

Forest Resources

Trees are cut down to make wood and paper products, such as your notebook. Trees also are cut for fuel and to clear land for agriculture, grazing, or building houses or highways.

Sometimes forests are cleared, as shown in **Figure 4.** **Deforestation** *is the removal of large areas of forests for human purposes.* Approximately 130,000 km^2 of tropical rain forests are cut down each year. For comparison, the area of the state of Florida is about 170,000 km^2. Tropical rain forests are home to an estimated 50 percent of all the species on Earth. Deforestation destroys habitats, which can lead to species' extinction.

Deforestation also can affect soil quality. Plant roots hold soil in place. Without these natural anchors, soil erodes away. In addition, deforestation affects air quality. Recall that trees remove carbon dioxide from the air when they undergo photosynthesis. When there are fewer trees on Earth, more carbon dioxide remains in the atmosphere. You will learn more about carbon dioxide in Lesson 4.

Figure 4 Deforestation causes dramatic changes in the appearance of a region.

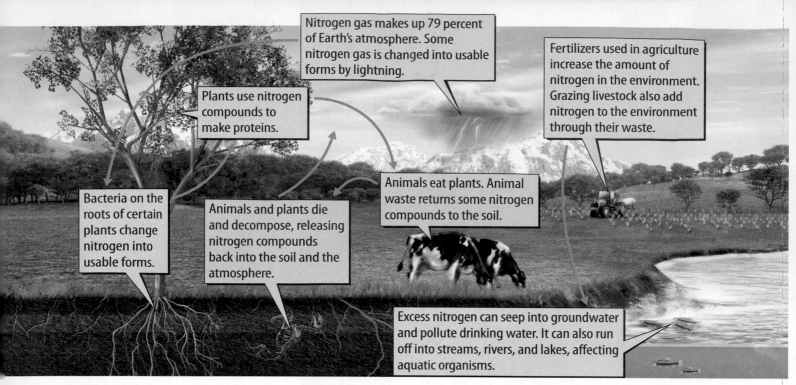

Nitrogen gas makes up 79 percent of Earth's atmosphere. Some nitrogen gas is changed into usable forms by lightning.

Plants use nitrogen compounds to make proteins.

Fertilizers used in agriculture increase the amount of nitrogen in the environment. Grazing livestock also add nitrogen to the environment through their waste.

Bacteria on the roots of certain plants change nitrogen into usable forms.

Animals and plants die and decompose, releasing nitrogen compounds back into the soil and the atmosphere.

Animals eat plants. Animal waste returns some nitrogen compounds to the soil.

Excess nitrogen can seep into groundwater and pollute drinking water. It can also run off into streams, rivers, and lakes, affecting aquatic organisms.

Figure 5 Agricultural practices can increase the amount of nitrogen that cycles through the ecosystem.

Active Reading 3. **Analyze** How does the use of fertilizers affect the environment?

Agriculture and the Nitrogen Cycle

It takes a lot of food to feed 6.7 billion people. To meet the food demands of the world's population, farmers often add fertilizers that contain nitrogen to soil to increase crop yields.

As shown in **Figure 5,** nitrogen is an element that naturally cycles through ecosystems. Living things use nitrogen to make proteins. When these living things die and decompose or produce waste, they release nitrogen into the soil or the atmosphere.

Although nitrogen gas makes up about 79 percent of Earth's atmosphere, most living things cannot use nitrogen in its gaseous form. Nitrogen must be converted into a usable form. Bacteria that live on the roots of certain plants convert atmospheric nitrogen to a form that is usable by plants. Modern agricultural practices include adding fertilizer that contains a usable form of nitrogen to soil.

Scientists estimate that human activities such as manufacturing and applying fertilizers to crops have doubled the amount of nitrogen cycling through ecosystems. Excess nitrogen can kill plants adapted to low nitrogen levels and affect organisms that depend on those plants for food. Fertilizers can seep into groundwater supplies, polluting drinking water. They can also run off into streams and rivers, affecting aquatic organisms.

Other Effects of Agriculture

Agriculture can impact soil quality in other ways, too. Soil erosion can occur when land is overfarmed or overgrazed. High rates of soil erosion can lead to desertification. **Desertification** *is the development of desertlike conditions due to human activities and/or climate change.* A region of land that undergoes desertification is no longer useful for food production.

Active Reading 4. **Recall** Underline the cause of desertification.

Figure 6 Some resources must be mined from the ground, such as phosphates, which are used in fertilizers, vitamins, soft drinks, and toothpaste.

Mining

Many useful rocks and minerals are removed from the ground by mining. For example, in Florida, phosphates are removed from the surface by digging a strip mine, such as the one shown in **Figure 6.** Coal and other in-ground resources also can be removed by digging underground mines.

Mines are essential for obtaining much-needed resources. However, digging mines disturbs habitats and changes the landscape. If proper regulations are not followed, water can be polluted by **runoff** that contains heavy metals from mines.

 5. **NGSSS Check Summarize** What are some consequences of using land as a resource? SC.7.E.6.6

_____

_____

_____

REVIEW VOCABULARY

runoff
the portion of precipitation that moves over land and eventually reaches streams, rivers, lakes, and oceans

Construction and Development

You have read about important resources that are found on or in land. But did you know that land itself is a resource? People use land for living space. Your home, your school, your favorite stores, and your neighborhood streets all are built on land.

Inquiry

SC.7.N.1.1, SC.7.N.1.3, SC.7.E.6.6

LAB STATION Try It!

MiniLab What happens when you mine? at connectED.mcgraw-hill.com

Apply It! After you complete the lab, answer these questions.

1. As you mined the "coal," how did you change your model landscape?

2. Where in Florida has mined land been restored?

Figure 7 Urban sprawl can lead to habitat destruction as forests are cut down to make room for housing developments such as these near Halifax, Nova Scotia.

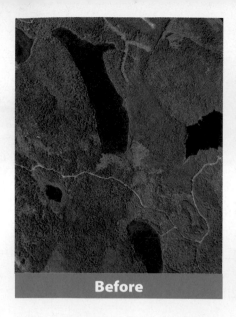
Before

After

Math Skills   MA.6.A.3.6

Use Percentages

Between 1960 and today, interstate highways increased from a total of 16,000 km to 47,000 km. What percent increase does this represent?

1. Subtract the starting value from the final value.

   47,000 km − 16,000 km = 31,000 km

2. Divide the difference by the starting value.

   $\dfrac{31,000 \text{ km}}{16,000 \text{ km}} = 1.94$

3. Multiply by 100 and add a % sign.

   $1.94 \times 100 = 194\%$

Practice

6. In 1950, the population of Florida was about 2,800,000. By 2000, it was nearly 16,000,000. What was the percent increase?

Urban Sprawl

In the 1950s, large tracts of rural land in the United States were developed as suburbs, residential areas on the outside edges of a city. When the suburbs became crowded, people moved farther out into the country. More open land was cleared for still more development. *The development of land for houses and other buildings near a city is called* **urban sprawl.** For example, the city of Miami has an area of 143 km^2, but the urban area of Miami is 2,891 km^2. The impacts of urban sprawl include habitat destruction, shown in **Figure 7,** and loss of farmland. Increased runoff also occurs, as large areas are paved for sidewalks and streets. An increase in runoff, especially if it contains sediments or chemical pollutants, can reduce the water quality of streams, rivers, and groundwater.

Roadways

Urban sprawl occurred at the same time as another trend in the United States—increased motor vehicle use. Only a small percentage of Americans owned cars before the 1940s. By 2005, there were 240 million vehicles for 295 million people, greatly increasing the need for roadways. In 1960, the United States had about 16,000 km of interstate highways. Today, the interstate highway system includes 47,000 km of paved roadways. Like urban sprawl, roadways increase runoff and disturb habitats.

Active Reading **7. Identify** (Circle) the two trends that triggered the need for more highways.

Recreation

Not all of the land used by people is paved and developed. People also use land for recreation. They hike, bike, ski, and picnic, among other activities. In urban areas, some of these activities take place in public parks. As you will learn later in this lesson, parks and other green spaces help decrease runoff.

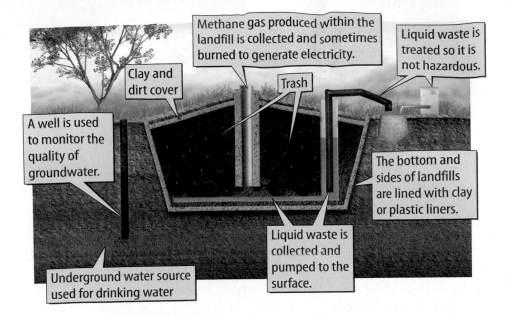

Methane gas produced within the landfill is collected and sometimes burned to generate electricity.

Clay and dirt cover

Trash

Liquid waste is treated so it is not hazardous.

A well is used to monitor the quality of groundwater.

The bottom and sides of landfills are lined with clay or plastic liners.

Underground water source used for drinking water

Liquid waste is collected and pumped to the surface.

Figure 8 Landfills are carefully constructed to prevent polution.

8. Examine How can the methane gas produced within a landfill be used?

_____

_____

Waste Management

On a typical day, each person in the United States generates about 2.1 kg of trash. That adds up to about 230 million metric tons per year! Where does all that trash go?

Landfills

About 31 percent of the trash is recycled and composted. About 14 percent is burned, and the remaining 55 percent is placed in landfills, such as the one shown in **Figure 8.** Landfills are areas where trash is buried. Landfills are another way that people use land.

A landfill is carefully designed to meet government regulations. Trash is covered by soil to keep it from blowing away. Special liners help prevent pollutants from leaking into soil and groundwater supplies.

 10. NGSSS Check Recall What is done to prevent the trash in landfills from polluting air, soil, and water? **SC.7.E.6.6**

_____

Hazardous Waste

Some trash cannot be placed in landfills because it contains harmful substances that can affect soil, air, and water quality. This trash is called hazardous waste. The substances in hazardous waste also can affect the health of humans and other living things.

Both industries and households generate hazardous waste. For example, hazardous waste from the medical industry includes used needles and bandages. Household hazardous waste includes used motor oil and batteries. The U.S. Environmental Protection Agency (EPA) works with state and local agencies to help people safely dispose of hazardous waste.

Active Reading **9. Report** What percentage of trash is recycled or

composted? _____ %

burned? _____ %

placed in landfills? _____ %

Active Reading

FOLDABLES LA.7.2.2.3

Use a sheet of notebook paper to make a horizontal two-tab concept map. Label and draw arrows as illustrated. Use the Foldable to identify positive and negative factors that have an impact on land.

Impacts on Land

+   −

Figure 9 Everglades National Park was established in 1947. It contains over 6,000 km² of wetlands and forests.

reclamation
from Latin *reclamare*, means "to call back"

Positive Actions

Human actions can have negative effects on the environment, but they can have positive impacts as well. Governments, society, and individuals can work together to reduce the impact of human activities on land resources.

Protecting Florida's Land

After Yellowstone National Park was established in 1847, other states began setting aside land for preservation. Florida contains 3 national forests, 150 state parks, and 10 national parks, including the Everglades National Park, which is shown in **Figure 9.** Some islands also have been protected as state parks or as part of the Gulf Islands National Seashore. Florida also contains 28 wildlife refuges, including Merritt Island, which is located near the Kennedy Space Center.

Reforestation and Reclamation

A forest is a complex ecosystem. With careful planning, it can be managed as a renewable resource. For example, trees can be select-cut. That means that only some trees in one area are cut down, rather than the entire forest. In addition, people can practice reforestation. **Reforestation** *involves planting trees to replace trees that have been cut or burned down.* Reforestation can keep a forest healthy or help reestablish a deforested area.

Mined land also can be made environmentally healthy through reclamation. **Reclamation** *is the process of restoring land disturbed by mining.* The before and after photos in **Figure 10** show that the mined area has been reshaped, covered with soil, and then replanted with trees and other vegetation.

Active Reading **11. Compare** How do reforestation and reclamation positively impact land?

_____

_____

Figure 10 As part of reclamation, grasses and trees were planted on this coal mine in Indiana.

Before

After

258 Chapter 7 • EXPLAIN

258 Chapter 7 • EXPLAIN

Green Spaces

In urban areas, much of the land is covered with parking lots, streets, buildings, and sidewalks. Many cities use green spaces to create natural environments in urban settings. Green spaces are areas that are left undeveloped or lightly developed. They include parks within cities and forests around suburbs. Green spaces, such as the park shown in **Figure 11,** provide recreational opportunities for people and shelter for wildlife. Green spaces also reduce runoff and improve air quality as plants remove excess carbon dioxide from the air.

How can you help?

Individuals can have a big impact on land-use issues by practicing the three Rs—reusing, reducing, and recycling. Reusing is using an item for a new purpose. For example, you might have made a bird feeder from a used plastic milk jug. Reducing is using fewer resources. You can turn off the lights when you leave a room to reduce your use of electricity.

Recycling is making a new product from a used product. Plastic containers can be recycled into new plastic products. Recycled aluminum cans are used to make new aluminum cans. Paper, shown in **Figure 11,** also can be recycled.

Figure 11 shows another way people can lessen their environmental impact on the land. The student in the bottom photo is composting food scraps into a material that is added to soil to increase its fertility. Compost is a mixture of decaying organic matter, such as leaves, food scraps, and grass clippings. It is used to improve soil quality by adding nutrients to soil. Composting and reusing, reducing, and recycling all help reduce the amount of trash that ends up in landfills.

 12. NGSSS Check Consider What can you do to help lessen your impact on the land?
SC.7.E.6.6

Figure 11 🔑 **13. Compose** Write captions for the photos below, relating each image to a positive impact on land.

Dead Zones

What causes lifeless areas in the ocean?

For thousands of years, people have lived on coasts, making a living by shipping goods or by fishing. Today, fisheries in the Gulf of Mexico provide jobs for thousands of people and food for millions more. Although humans and other organisms depend on the ocean, human activities can harm marine ecosystems. Scientists have been tracking dead zones in the ocean for several decades. They believe that these zones are a result of human activities on land.

A large dead zone in the Gulf of Mexico forms every year when runoff from spring and summer rain in the Midwest drains into the Mississippi River. The runoff contains nitrogen and phosphorous from fertilizer, animal waste, and sewage from farms and cities. This nutrient-rich water flows into the gulf. Algae feed on excess nutrients and multiply rapidly, creating an algal bloom. The results of the algal bloom are shown below.

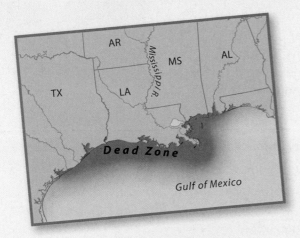

Some simple changes in human activity can help prevent dead zones. People upstream from the Gulf can decrease the use of fertilizer and apply it at times when it is less likely to be carried away by runoff. Picking up or containing animal waste can help, too. Also, people can modernize and improve septic and sewage systems. How do we know these steps would work? Using them has already restored life to dead zones in the Great Lakes!

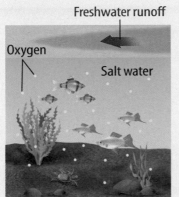

1 River water flows into the Gulf of Mexico.

2 After the algal bloom, dead algae sink to the ocean floor.

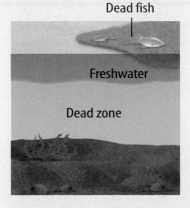

3 Decomposing algae deplete the water's oxygen, killing other organisms.

It's Your Turn

RESEARCH AND REPORT Earth's oceans contain about 150 dead zones. Choose three. Plot them on a map and write a report about what causes each dead zone. **LA.7.4.2.2**

Impacts on the ATMOSPHERE

ESSENTIAL QUESTIONS

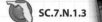

What are some types of air pollution?

How are global warming and the carbon cycle related?

How does air pollution affect human health?

What actions help prevent air pollution?

Vocabulary

photochemical smog p. 271

acid precipitation p. 272

particulate matter p. 272

global warming p. 273

greenhouse effect p. 274

Air Quality Index p. 275

Florida NGSSS

LA.7.2.2.3 The student will organize information to show understanding (e.g., representing main ideas within text through charting, mapping, paraphrasing, summarizing, or comparing/contrasting);

SC.7.E.6.6 Identify the impact that humans have had on Earth, such as deforestation, urbanization, desertification, erosion, air and water quality, changing the flow of water.

SC.7.N.1.3 Distinguish between an experiment (which must involve the identification and control of variables) and other forms of scientific investigation and explain that not all scientific knowledge is derived from experimentation.

 inquiry Launch Lab

SC.7.N.1.3

20 minutes

Where's the air?

In 1986, an explosion at a nuclear power plant in Chernobyl, Russia, sent radioactive pollution 6 km into the atmosphere. Within three weeks, the pollution had reached Italy, Finland, Iceland, and North America.

Procedure

1. Read and complete a lab safety form.

2. With your group, move to your assigned area of the room.

3. Lay out **sheets of paper** to cover the table.

4. When the **fan** starts blowing, observe whether water droplets appear on the paper. Record your observations.

5. Lay out another set of paper sheets and record your observations when the fan blows in a different direction.

Data and Observations

Think About This

1. Did the water droplets reach your location? Why or why not?

2. How is the movement of air and particles by the fan similar to the movement of the pollution from Chernobyl? How does the movement differ?

3. **Key Concept** How do you think the health of a person in Iceland could be affected by the explosion in Chernobyl?

Inquiry **Why wear a mask?**

1. In some areas of the world, people wear masks to help protect themselves against high levels of air pollution. Where does this pollution come from? How do you think air pollution affects human health and the environment?

_____

_____

_____

_____

_____

_____

_____

Importance of Clean Air

Your body, and the bodies of other animals, uses oxygen in air to produce some of the energy it needs. Many organisms can survive for only a few minutes without air. But the air you breathe must be clean or it can harm your body.

Types of Air Pollution

Human activities can produce pollution that enters the air and affects air quality. Types of air pollution include smog, acid precipitation, particulate matter, chlorofluorocarbons (CFCs), and carbon monoxide.

Smog

The brownish haze in the sky in **Figure 16** is photochemical smog. **Photochemical smog** _is caused when nitrogen and carbon compounds in the air react in sunlight._ Nitrogen and carbon compounds are released when fossil fuels are burned to provide energy for vehicles and power plants. These compounds react in sunlight and form other substances. One of these substances is ozone. Ozone high in the atmosphere helps protect living things from the Sun's ultraviolet radiation. However, ozone close to Earth's surface is a major component of smog.

Active Reading 2. **Distinguish** What is one benefit of ozone?

_____

_____

Figure 16 Burning fossil fuels releases compounds that can react in sunlight and form smog.

Acid Precipitation

Another form of pollution that occurs as a result of burning fossil fuels is acid precipitation. **Acid precipitation** *is rain or snow that has a lower pH than that of normal rainwater.* The pH of normal rainwater is about 5.6. Acid precipitation forms when gases containing nitrogen and sulfur react with water, oxygen, and other chemicals in the atmosphere. Acid precipitation falls into lakes and ponds or onto the ground. It makes the water and the soil more acidic. Many living things cannot survive if the pH of water or soil becomes too low. The trees shown in **Figure 17** have been affected by acid precipitation.

Particulate Matter

The mix of both solid and liquid particles in the air is called **particulate matter.** Solid particles include smoke, dust, and dirt. These particles enter the air from natural processes, such as volcanic eruptions and forest fires. Human activities, such as burning fossil fuels at power plants and in vehicles, also release particulate matter. Inhaling particulate matter can cause coughing, difficulty breathing, and other respiratory problems.

CFCs

Ozone in the upper atmosphere absorbs harmful ultraviolet (UV) rays from the Sun. Using products that contain CFCs, such as air conditioners and refrigerators made before 1996, affects the ozone layer. CFCs react with sunlight and destroy ozone molecules. As a result, the ozone layer thins and more UV rays reach Earth's surface. Increased skin cancer rates have been linked with an increase in UV rays.

Carbon Monoxide

Carbon monoxide is a gas released from vehicles and industrial processes. Forest fires also release carbon monoxide into the air. Wood-burning and gas stoves are sources of carbon monoxide indoors. Breathing carbon monoxide reduces the amount of oxygen that reaches the body's tissues and organs.

Figure 17

Active Reading **3. Summarize** How have these trees been affected by air pollution?

_____

_____

_____

_____

Active Reading

4. NGSSS Check Point Out (Circle) types of air pollution. SC.7.E.6.6

Active Reading

FOLDABLES® LA.7.2.2.3

Make a two-tab book. Label the tabs as illustrated. Use your Foldable to record factors that increase or decrease air pollution.

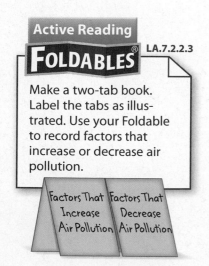

Factors That Increase Air Pollution | Factors That Decrease Air Pollution

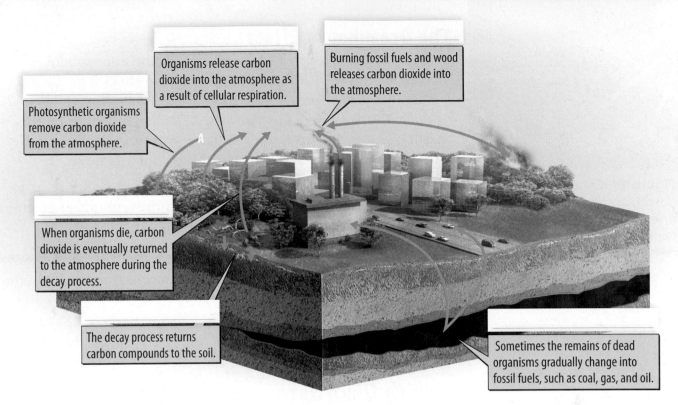

Photosynthetic organisms remove carbon dioxide from the atmosphere.

Organisms release carbon dioxide into the atmosphere as a result of cellular respiration.

Burning fossil fuels and wood releases carbon dioxide into the atmosphere.

When organisms die, carbon dioxide is eventually returned to the atmosphere during the decay process.

The decay process returns carbon compounds to the soil.

Sometimes the remains of dead organisms gradually change into fossil fuels, such as coal, gas, and oil.

Global Warming and the Carbon Cycle

Air pollution affects natural cycles on Earth. For example, burning fossil fuels for electricity, heating, and transportation releases substances that cause acid precipitation. Burning fossil fuels also releases carbon dioxide into the atmosphere, as shown in **Figure 18.** An increased concentration of carbon dioxide in the atmosphere can lead to **global warming**, *an increase in Earth's average surface temperature.* Earth's temperature has increased about 0.7°C over the past 100 years. Scientists estimate it will rise an additional 1.8 to 4.0°C over the next 100 years. Even a small increase in Earth's average surface temperature can cause widespread problems.

Effects of Global Warming

Warmer temperatures can cause ice to melt, making sea levels rise. Higher sea levels can cause flooding along coastal areas. In addition, warmer ocean waters might lead to an increase in the intensity and frequency of storms.

Global warming also can affect the kinds of living things found in ecosystems. Some hardwood trees, for example, do not thrive in warm environments. These trees will no longer be found in some areas if temperatures continue to rise.

5. NGSSS Check Relate How are global warming and the carbon cycle related? SC.7.E.6.6

_____

_____

Figure 18 🔑 Some human activities can increase the amount of carbon dioxide in the atmosphere.

Active Reading 6. Organize Label each process in the carbon cycle as a process that either uses carbon or releases carbon.

WORD ORIGIN

particulate
from Latin *particula*, means "small part"

Table 1 Air Quality Index

Ozone Concentration (parts per million)	Air Quality Index Values	Air Quality Description	Preventative Actions
0.0 to 0.064	0 to 50	good	No preventative actions needed.
0.065 to 0.084	51 to 100	moderate	Highly sensitive people should limit prolonged outdoor activity.
0.085 to 0.104	101 to 150	unhealthy for sensitive groups	Sensitive people should limit prolonged outdoor activity.
0.105 to 0.124	151 to 200	unhealthy	All groups should limit prolonged outdoor activity.
0.125 to 0.404	201 to 300	very unhealthy	Sensitive people should avoid outdoor activity. All other groups should limit outdoor activity.

Measuring Air Quality

Some pollutants, such as smoke from forest fires, are easily seen. Other pollutants, such as carbon monoxide, are invisible. How can people know when levels of air pollution are high?

The EPA works with state and local agencies to measure and report air quality. *The **Air Quality Index** (AQI) is a scale that ranks levels of ozone and other air pollutants.* Study the AQI for ozone in **Table 1.** It uses color codes to rank ozone levels on a scale of 0 to 300. Although ozone in the upper atmosphere blocks harmful rays from the Sun, ozone that is close to Earth's surface can cause health problems, including throat irritation, coughing, and chest pain. The EPA cautions that no one should do physical activities outside when AQI values reach 300.

Inquiry
LAB STATION
Try It!
SC.7.E.6.6

MiniLab *What is in the air?* at
connectED.mcgraw-hill.com

Apply It! After you complete the lab, answer these questions.

1. What is the AQI value for your area in Florida today?

2. How would you describe the air quality?

3. How will the air quality affect your daily activities?

Hybrid car

Solar car

Figure 21 Energy-efficient and renewable-energy vehicles help reduce air pollution.

Positive Actions

Countries around the world are working together to reduce air pollution. For example, 190 countries, including the United States, have signed the Montreal Protocol to phase out the use of CFCs. Levels of CFCs have since decreased. The Kyoto Protocol aims to reduce emissions of greenhouse gases. Currently, 184 countries have accepted the agreement.

National Initiatives

In the United States, the Clean Air Act sets limits on the amount of certain pollutants that can be released into the air. Some states, such as California, have established their own emissions standards for motor vehicles. Other states, such as Florida, have adopted or plan to adopt standards similar to California's in an attempt to reduce air pollution.

Since the Clean Air Act was passed in 1970, amounts of carbon monoxide, ozone near Earth's surface, and acid precipitation-producing substances have decreased by more than 50 percent. Toxins from industrial factories have gone down by 90 percent.

Cleaner Energy

Using renewable energy resources such as solar power, wind power, and geothermal energy to heat homes helps reduce air pollution. Recall that renewable resources are resources that can be replaced by natural processes in a relatively short amount of time. People also can invest in more energy-efficient appliances and vehicles. The hybrid car shown in **Figure 21** uses both a battery and fossil fuels for power. It is more energy efficient and emits less pollution than vehicles that are powered by fossil fuels alone. The solar car shown in **Figure 21** uses only the Sun's energy for power.

Active Reading **9. Compare** How do hybrid cars and solar cars help reduce air pollution?

_____

_____

How can you help?

Reducing energy use means that fewer pollutants are released into the air. You can turn the thermostat down in the winter and up in the summer to save energy. You can walk to the store or use public transportation. Each small step you take to conserve energy helps improve air, water, and soil quality.

10. NGSSS Check Recommend How can people help prevent air pollution? SC.7.E.6.6

_____

Active Reading

FOLDABLES® Chapter Project

Assemble your lesson Foldables as shown to make a Chapter Project. Use the project to review what you have learned in this chapter.

Environmental Impacts

Use Vocabulary

1 Use the term *carrying capacity* in a sentence.

_____

_____

2 Distinguish between desertification and deforestation.

_____

_____

3 Planting trees to replace logged trees is called

_____

4 Distinguish between point-source and nonpoint-source pollution.

_____

_____

_____

5 Define *greenhouse effect* in your own words.

_____

_____

6 Solid and liquid particles in the air are called

_____

Link Vocabulary and Key Concepts

Use vocabulary terms from the previous page to complete the concept map.

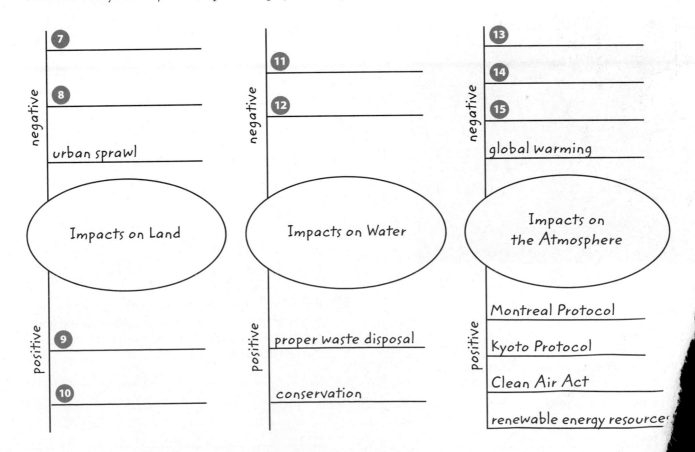

negative

7 _____

8 _____

urban sprawl

Impacts on Land

positive

9 _____

10 _____

negative

11 _____

12 _____

Impacts on Water

positive

proper waste disposal

conservation

negative

13 _____

14 _____

15 _____

global warming

Impacts on the Atmosphere

positive

Montreal Protocol

Kyoto Protocol

Clean Air Act

renewable energy resources

Fill in the correct answer choice.

🔑 Understand Key Concepts

1 Which is a population? **LA.7.2.2.3**
- Ⓐ all the animals in a zoo
- Ⓑ all the living things in a forest
- Ⓒ all the people in a park
- Ⓓ all the plants in a meadow

2 Which had the greatest influence on the growth of the human population? **SC.7.E.6.6**
- Ⓐ higher death rates
- Ⓑ increased marriage rates
- Ⓒ medical advances
- Ⓓ widespread disease

3 What percentage of species on Earth live in tropical rain forests? **SC.7.E.6.6**
- Ⓐ 10 percent
- Ⓑ 25 percent
- Ⓒ 50 percent
- Ⓓ 75 percent

4 What process is illustrated in the diagram below? **SC.7.E.6.6**

Newly planted trees

- Ⓐ desertification
- Ⓑ recycling
- Ⓒ reforestation
- Ⓓ waste management

Which could harm human health? **SC.7.E.6.6**
- Ⓐ compost
- Ⓑ hazardous waste
- Ⓒ nitrogen
- Ⓓ reclamation

Critical Thinking

6 **Decide** Rates of human population growth are higher in developing countries than in developed countries. Yet people in developed countries use more resources than those in developing countries. Should international efforts focus on reducing population growth or reducing resource use? Explain. **SC.7.E.6.6**

7 **Relate** How does the carrying capacity for a species help regulate its population growth? **SC.7.E.6.6**

8 **Assess** your personal impact on the environment today. Include both positive and negative impacts on soil, water, and air. **SC.7.E.6.6**

9 **Infer** How does deforestation affect levels of carbon in the atmosphere? **SC.7.E.6.6**

10 **Role-Play** Suppose you are a soil expert advising a farmer on the use of fertilizers. What would you tell the farmer about the environmental impact of the fertilizers? **SC.7.E.6.6**

11 **Create** Use the data below to create a circle graph showing waste disposal methods in the United States. SC.7.E.6.6

Waste Disposal Methods—United States	
Method	**Percent of Waste Disposed**
Landfill	55%
Recycling/composting	31%
Incineration	14%

Writing in Science

12 **Compose** On a separate piece of paper, compose a letter to a younger student to help him or her understand air pollution. The letter should identify the different kinds of pollution and explain their causes. SC.7.E.6.6

Big Idea Review

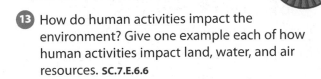

13 How do human activities impact the environment? Give one example each of how human activities impact land, water, and air resources. SC.7.E.6.6

_____

_____

_____

14 What positive actions can people take to reduce or reverse negative impacts on the environment? SC.7.E.6.6

_____

_____

_____

Math Skills   MA.6.A.3.6

Use Percentages

15 Between 1960 and 1990, the number of people per square mile in the United States grew from 50.7 people to 70.3 people. What was the percent change?

16 Between 1950 and 1998, the rural population in the United States decreased from 66.2 million to 53.8 million people. What was the percent change in rural population?

17 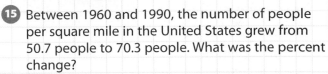 During the twentieth century, the population of Florida increased from 530,000 people to 16 million people. What was the percent change during the century?

Florida NGSSS

Benchmark Practice

Fill in the correct answer choice.

Multiple Choice

1 Which action can help restore land that has been disturbed by mining? SC.7.E.6.6

Ⓐ deforestation

Ⓑ desertification

Ⓒ preservation

Ⓓ reclamation

2 Which is a consequence of deforestation? SC.7.E.6.6

Ⓕ Animal habitats are destroyed.

Ⓖ Carbon in the atmosphere is reduced.

Ⓗ Soil erosion is prevented.

Ⓘ The rate of extinction is slowed.

Use the graph below to answer question 3.

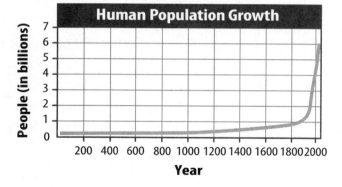

3 What can be inferred about the population over the last 200 years? SC.7.E.6.6

Ⓐ Land use has remained the same.

Ⓑ Population increased sharply.

Ⓒ Deforestation has decreased.

Ⓓ Water consumption decreased.

4 Which accounts for the least water use in the United States? SC.7.E.6.6

Ⓕ electricity-generating power plants

Ⓖ irrigation of agricultural crops

Ⓗ mines, livestock, and aquaculture

Ⓘ public supply, including houses

5 Which is a point source of water pollution? SC.7.E.6.6

Ⓐ discharge pipes

Ⓑ runoff from farms

Ⓒ runoff from construction sites

Ⓓ runoff from urban areas

6 Which air pollutant contains ozone? SC.7.E.6.6

Ⓕ acid precipitation

Ⓖ carbon monoxide

Ⓗ CFCs

Ⓘ smog

Use the figure below to answer question 7.

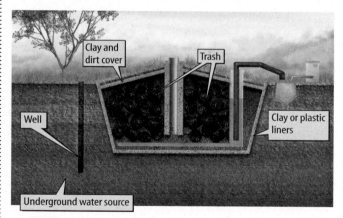

7 What is the function of the well in the figure above? SC.7.E.6.6

Ⓐ to generate electricity

Ⓑ to monitor quality of groundwater

Ⓒ to prevent pollution of nearby land

Ⓓ to treat hazardous water

8 Which action helps prevent water pollution? SC.7.E.6.6

Ⓕ pouring motor oil on the ground

Ⓖ putting hazardous wastes in the trash

Ⓗ using fertilizers when gardening

Ⓘ using vinegar when cleaning

Multiple Choice *Bubble the correct answer.*

Use the graph below to answer questions 1 and 2.

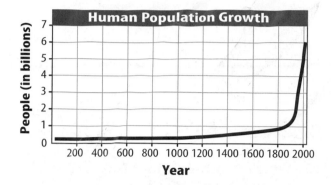

1. Based on the graph above, around what year did technological advances lead to sharply increased human population growth? **SC.7.E.6.6**

- (A) 1200
- (B) 1400
- (C) 1700
- (D) 1900

2. The graph above shows human population growth. How would the line in the graph above most likely differ if the Industrial Revolution had not taken place? **SC.7.E.6.6**

- (F) The line between 1600 and 1800 would dip sharply.
- (G) The line between 1600 and 1800 would have a much sharper increasing angle.
- (H) The line between 1800 and 2000 would look more like it does between 1600 and 1800.
- (I) The line between 1800 and 2000 would point sharply downward instead of upward.

Use the graph below to answer questions 3 and 4.

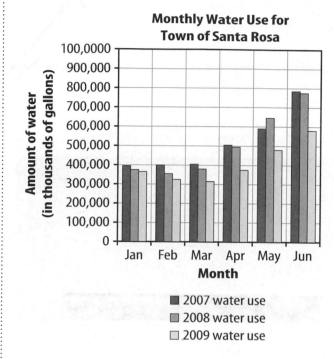

3. In the graph above, which month in 2009 shows the least amount of water used? **SC.7.E.6.6**

- (A) January
- (C) April
- (B) March
- (D) June

4. How would the graph above differ if the population of Santa Rosa had increased drastically during 2008? **SC.7.E.6.6**

- (F) The bars for 2008 would be higher at the beginning of the year.
- (G) The bars for 2008 would increase, but the bars for 2009 would stay the same.
- (H) The bars for 2009 would be much higher than they are now.
- (I) The bars for both 2008 and 2009 would be lower than they are now.

Copyright © Glencoe/McGraw-Hill, a division of The McGraw-Hill Companies, Inc.

9 What results from an increase of nitrogen, sulfur, and other pollutants reacting with water and oxygen in the air? **SC.7.E.6.6**

Ⓐ particulate matter

Ⓑ acid precipitation

Ⓒ photochemical smog

Ⓓ global warming

Use the figure below to answer question 10.

10 Which term describes what is shown in the figure above? **SC.7.E.6.6**

Ⓕ acid precipitation

Ⓖ global warming

Ⓗ greenhouse effect

Ⓘ urban sprawl

11 Which results in a human impact to Earth? **SC.7.E.6.6**

Ⓐ reclamation

Ⓑ reforestation

Ⓒ urban sprawl

Ⓓ water conservation

Use the figure below to answer questions 12 and 13.

12 Which is NOT directly related to the mining of coal? **SC.7.E.6.6**

Ⓕ sediment filling streams

Ⓖ carbon dioxide entering the air

Ⓗ acidification of lakes and streams

Ⓘ contamination of groundwater drinking supplies

13 Strip-mining operations often begin with which of the following? **SC.7.E.6.6**

Ⓐ deforestation

Ⓑ urbanization

Ⓒ desertification

Ⓓ reclamation

14 Acid rain has what affect on the pH of natural lakes and streams? **SC.7.E.6.6**

Ⓕ It increases the pH.

Ⓖ It decreases the pH

Ⓗ It stabilizes the pH.

Ⓘ It doesn't affect the pH.

NEED EXTRA HELP?

If You Missed Question...	1	2	3	4	5	6	7	8	9	10	11	12	13	14
Go to Lesson...	2	2	1	3	3	4	2	3	4	2	2	4	4	4

Benchmark Mini-Assessment Chapter 7 • Lesson 2

Multiple Choice *Bubble the correct answer.*

Use the image below to answer questions 1 through 4.

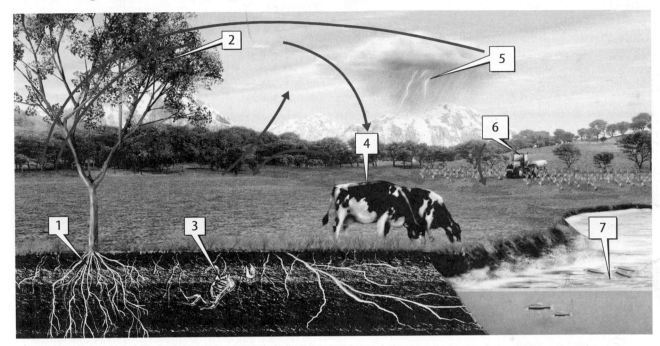

1. The figure above shows the nitrogen cycle. Which of the following steps involves decomposers? **SC.7.E.6.6**

- (A) 1
- (B) 2
- (C) 3
- (D) 5

2. A bacteriophage, which is a virus that infects bacteria, is accidentally released into the environment. Which step of the nitrogen cycle would be most affected? **SC.7.E.6.6**

- (F) 1
- (G) 3
- (H) 4
- (I) 5

3. In the figure, which number indicates the area that would be most negatively affected by the use of excess fertilizers? **SC.7.E.6.6**

- (A) 2
- (B) 3
- (C) 5
- (D) 7

4. What takes place during step 2 of the nitrogen cycle? **SC.7.E.6.6**

- (F) Plants release nitrogen into the air.
- (G) Plants use nitrogen to make proteins.
- (H) Plants break down nitrogen compounds, releasing them into the environment.
- (I) Plants convert nitrogen to a form that can be used by other organisms.

Copyright © Glencoe/McGraw-Hill, a division of The McGraw-Hill Companies, Inc.

Benchmark Mini-Assessment   Chapter 7 • Lesson 3

Multiple Choice *Bubble the correct answer.*

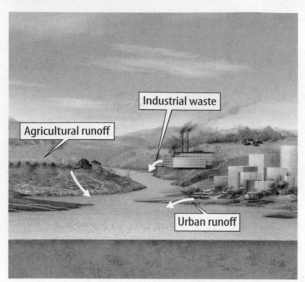

1. The image above shows various human activities that produce pollution. Which statement below is true? **SC.7.E.6.6**

    (A) They are all nonpoint sources of pollution.

    (B) They are all point sources of pollution.

    (C) Only industrial waste is a point source.

    (D) Only urban runoff is a point source.

2. Which of the following is an example of point-source pollution? **SC.7.E.6.6**

    (F) A container ship strikes a bridge, releasing fuel into the water.

    (G) Dirt washes away from a housing division during construction.

    (H) Fertilizer and pesticides run off fields during a rainstorm.

    (I) A flood overflows the underground sewage system of a city.

Use the graph below to answer questions 3 and 4.

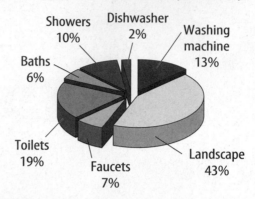

Typical Water Use by a Family

3. The circle graph above shows how Elena's family uses water. Which change would produce the greatest reduction in water use in Elena's household? **SC.7.E.6.6**

    (A) They all reduce their shower time.

    (B) They stop using their dishwasher.

    (C) They change their washing machine to a more efficient model.

    (D) They use plants in their yard that do not need much water to survive.

4. Most, but not all, of the water used in a house is treated in a plant to remove harmful chemicals. Based on the circle graph above, which form of water use might contribute the most to nonpoint-source pollution? **SC.7.E.6.6**

    (F) baths

    (G) faucets

    (H) landscape

    (I) showers

Copyright © Glencoe/McGraw-Hill, a division of The McGraw-Hill Companies, Inc.

Multiple Choice *Bubble the correct answer.*

Use the figure below to answer questions 1 and 2.

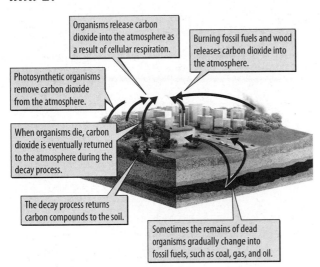

Use the graph below to answer questions 3 and 4.

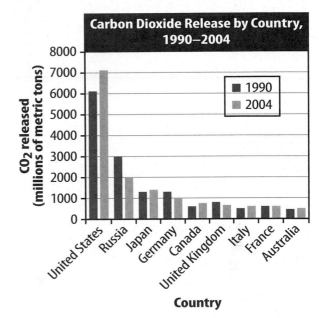

1. Based on the image above, how would planting trees affect the carbon cycle? SC.7.E.6.6

   Ⓐ The trees would eventually become fossil fuels.

   Ⓑ The trees would release carbon compounds into the soil.

   Ⓒ The trees would remove more carbon from the atmosphere.

   Ⓓ The trees would support decomposers, adding carbon to the atmosphere.

2. Which part of the carbon cycle shown in the figure above contributes the most to the greenhouse effect? SC.7.E.6.6

   Ⓕ Burning fossil fuels and wood releases carbon dioxide into the atmosphere.

   Ⓖ The decay process returns carbon dioxide to the soil.

   Ⓗ Photosynthetic organisms remove carbon dioxide from the atmosphere.

   Ⓘ The remains of dead organisms gradually change into fossil fuels.

3. In the graph above, which country decreased their carbon dioxide emissions by the greatest percentage? SC.7.E.6.6

   Ⓐ Canada

   Ⓑ France

   Ⓒ Germany

   Ⓓ Russia

4. Based on the graph above, which statement is true? SC.7.E.6.6

   Ⓕ France released the smallest amount of greenhouse gases.

   Ⓖ If the United States halved their emissions, they would release fewer greenhouse gases than Russia.

   Ⓗ Japan released more greenhouse gases than any other country except the United States.

   Ⓘ The United States released the largest amount of greenhouse gases.

Copyright © Glencoe/McGraw-Hill, a division of The McGraw-Hill Companies, Inc.

Notes

Notes

1660
Robert Hooke publishes the wave theory of light, comparing light's movement to that of waves in water.

1705
Francis Hauksbee experiments with a clock in a vacuum and proves that sound cannot travel without air.

1820
Danish physicist Hans Christian Ørsted publishes his discovery that an electric current passing through a wire produces a magnetic field.

1878
Thomas Edison develops a system to provide electricity to homes and businesses using locally generated and distributed direct current (DC) electricity.

1882
Thomas Edison develops and builds the first electricity-generating plant in New York City, which provides 110 V of direct current to 59 customers in lower Manhattan.

1883
The first standardized incandescent electric lighting system using overhead wires begins service in Roselle, New Jersey.

1890s
Physicist Nikola Tesla introduces alternating current (AC) by inventing the alternating current generator, allowing electricity to be transmitted at higher voltages over longer distances.

1947
Chuck Yeager becomes the first pilot to travel faster than the speed of sound.

? Inquiry
Visit ConnectED for this unit's STEM activity.

Graphs

Have you ever felt a shock from static electricity? That electric energy is similar to the energy in lightning, such as in **Figure 1,** only smaller. Scientists investigate what causes lightning and where it will occur. They use graphs to learn about lightning in different places and at different times. A **graph** is a type of chart that shows relationships between variables. Graphs organize and summarize data in a visual way. Three of the most common graphs are circle line, bar graphs, and circle graphs.

Types of Graphs

Line Graphs

A line graph is used when you want to analyze how a change in one variable affects another variable. This line graph shows how the average number of lightning flashes changes over time in Illinois. Time is plotted on the x-axis. The average number of lightning flashes is plotted on the y-axis. Each data point indicates the average number of flashes recorded during that hour. A line connects the data points to help determine any trends.

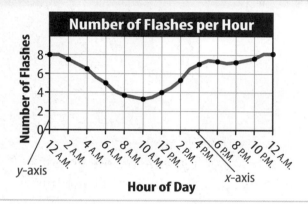

Bar Graphs

When you want to compare amounts in different categories, you use a bar graph. The horizontal axis often contains categories instead of numbers. This bar graph shows the average number of lightning flashes that occur in different states. On average, about 9.8 lightning flashes strike each square kilometer of land in Florida every year. Florida has more lightning flashes per square kilometer than all other states shown.

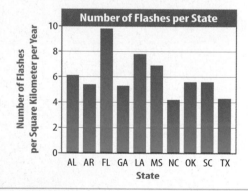

Circle Graphs

To show how the parts of something relate to the whole, use a circle graph. This circle graph shows the average percentage of lightning flashes each U.S. region receives in a year.

Active Reading **1. Complete** Fill in the blanks to complete the paragraph below.
The graph shows that the _____ region receives about 14 percent of all lightning flashes that strike each year. You can also determine that the _____ receives the most lightning in a given year.

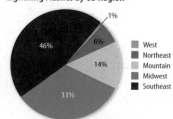

Lightning Flashes by US Region

Active Reading **2. Infer** Determine in which region of the U.S. Florida is located. How does the Florida information from the bar graph above relate to the circle graph data?

_____

_____

Line Graphs and Trends

Suppose you are planning a picnic in an area that experiences quite a bit of lightning. When would be the safest time to go? First, you gather data about the average number of lightning flashes per hour. Next, you plot the data on a line graph and analyze trends. Trends are patterns in data that help you find relationships among the data and make predictions.

Active Reading 3. **Determine** Highlight the definition of *trend* as used with a graph.

Follow the orange line on the line graph from 12 A.M. to 10 A.M. in **Figure 2.** Notice that the line slopes downward, indicated by the green arrow. A downward slope means that as measurements on the *x*-axis increase, measurements on the *y*-axis decrease. So, as time passes from 12 A.M. to 10 A.M., the number of lightning flashes decreases.

Follow the orange line on the graph from 12 P.M. to 5 P.M. Notice that the line slopes upward, indicated by the blue arrow. An upward slope means that as the measurements on the *x*-axis increase, the measurements on the *y*-axis also increase. So, as time passes from 12 P.M. to 5 P.M., the number of lightning flashes increases.

Active Reading 4. **Predict** Analyze the line graph to determine when you would have the least risk of lightning during your picnic.

▲ **Figure 1** Scientists study lightning to get a better understanding of what causes it and to predict when it will occur.

Inquiry **⊙LAB STATION** MA.6.A.3.6 **Try It!**

MiniLab *When does lightning strike?* at connectED.mcgraw-hill.com

Apply It!
After you complete the lab, analyze and answer the following questions.

1. **Construct** Create a double line graph of data from Illinois, found on the previous page, and data from Florida found in the lab.

2. **Predict** Compare the trends for the different states as shown on your graph.

Figure 2 The slope of a line in a line graph shows the relationship between the variable on the *x*-axis and the variable on the *y*-axis. ▼

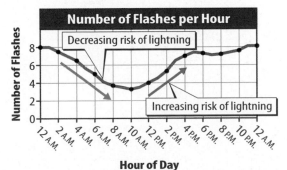

Number of Flashes per Hour

Decreasing risk of lightning

Increasing risk of lightning

Number of Flashes

12 A.M. 2 A.M. 4 A.M. 6 A.M. 8 A.M. 10 A.M. 12 P.M. 2 P.M. 4 P.M. 6 P.M. 8 P.M. 10 P.M. 12 A.M.

Hour of Day

Notes

Energy Transfers and TRANSFORMATIONS

ESSENTIAL QUESTIONS

 What is the law of conservation of energy?

 How is energy transformed and transferred?

 What are renewable and nonrenewable energy resources?

Vocabulary

law of conservation of energy p. 308

energy transfer p. 309

energy transformation p. 309

work p. 309

open system p. 311

closed system p. 311

renewable energy resource p. 312

nonrenewable energy resource p. 314

Florida NGSSS

SC.7.N.1.1 Define a problem from the seventh grade curriculum, use appropriate reference materials to support scientific understanding, plan and carry out scientific investigation of various types, such as systematic observations or experiments, identify variables, collect and organize data, interpret data in charts, tables, and graphics, analyze information, make predictions, and defend conclusions.

SC.7.P.11.2 Investigate and describe the transformation of energy from one form to another.

SC.7.P.11.3 Cite evidence to explain that energy cannot be created nor destroyed, only changed from one form to another.

MA.6.A.3.6 Construct and analyze tables, graphs, and equations to describe linear functions and other simple relations using both common language and algebraic notation.

LA.7.2.2.3 The student will organize information to show understanding (e.g., representing main ideas within text through charting, mapping, paraphrasing, summarizing, or comparing/contrasting);

Inquiry **Launch Lab** SC.7.N.1.1

20 minutes

How does a flashlight work?

If the lights go out, you might turn on a flashlight. You know that when you flip the switch, the light will go on. What happens? How does the flashlight work?

Procedure

1. Read and complete a lab safety form.

2. Examine a **flashlight.** List the parts that you can see. Predict the types of energy involved in the operation of the flashlight.

3. Use the switch to turn the flashlight on. What do you think happened inside the flashlight to produce the light? Write your ideas in the Data and Observation section.

4. Take the flashlight apart. Discuss the kinds of energy involved in producing light.

Data and Observations

Think About This

1. Was light the only type of energy produced? Why or why not?

2. **Key Concept** Describe the different types of energy involved in a flashlight. Draw a sequence diagram showing how each form of energy changes to the next form.

Warm and Cozy?

1. This penguin chick lives in one of the coldest places on Earth—Antarctica. The chick is standing on its parent's feet to insulate its feet from the ice. This helps prevent thermal energy of the chick's body from transferring to the ice. The chick cuddles with its parent to absorb thermal energy from its parent's body. Why do a penguin's feet need to be insulated from ice?

_____

_____

_____

_____

Active Reading **2. Organize** Each time you read a heading on the page, make a flash card with that heading. Then write the main idea of the information under the heading of the card.

Law of Conservation of Energy

Think about turning on the flashlight in the Launch Lab. *The law of conservation of energy says that energy can be transformed from one form to another, but it cannot be created or destroyed.* In the flashlight shown in **Figure 10,** chemical energy of the battery is transformed to electric energy (moving electrons) that moves through the contact strip to the bulb. The electric energy is transformed into radiant energy and thermal energy in the light-bulb. The law of conservation of energy indicates that the amount of radiant energy that shines out of the flashlight cannot be greater than the chemical energy stored in the battery.

 3. NGSSS Check **Recognize** Underline the law of conservation of energy. SC.7.P.11.2

The amount of radiant energy given off by the flashlight is less than the chemical energy in the battery. Where is the missing energy? As you read this lesson, you will learn that in every energy transformation, some energy transfers to the environment.

Figure 10 Several energy changes occur in a flashlight.

Active Reading **4. Name** Fill in the blanks to understand the energy changes in a flashlight.

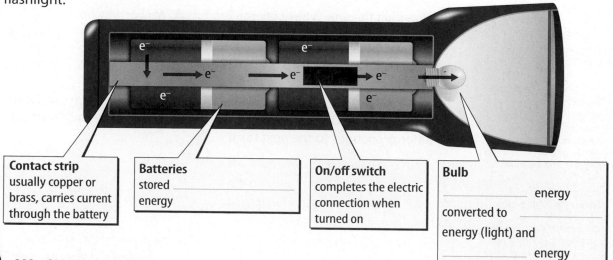

Contact strip
usually copper or brass, carries current through the battery

Batteries
stored _____ energy

On/off switch
completes the electric connection when turned on

Bulb
_____ energy converted to _____ energy (light) and _____ energy

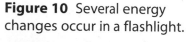

Chemical energy is transformed to mechanical energy.

Mechanical energy is transferred to the tennis ball.

Figure 11 Energy transfers and transformations take place when the tennis player hits the ball.

Active Reading **5. Identify** What is one energy transformation that occurs as the ball moves through the air?

_____

_____

_____

Energy Transfer

What happens when the tennis player in **Figure 11** hits the ball with the racket? The mechanical energy of the racket changes the movement of the ball, and the ball's mechanical energy increases. *When energy moves from one object to another without changing form, an* **energy transfer** *occurs.* The tennis racket transfers mechanical energy to the tennis ball.

Energy Transformation

Where does the mechanical energy in the tennis player's racket come from? Chemical energy stored in the player's muscles changes to mechanical energy when she swings her arm. *When one form of energy is converted to another form of energy, an* **energy transformation** *occurs.*

 6. NGSSS Check **Identify** What is an energy transfer and an energy transformation that occurs when someone plays a guitar? **SC.7.P.11.2**

_____

_____

Energy and Work

You might be thinking that reading about energy is a lot of work. But to a scientist, it's not work at all. To a scientist, **work** *is the transfer of energy that occurs when a force makes an object move in the direction of the force. Work is only being done while the force is acting on the object.* As the tennis player swings the racket, the racket applies a force to the ball for about a meter. Although the ball moves 10 m, work is done by the racket only during the time the racket applies a force to the ball. When the ball separates from the racket, the racket no longer does work.

Suppose the tennis player is standing still before she serves the ball. She is using her muscles to hold the ball. Is she doing work on the ball? No; because the ball is not moving, she is not doing work. If a force does not make an object move in the direction of the force, it does no work on the object.

Math Skills MA.6.A.3.6

Use a Formula
The amount of work done on an object is calculated using the formula $W = F \times d$, where W = work, F = the force applied to the object, and d = the distance the force moves the object. For example, a student slides a library book across a table. The student pushes the book with a force of **8.5 newtons (N)** a distance of **0.3 m**. The book slides a total distance of 1 m. How much work is done on the book?

$W = F \times d$

$W = 8.5 \text{ N} \times 0.30 \text{ m} = 2.55 \text{ N·m} = 2.6 \text{ J}$

Note: 1 N·m = 1 J, so 2.6 J of work was done on the book.

Practice
7. A student lifts a backpack straight up with a force of 53.5 N for a distance of 0.65 m. How much work is done on the backpack? Use the space below for your calculations.

Inefficiency of Energy Transformations

When a tennis player hits a ball with a racket, most of the mechanical energy of the racket transfers to the ball, but not all of it. You know when a ball hits a racket because you can hear a sound. Some of the mechanical energy of the racket is transformed to sound energy. In addition, some of the mechanical energy of the racket is transformed to thermal energy. The temperature of the racket, the ball, and the air surrounding both objects increases slightly. Anytime there is an energy transformation or energy transfer, some energy is transformed into thermal energy.

Active Reading

8. Explain What makes an energy transformation inefficient?

_____

_____

_____

Active Reading

9. Summarize Explain the energy transformations that occur when a tennis racket hits a tennis ball using the graphic organizer below.

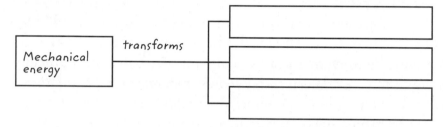

Recall the flashlight at the beginning of the lesson. The transformation of chemical energy of the battery to radiant energy from the lightbulb is inefficient, too. As the electric energy moves through the circuit, some electric energy transforms to thermal energy. When electric energy transforms to radiant energy in the lightbulb, more energy transforms to thermal energy. In some flashlights, the bulb is warm to the touch.

Recall that the law of conservation of energy says that energy cannot be created or destroyed. When scientists say that energy transformations are inefficient, they do not mean that energy is destroyed. Energy transformations are inefficient because not all the energy that is transformed to another form of energy is usable.

Inquiry

LAB STATION Try It!

SC.7.N.1.1, SC.7.P.11.2

MiniLab *How can you transfer energy?* at connectED.mcgraw-hill.com

Apply It! After you complete the lab, answer these questions.

1. Describe How can you make your least effective method of moving the cork more efficient?

2. Explain How is work the transfer of energy?

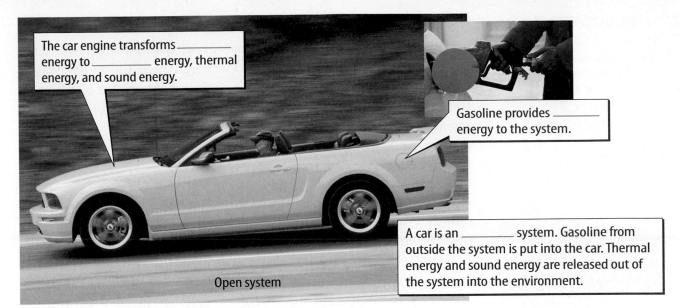

The car engine transforms _____ energy to _____ energy, thermal energy, and sound energy.

Gasoline provides _____ energy to the system.

A car is an _____ system. Gasoline from outside the system is put into the car. Thermal energy and sound energy are released out of the system into the environment.

Open system

Open Systems

In Lesson 1, you read that scientists often study the energy of systems. A car, as shown in **Figure 12,** is a system. Chemical energy of the fuel is transformed to mechanical energy of the moving car. Because energy transformations are inefficient, some of the chemical energy transforms to thermal energy and sound energy that are released to the environment. An **open system,** such as a car engine, *is a system that exchanges matter or energy with the environment.*

Closed Systems

Can you think of a system that does not exchange energy with the environment? What about a flashlight? A flashlight releases radiant energy and thermal energy into the environment. What about your body? You eat food, which contains chemical energy and comes from the environment. Your body also releases several types of energy into the environment, including thermal energy, mechanical energy, and sound energy. *A* **closed system** *is a system that does not exchange matter or energy with the environment.* In reality, there are no closed systems. Every physical system transfers some energy to or from its environment. Scientists use the idea of a closed system to study and model the movement of energy.

Energy Transformations and Electric Energy

You probably have heard someone say, "turn off the lights, you're wasting energy." This form of energy is electric energy. Most appliances you use every day require electric energy. Where does this energy come from?

Figure 12 A car is a system. Gasoline is an input. Thermal energy and sound energy are outputs.

Active Reading 10. **Write** Fill in the missing information in the boxes.

Active Reading 11. **Review** Underline the definitions of open system and closed system.

WORD ORIGIN
system
from Greek *systema*, means "whole made of several parts"

Changes Between Liquids and Gases

What happens when ice melts? As thermal energy transfers to the ice, the particles move faster and faster. The average kinetic energy of the water particles that make up ice increases and the ice melts. The temperature of the ice continues to increase until it reaches 100°C. At 100°C, water begins to vaporize. **Vaporization** *is the change of state from a liquid to a gas*. While the water is changing state—from a liquid to a gas—the kinetic energy of the particles remains constant.

Liquids vaporize in two ways—boiling and evaporation. Vaporization that occurs within a liquid is called boiling. Vaporization that occurs at the surface of a liquid is called evaporation. Have you heard the term *water vapor?* The gaseous state of a substance that is normally a liquid or a solid at room temperature is called vapor. Because water is liquid at room temperature, its gaseous state is referred to as water vapor.

The reverse process also can occur. Removing thermal energy from a gas changes it to a liquid. The change of state from a gas to a liquid is condensation. For example, sometimes at night, water vapor condenses on grass and is called dew.

Active Reading **11. Identify** What are the two ways liquid can vaporize?

_____

_____

Changes Between Solids and Gases

Usually, water transforms from a solid to a liquid and then to a gas as it absorbs thermal energy. However, this is not always the case. On cold winter days, ice often changes directly to water vapor without passing through the liquid state. **Sublimation** is the change of state that occurs when a solid changes to a gas without passing through the liquid state. Dry ice, or solid carbon dioxide, sublimes as shown in **Figure 24**. Dry ice is used to keep foods frozen when they are shipped.

When thermal energy is removed from some materials, they undergo deposition. Deposition is the change of state from a gas directly to a solid without passing through the liquid state. Water vapor undergoes deposition when it freezes and forms frost, as shown in **Figure 24**.

SCIENCE USE V. COMMON USE

sublime

Science Use to change from a solid state to a gas state without passing through the liquid state

Common Use inspiring awe; supreme, outstanding, or lofty in thought or language

Figure 24 The solid dry ice is changing directly to a gas by sublimation (left). Water vapor in the air changed directly to the solid ice crystals on the plant leaves by deposition (right).

Active Reading **12. Compare** How are sublimation and deposition related?

_____

_____

Figure 25 🔑 The color variations of this thermogram show the temperature variations in the pan and stove burner. The temperature scale is from white (warmest) through red, yellow, green, cyan, blue, and black (coolest).

Active Reading 13. **Explain** Why are the handles black?

Conductors and Insulators

When you put a metal pan on a burner, the pan gets very hot. If the pan has a handle made of wood or plastic, such as the one in **Figure 25,** the handle stays cool. Why doesn't the handle get hot like the pan as a result of thermal conduction?

The metal that makes up the pan is a **thermal conductor,** *a material in which thermal energy moves quickly.* The atoms that make up thermal conductors have electrons that are free to move, transferring thermal energy easily. The material that makes up the pan's handles is a **thermal insulator,** *a material in which thermal energy moves slowly.* The electrons in thermal insulators are held tightly in place and do not transfer thermal energy easily.

Active Reading 14. **Contrast** How do thermal conductors differ from thermal insulators?

Inquiry LAB STATION Try It!

SC.7.P.11.1, SC.7.P.11.4

MiniLab *What affects the transfer of thermal energy?* at connectED.mcgraw-hill.com

Apply It! After you complete the lab, answer these questions.

1. **Examine** Contrast the ability of materials to transfer thermal energy in thermal conductors and thermal insulators.

2. **Compose** Describe the thermal energy transfers between particles that occur when you overheat soup for lunch, and then drop an ice cube in it to cool it off.

Visual Summary

The kinetic molecular theory explains how particles move in matter.

Thermal energy is transferred in various ways by particles and waves.

Materials vary in how well they conduct thermal energy.

Inquiry SC.7.N.1.1, SC.7.N.1.3, SC.7.P.11.2

LAB STATION **Try It!**

Skill Lab *Power Device with a Potato* at connectED.mcgraw-hill.com

Use Vocabulary

1 **Define** *temperature* in your own words.

_____

_____

2 **Explain** how heat is related to thermal energy.

_____

_____

Understand Key Concepts

3 **Summarize** the kinetic molecular theory.

_____

_____

_____

4 Which is NOT a way in which thermal energy is transferred?

Ⓐ conduction   Ⓒ radiation

Ⓑ convection   Ⓓ sublimation

5 **Differentiate** between a cloth safety belt and a metal buckle in terms of thermal conductors and insulators.

_____

_____

Interpret Graphics

6 **Summarize** Fill in the graphic organizer below showing the state-of-matter changes as thermal energy is added to ice. LA.7.2.2.3

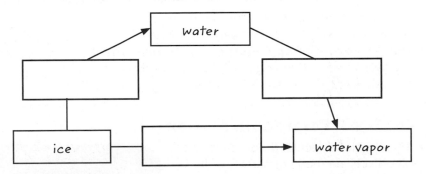

Critical Thinking

7 **Compare** You hold a 65°C cup of cocoa. Your hand is 37°C and the outside air is 6°C. Describe the flow of thermal energy. SC.7.P.11.4

_____

_____

_____

Chapter 8 Study Guide

 Think About It! Energy is involved in all physical processes; it is transferred when it moves from object to object without changing form. Energy is transformed when it is converted to another form. Energy is conserved during transfers and transformations.

🔑 Key Concepts Summary

Vocabulary

LESSON 1 Forms of Energy

- **Potential energy** is stored energy, and **kinetic energy** is energy of motion.
- Both **mechanical energy** and **thermal energy** involve kinetic energy and potential energy. Mechanical energy is the sum of the kinetic energy and the potential energy in a system of objects. Thermal energy is the sum of the kinetic energy and the potential energy in a system of particles.
- **Sound energy** and **radiant energy** are carried by waves.

energy p. 299

potential energy p. 299

chemical energy p. 300

nuclear energy p. 300

kinetic energy p. 301

electric energy p. 301

mechanical energy p. 302

thermal energy p. 302

wave p. 303

sound energy p. 303

radiant energy p. 304

LESSON 2 Energy Transfers and Transformations

- The **law of conservation of energy** says that energy can be transformed from one form to another, but it cannot be created or destroyed.
- Energy is transformed when it is converted from one form to another. It is transferred when it moves from one object to another.
- **Renewable energy resources** are resources that are replaced as fast as, or faster than they are used. **Nonrenewable energy resources** are resources that are available in limited quantities or are used faster than they can be replaced.

law of conservation of energy p. 308

energy transfer p. 309

energy transformation p. 309

work p. 309

open system p. 311

closed system p. 311

renewable energy resource p. 312

nonrenewable energy resource p. 314

LESSON 3 Particles in Motion

- The kinetic molecular theory says that all objects are made of particles; all particles are in constant, random motion; and the particles collide with each other and with the walls of their container.
- Thermal energy is transferred by **conduction, radiation,** and **convection.**
- A **thermal conductor** transfers thermal energy easily and a **thermal insulator** does not transfer thermal energy easily.

temperature p. 319

heat p. 320

conduction p. 321

radiation p. 321

convection p. 321

vaporization p. 323

thermal conductor p. 324

thermal insulator p. 324

FOLDABLES® Chapter Project

Assemble your lesson Foldables as shown to make a Chapter Project. Use the project to review what you have learned in this chapter.

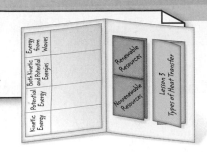

Use Vocabulary

1 Compare and contrast sound energy and radiant energy.

_____

_____

2 Explain why chemical energy and nuclear energy are both considered potential energy.

_____

_____

3 Describe how an open system differs from a closed system.

_____

_____

4 The energy of moving electrons is _____

_____ .

5 The energy carried by electromagnetic waves is

_____ .

6 Define *work* in your own words.

_____

7 Define *conduction* in your own words.

_____

Link Vocabulary and Key Concepts

Use vocabulary terms from the previous page to complete the concept map.

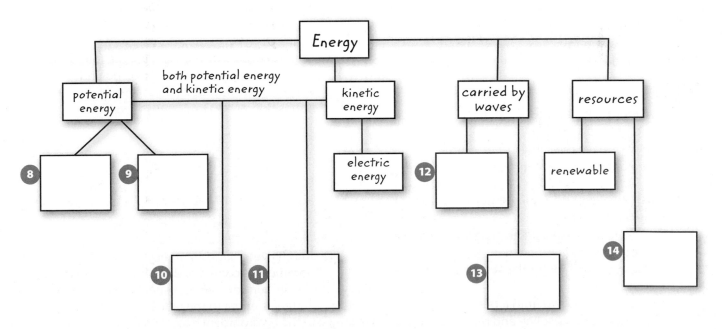

Fill in the correct answer.

🔑 Understand Key Concepts

1 What type of energy does the statue have? SC.7.P.11.3

- Ⓐ electric energy
- Ⓑ mechanical energy
- Ⓒ sound energy
- Ⓓ thermal energy

2 Which is a form of energy that cannot be stored? SC.7.P.11.3

- Ⓐ chemical energy
- Ⓑ gravitational potential energy
- Ⓒ nuclear energy
- Ⓓ sound energy

3 Waves can transfer energy, but they SC.7.P.10.3

- Ⓐ cannot carry sounds.
- Ⓑ do not move matter.
- Ⓒ always have the same wavelength.
- Ⓓ are unable to move through empty space.

4 Which involves ONLY an energy transfer? SC.7.P.11.2

- Ⓐ A boy turns on an electric toaster to warm a piece of bread.
- Ⓑ A can of juice cools off in a cooler on a hot summer day.
- Ⓒ A cat jumps down from a tree branch.
- Ⓓ A truck burns gasoline and moves 20 km.

Critical Thinking

5 **Identify** all the different forms of energy and all the energy transformations that you see in the picture below. SC.7.P.11.2

6 **Compare** the energy transformations that take place in the human body to the energy transformations that take place in a gasoline-powered car. SC.7.P.11.2

7 **Compare and contrast** each of the following terms: melting and freezing, boiling and evaporation, and sublimation and deposition. SC.7.P.11.1

8 **Judge** Determine whether the following statement is correct and explain your reasoning: The amount of chemical energy in a flashlight's battery is equal to the radiant energy transferred to the environment. SC.7.P.11.3

9 **Evaluate** Using what you read about thermal energy, explain why sidewalks are built as panels with space between them. **LA.7.2.2.3**

10 **Explain** Convection space heaters are small appliances that sit on the floor. Explain how they can heat an entire room. **SC.7.P.11.4**

Writing in Science

11 **Write** On a separate sheet of paper, write a short explanation to a friend explaining the following scenario: Most air-conditioned rooms are set to a temperature of about 22°C. Human body temperature is 37°C. Why don't people in an air-conditioned room come to thermal equilibrium with the room? **SC.7.P.11.4**

Big Idea Review

12 Describe at least four energy transfers or energy transformations that occur in your body? **SC.7.P.11.4**

13 The photo below shows an electrical power plant that uses coal to generate electricity. What type of energy resource is coal? What are the advantages and disadvantages of using coal to generate electricity? **SC.7.P.11.2**

Math Skills   MA.6.A.3.6

Use a Formula

14 A child pulls a toy wagon with a force of 25.0 N for a distance of 8.5 m. How much work did the child do on the wagon?

15 A man pushes a box with a force of 75.0 N across a 12.0 m loading dock. How much work did he do on the box?

Fill in the correct answer choice.

Multiple Choice

1 Which is NOT a method by which heat is transferred from a warmer object to a cooler one? SC.7.P.11.4

Ⓐ conduction

Ⓑ convection

Ⓒ insulation

Ⓓ radiation

2 A rock falls from a ledge to the ground. What energy conversion occurs during this process? SC.7.P.11.2

Ⓕ Potential energy is converted into kinetic energy.

Ⓖ Kinetic energy is converted into potential energy.

Ⓗ Thermal energy is converted into mechanical energy.

Ⓘ Chemical energy is converted into mechanical energy.

3 Which energy type is transferred from one object at a high temperature to another at a lower temperature? SC.7.P.11.4

Ⓐ chemical

Ⓑ electrical

Ⓒ heat

Ⓓ mechanical

4 Which is a statement of the law of conservation of energy? SC.7.P.11.2

Ⓕ Energy never can be created or destroyed.

Ⓖ Energy can be created or destroyed.

Ⓗ Energy can be destroyed, but never created.

Ⓘ Energy can be created and destroyed when it changes form.

5 Ice waters melt to form liquid water. The water spreads out to match the shape of its container. After heating on the stove, the water begins to boil and change into steam. Which statement is true about this process? SC.7.P.11.1

Ⓐ The average kinetic energy of the water molecules decrease as time passes.

Ⓑ The average kinetic energy of the water molecules increases as time passes.

Ⓒ The average kinetic energy of the water molecules does not change as time passes.

Ⓓ The average kinetic energy of the water molecules increases and then decreases as time passes.

Use the figure to answer questions 6 and 7.

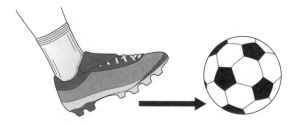

6 The figure above shows someone kicking a soccer ball. Which energy transformation occurs to make the foot move? SC.7.P.11.2

Ⓕ chemical energy to mechanical energy

Ⓖ mechanical energy to chemical energy

Ⓗ mechanical energy to mechanical energy

Ⓘ thermal energy to mechanical energy

7 Which energy transfer occurs to make the ball move? SC.7.P.11.2

Ⓐ chemical energy from ball to foot

Ⓑ chemical energy from foot to ball

Ⓒ mechanical energy from ball to foot

Ⓓ mechanical energy from foot to ball

8 A portable radio transforms chemical energy in batteries into electric energy and then into sound energy. However, not all of the energy from the batteries is converted to sound energy. Which describes how a portable radio still upholds the law of conservation of energy? SC.7.P.11.3

Ⓕ Some energy is destroyed due to inefficiency.

Ⓖ Some energy goes back into the batteries to be used later.

Ⓗ Some energy is lost to the surroundings as thermal energy.

Ⓘ Some energy is lost to the surroundings as chemical energy.

Use the figure to answer question 9.

9 The figure above shows a pan of water being heated on a stove. Which statement is true? SC.7.P.11.1

Ⓐ The pan and the flame are a closed system.

Ⓑ The natural gas is undergoing an energy transfer.

Ⓒ Thermal energy is not transferred, and the temperature remains constant.

Ⓓ This process results in a temperature change and possibly a change of state.

10 Which describes how the total amount of energy changes during an energy transformation? SC.7.P.11.3

Ⓕ It increases.

Ⓖ It decreases.

Ⓗ It stays the same.

Ⓘ It depends on the form of energy being transferred.

Use the figure to answer question 11.

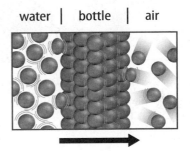

11 Imagine putting a warm bottle of water into a refrigerator. The figure models the particles that make up the water, the bottle, and the air. What happens to the temperature of the water? SC.7.P.11.4

Ⓐ The water molecules lose kinetic energy, which causes the average kinetic energy (temperature) to decrease.

Ⓑ The water molecules gain kinetic energy, which causes the average kinetic energy (temperature) to decrease.

Ⓒ The water molecules lose kinetic energy, which causes the average kinetic energy (temperature) to increase.

Ⓓ The water molecules gain kinetic energy, which causes the average kinetic energy (temperature) to increase.

NEED EXTRA HELP?

If You Missed Question...	1	2	3	4	5	6	7	8	9	10	11
Go to Lesson...	3	1	3	3	3	2	2	2	2	3	3

Multiple Choice *Bubble the correct answer.*

Use the image below to answer questions 1 and 2.

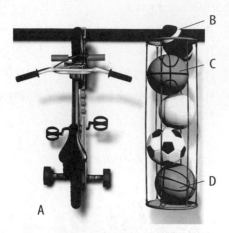

1. Which correctly describes gravitational potential energy of objects in the figure above? **SC.7.P.10.3**

   Ⓐ A = B

   Ⓑ B = C

   Ⓒ A < D

   Ⓓ C > D

2. If all of the items in the figure above have the same mass, which two items will have the same gravitational potential energy? **SC.7.P.10.3**

   Ⓕ A and B

   Ⓖ A and D

   Ⓗ B and C

   Ⓘ C and D

Use the table below to answer questions 3 and 4.

Kinetic Energy of Three Bowling Balls

Ball	Mass	Speed	Kinetic energy
A	4.0 kg	0 m/s	0 J
B	4.0 kg	8.0 m/s	130 J
C	5.0 kg	8.0 m/s	160 J

3. Based on the table above, which statement regarding kinetic energy is true? **SC.7.P.10.3**

   Ⓐ The kinetic energy of an object is calculated by multiplying mass times speed.

   Ⓑ The objects at rest will have more kinetic energy than the objects in motion.

   Ⓒ Objects that have the same mass have the same kinetic energy no matter how fast the object moves.

   Ⓓ When objects travel at the same speed, the more massive object has more kinetic energy.

4. Which correctly relates the kinetic energy of the balls above? **SC.7.P.10.3**

   Ⓕ A = B

   Ⓖ B = C

   Ⓗ A > C

   Ⓘ B < C

Copyright © Glencoe/McGraw-Hill, a division of The McGraw-Hill Companies, Inc.

Multiple Choice *Bubble the correct answer.*

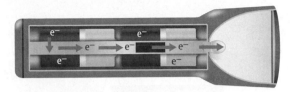

1. In the figure above, what is the first energy conversion that occurs? **SC.7.P.11.2**

   (A) chemical energy → electric energy

   (B) chemical energy → radiant energy

   (C) radiant energy → thermal energy

   (D) thermal energy → electric energy

2. During a soccer game, a player kicks the winning goal. When is work being done by the player? **SC.7.P.11.2**

   (F) when the ball flies through the air

   (G) when the player moves toward the ball

   (H) when the player's foot is in contact with the ball

   (I) when the player's foot continues to move after touching the ball

Use the table below to answer questions 3 and 4.

Electric Energy Net Generation by Resource (as of 2007)

Resource	Percentage
coal	48.5
natural gas	21.6
uranium	19.4
hydroelectric	5.8
petroleum	1.6
biomass	about 1.0
geothermal	<1.0
solar and other	<1.0
wind	<1.0
other gases	0.3

3. Based on the table above, which renewable resource generates the greatest amount of energy? **SC.7.P.11.3**

   (A) biomass

   (B) coal

   (C) hydroelectric

   (D) uranium

4. What percentage of energy generation is produced by fossil fuels? **SC.7.P.11.2**

   (F) 19.4 %

   (G) 48.5 %

   (H) 50.1 %

   (I) 71.7 %

Copyright © Glencoe/McGraw-Hill, a division of The McGraw-Hill Companies, Inc.

Multiple Choice *Bubble the correct answer.*

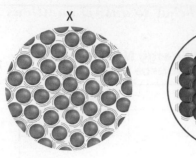

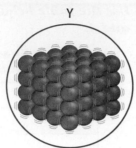

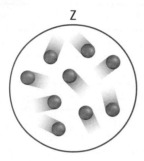

1. What does the molecular structure of Y show? **SC.7.P.11.1**

(A) gas particles

(B) liquid particles

(C) plasma particles

(D) solid particles

2. The measure of the average kinetic energy of the particles in a material is known as **SC.7.P.11.1**

(F) radiation.

(G) temperature.

(H) heat transfer.

(I) thermal expansion.

3. Which correctly identifies a type of heat transfer shown in the figure above? **SC.7.P.11.4**

(A) A = radiation

(B) A = convection

(C) B = conduction

(D) C = radiation

4. During what process does a liquid change state from a liquid to a gas? **SC.7.P.11.1**

(F) condensation

(G) deposition

(H) sublimation

(I) vaporization

Copyright © Glencoe/McGraw-Hill, a division of The McGraw-Hill Companies, Inc.

Notes

Name _____ Date _____ Class _____

Name _____ Date _____ Class _____

Getting the Ball Back

Two brothers are playing soccer on the beach. One brother kicks the ball really hard and the ball lands in the water, about 50 meters from the beach. They wonder if the ball will float back to the beach. This is what they said:

Todd: Waves carry objects as they travel through water. If we wait, the waves will move the ball back onto the beach

Brian: Waves don't carry things as they travel through water. I think we need to swim out and get the ball.

Who do you agree with?

Explain why you agree using ideas about waves.

WAVES

The Big Idea

Think About It!

How do waves travel through matter?

South of Cocoa Beach is "Surf City." That's where you will find the best surfers and some of Florida's best waves to ride. Waves are actually energy moving through matter. Think about the amount of energy this wave must be carrying.

1 What do you think caused this giant wave?

_____

_____

_____

2 Do you think this is the only large wave in the area?

_____

_____

_____

_____

3 How do you think this wave moves through water?

_____

_____

_____

_____

Get Ready to Read

What do you think about energy and wave motion?

Before you read, decide if you agree or disagree with each of these statements. As you read this chapter, see if you change your mind about any of the statements.

	AGREE	DISAGREE
1 Waves carry matter as they travel from one place to another.	☐	☐
2 Sound waves can travel where there is no matter.	☐	☐
3 Waves that carry more energy cause particles in a material to move a greater distance.	☐	☐
4 Sound waves travel fastest in gases.	☐	☐
5 When light waves strike a mirror, they change direction.	☐	☐
6 Light waves travel at the same speed in all materials.	☐	☐

What are WAVES?

ESSENTIAL QUESTIONS

 What is a wave?

 How do different types of waves make particles of matter move?

 Can waves travel through empty space?

Vocabulary

wave p. 341

mechanical wave p. 343

medium p. 343

transverse wave p. 343

crest p. 343

trough p. 343

longitudinal wave p. 344

compression p. 344

rarefaction p. 344

electromagnetic wave p. 347

Florida NGSSS

LA.7.2.2.3 The student will organize information to show understanding (e.g., representing main ideas within text through charting, mapping, paraphrasing, summarizing, or comparing/contrasting);

SC.7.P.10.1 Illustrate that the sun's energy arrives as radiation with a wide range of wavelengths, including infrared, visible, and ultraviolet, and that white light is made up of a spectrum of many different colors.

SC.7.N.1.3 Distinguish between an experiment (which must involve the identification and control of variables) and other forms of scientific investigation and explain that not all scientific knowledge is derived from experimentation.

Inquiry Launch Lab SC.7.N.1.3

20 minutes

How can you make waves?

Oceans, lakes, and ponds aren't the only places you can find waves. Can you create waves in a cup of water?

Procedure

1. Read and complete a lab safety form.
2. Add **water** to a **clear plastic cup** until it is about two-thirds full. Place the cup on a **paper towel.**
3. Explore ways of producing water waves by touching the cup. Do not move the cup.
4. Explore ways of producing water waves without touching the cup. Do not move the cup.

Data and Observations

Think About This

1. **Describe** How did the water's surface change when you produced water waves in the cup?

2. **Key Concept** **Explain** What did the different ways of producing water waves have in common?

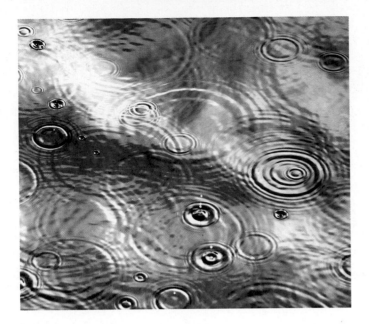

1. If you don't like rain, try living in Key West, where the average monthly rainfall is under 7.5 cm. Or maybe you like rain and enjoy watching raindrops falling into a pool of water. Why do raindrops form patterns of circles on the water's surface? The circles are small waves that spread out from where the raindrops hit the water.

• What controls how fast a wave moves across a body of water?

_____

_____

• How do waves transfer energy?

_____

_____

What are waves?

Imagine a warm summer day. You are floating on a raft in the middle of a calm pool. Suddenly, a friend does a cannonball dive into the pool. You probably know what happens next—you are no longer resting peacefully on your raft. Your friend's dive causes you to start bobbing up and down on the water. You might notice that after you stop moving up and down, you haven't moved forward or backward in the pool.

Why did your friend's dive make you move up and down? Your friend created waves by jumping into the pool. *A* **wave** *is a disturbance that transfers energy from one place to another without transferring matter.* You moved up and down because these waves transferred **energy.**

Active Reading   **2. Identify** Highlight the definition of a wave.

A Source of Energy

The photo on this page shows the waves produced as raindrops fall into a pond. The falling raindrops are the sources of energy for these water waves. Waves transfer energy away from the source of the energy. **Figure 1** shows how light waves spread out in all directions away from a flame. The burning wax is the energy source for these light waves.

Active Reading   **3. Illustrate** On the photo at the top of this page, draw a red dot at the source of one of the waves. Then, draw several arrows on the wave to show the direction(s) the wave travels.

Active Reading
FOLDABLES LA.7.2.2.3

Make a two-tab book and label it as shown. As you read the lesson, use your book to organize information about mechanical and electromagnetic waves.

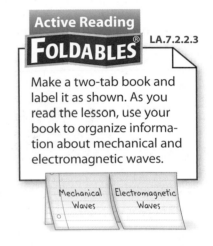

Mechanical Waves | Electromagnetic Waves

Figure 1 All waves, such as light waves, spread out from the energy source that produces the waves.

Wave Speed

Different types of waves travel at different speeds. For example, light waves from a lightning flash travel almost 1 million times faster than the sound waves you hear as thunder.

Wave Speed Through Different Materials

The same type of waves travel at different speeds in different materials. Mechanical waves, such as sound waves, usually travel fastest in solids and slowest in gases, as shown in **Table 2.** Mechanical waves also usually travel faster as the temperature of the medium increases. Unlike mechanical waves, electromagnetic waves move fastest in empty space and slowest in solids.

Calculating Wave Speed

You can calculate the speed of a wave by multiplying its wavelength and its frequency together, as shown below. The symbol for wavelength is λ, which is the Greek letter *lambda*.

Wave Speed Equation

wave speed (in m/s) = **frequency** (in Hz) × **wavelength** (in m)

$$s = f\lambda$$

When you multiply wavelength and frequency, the result has units of m × Hz. This equals m/s—the unit for speed.

Table 2 Speed of Sound Waves in Different Materials

Material	Wave Speed (m/s)
Gases (0°C)	
Oxygen	316
Dry air	331
Liquids (25°C)	
Ethanol	1,207
Water	1,500
Solids	
Ice	3,850
Aluminum	6,420

Math Skills   **Use a Simple Equation**   MA.6.A.3.6

Solve for Wave Speed Florida's state bird, the northern mockingbird, can imitate the songs of other birds. A typical songbird produces sound waves in the range of 8,000 Hz, with wavelengths in the range of 0.041 m. How fast do these sound waves travel?

1. This is what you know: frequency: f = 8,000 Hz

                               wavelength: λ = 0.041 m

2. This is what you need to find: wave speed: s

3. Use this formula: $s = f\lambda$

4. Substitute: s = **8,000 Hz** × **0.041 m** = 328 Hz × m

    the values for f and λ into the formula and multiply

5. Convert units: (**Hz**) × (**m**) = (1/s) × (m) = m/s

Answer: The wave speed is 328 m/s.

Practice
8. What is the speed of a wave that has a frequency of 8,500 Hz and a wavelength of 1.5 m?

Visual Summary

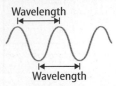

Amplitude

Amplitude

The amplitude of a transverse wave is the maximum distance that the wave moves from its rest position.

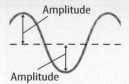

Wavelength

Wavelength

The wavelength of a transverse wave is the distance from one point on a wave to the same point on the next wave, such as from crest to crest or from trough to trough.

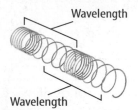

Wavelength

Wavelength

The wavelength of a longitudinal wave is the distance from one point on a wave to the nearest point just like it, such as from compression to compression or from rarefaction to rarefaction.

Inquiry MA.6.5.6.2, SC.7.N.1.6

Try It!
⊕LAB STATION

Skill Lab *How are the properties of waves related?* at connectED.mcgraw-hill.com

Use Vocabulary

1 For a transverse wave, the _____ depends on the distance from the rest position to a crest or a trough.

2 The unit for the _____ of a wave is the Hz, which means "per second."

Understand Key Concepts 🔑

3 **Compare** Which has the greatest speed? Explain your answer.

1) a sound wave from a vibrating piano string

2) a sound wave created by a boat anchor striking an underwater rock? SC.7.P.10.3

_____

_____

4 In which medium would an electromagnetic wave travel the fastest? SC.7.P.10.3

(A) air (C) vacuum

(B) granite (D) water

Interpret Graphics

5 **Determine** which wave carries the greater amount of energy. Explain.

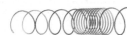 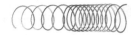

_____

Critical Thinking

6 **Infer** A loudspeaker produces sound waves that change in wavelength from 1.0 m to 1.5 m. If the wave speed is constant, how did the vibration of the loudspeaker change? Explain.

_____

_____

Math Skills MA.6.A.3.6

7 **Use a Simple Equation** A water wave has a frequency of 10 Hz and a wavelength of 150 m. What is the wave speed?

Connect Key Concepts

1 Organize Fill in the chart to organize information about wave speed.

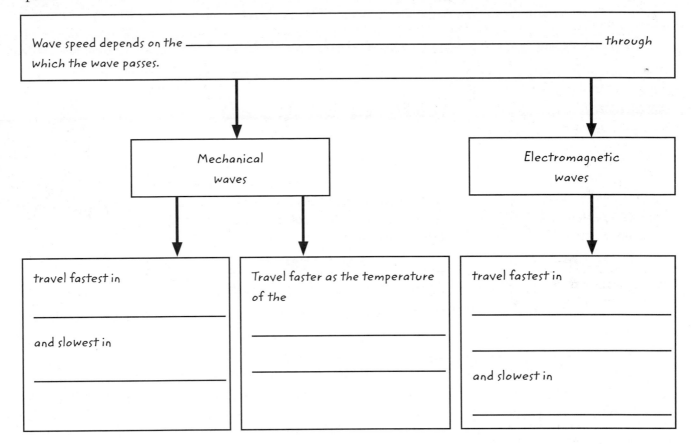

Wave speed depends on the _____ through which the wave passes.

Mechanical waves

Electromagnetic waves

travel fastest in

_____

and slowest in

_____

Travel faster as the temperature of the

_____

_____

travel fastest in

_____

_____

and slowest in

_____

2 Evaluate wave speed by completing the following formula.

Wave speed (in m/s) = _____ × _____

3 Contrast the amplitude and energy of the sound waves you make when you shout across a room with the sound waves you make when you speak softly.

_____

_____

_____

_____

Wave INTERACTIONS

Vocabulary

absorption p. 360

transmission p. 360

reflection p. 360

law of reflection p. 361

refraction p. 362

diffraction p. 362

interference p. 363

Florida NGSSS

LA.7.2.2.3 The student will organize information to show understanding (e.g., representing main ideas within text through charting, mapping, paraphrasing, summarizing, or comparing/contrasting);

SC.7.P.10.2 Observe and explain that light can be reflected, refracted, and/or absorbed.

SC.7.N.1.1 Define a problem from the seventh grade curriculum, use appropriate reference materials to support scientific understanding, plan and carry out scientific investigation of various types, such as systematic observations or experiments, identify variables, collect and organize data, interpret data in charts, tables, and graphics, analyze information, make predictions, and defend conclusions.

SC.7.N.1.3 Distinguish between an experiment (which must involve the identification and control of variables) and other forms of scientific investigation and explain that not all scientific knowledge is derived from experimentation.

SC.7.N.1.6 Explain that empirical evidence is the cumulative body of observations of a natural phenomenon on which scientific explanations are based.

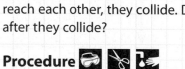

Inquiry Launch Lab SC.7.N.1.6

20 minutes

What happens in wave collisions?

You might have seen ripples on water spreading out from different points. As the water waves reach each other, they collide. Do waves change after they collide?

Procedure

1. Read and complete a lab safety form.

2. Stretch a **metal coiled spring toy** about 30–40 cm between you and a partner.

3. Make a wave by grabbing about five coils at one end and then releasing them. Record your observations.

4. Make waves at both ends of the spring with your partner. Make waves that appear much different from each other so you can distinguish them easily. Then release them at the same time. Observe and record how each wave moves before, during, and after the collision.

Data and Observations

Think About This

1. **Describe** Compare how the two waves moved after the coils were released.

2. **Explain** How were the two waves affected by their collision?

1. Have you ever watched two waves bump into each other? If so, you might have noticed that the shapes of the waves changed. How do the shapes of waves change when they interact?

_____

_____

_____

_____

Interaction of Waves with Matter

Have you seen photos, like the one shown in **Figure 12,** of objects in space taken with the *Hubble Space Telescope?* The *Hubble* orbits Earth collecting light waves before they enter Earth's atmosphere. Photos taken with the *Hubble* are clearer than photos taken with telescopes on Earth's surface. This is because light waves strike the telescope before they interact with matter in Earth's atmosphere.

Waves interact with matter in several ways. Waves can be reflected by matter, or they can change direction when they travel from one material to another. In addition, as waves pass through matter, some of the energy they carry can be transferred to matter. For example, the energy from sound waves can be transferred to soft surfaces, such as the padded walls in movie theaters. Waves also interact with each other. When two different waves overlap, a new wave forms. The new wave has different properties from either original wave.

Active Reading **2. Compare** Complete the chart below to explain the interactions of waves..

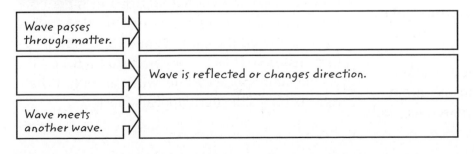

Wave passes through matter.	
	Wave is reflected or changes direction.
Wave meets another wave.	

Figure 12 This *Hubble Space Telescope* photo shows a giant cloud of dust and gas called NGC 3603. Many stars are forming in this cloud.

Figure 13 Waves can be absorbed, transmitted, or reflected by matter.

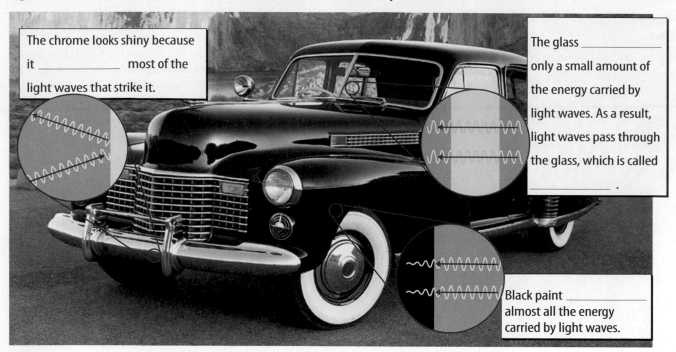

The chrome looks shiny because it _____ most of the light waves that strike it.

The glass _____ only a small amount of the energy carried by light waves. As a result, light waves pass through the glass, which is called _____.

Black paint _____ almost all the energy carried by light waves.

 Active Reading **3. Label** Fill in the blanks above with the correct type of wave interaction.

Absorption

When you shout, you create sound waves. As these waves travel in air, some of their energy transfers to particles in the air. As a result, the energy the waves carry decreases as they travel through matter. **Absorption** *is the transfer of energy by a wave to the medium through which it travels.* The amount of energy absorbed depends on the type of wave and the material in which it moves.

Active Reading **4. Consider** Why does the energy carried by sound waves decrease as the sound waves travel through air?

_____

_____

_____

_____

Absorption also occurs for electromagnetic waves. All materials absorb electromagnetic waves, although some materials absorb more electromagnetic waves than others. Darker materials, such as tinted glass, absorb more visible light waves than lighter materials, such as glass that is not tinted.

Absorption occurs at the surface of the car in **Figure 13.** The car's black paint absorbs much of the energy carried by light waves.

Transmission

Why can you see through a sheet of clear plastic wrap but not through a sheet of black construction paper? When visible light waves reach the paper, almost all their energy is absorbed. As a result, no light waves pass through the paper. However, light waves pass through the plastic wrap because the plastic absorbs only a small amount of the wave's energy. **Transmission** *is the passage of light through an object,* such as the windows in **Figure 13.**

Reflection

When waves reach the surface of materials, they can also be reflected. **Reflection** *is the bouncing of a wave off a surface.* Reflection causes the chrome on the car in **Figure 13** to appear grey instead of black. An object that reflects all visible light appears white, while an object that reflects no visible light appears black.

All types of waves, including sound waves, light waves, and water waves, can reflect when they hit a surface. Light waves reflect when they reach a mirror. Sound waves reflect when they reach a wall. Reflection causes waves to change direction. When you drop a basketball at an angle, it bounces up at the same angle but in the opposite direction. When waves reflect from a surface, they change direction like a basketball bouncing off a surface.

 5. NGSSS Check List Give three examples of how waves interact with matter that have not been discussed in the text.

_____

_____

_____

The Law of Reflection

The direction of a wave that hits a surface and the direction of the reflected wave are related. As shown in **Figure 14,** an imaginary line, perpendicular to a surface, is called a **normal.** The angle between the direction of the incoming wave and the normal is the angle of incidence. The angle between the direction of the reflected wave and the normal is the angle of reflection. According to the **law of reflection,** *when a wave is reflected from a surface, the angle of reflection is equal to the angle of incidence.*

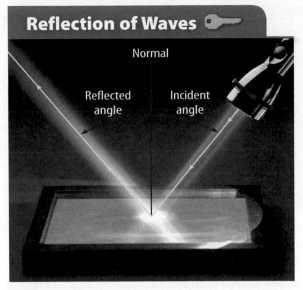

Reflection of Waves 🔑

Normal

Reflected angle

Incident angle

Figure 14 All waves obey the law of reflection. According to the law of reflection, the incident angle equals the reflected angle.

Active Reading **6. Label** Identify the incoming wave, reflected wave, incidence angle, and reflected angle in **Figure 14.**

SCIENCE USE V. COMMON USE

normal
Science Use perpendicular to or forming a right angle with a line or plane

Common Use conforming to a standard or common

Inquiry SC.7.P.10.2, SC.7.N.1.1

LAB STATION **Try It!**

MiniLab *How can reflection be used?* at connectED.mcgraw-hill.com

Apply It! After you finish the lab, complete the following activity.

1. Recall the law of reflection. Describe the path of the light as it traveled from the small object to your eye.

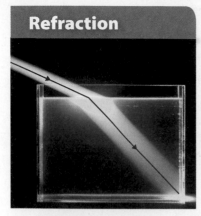

Refraction

Figure 15 The beam of light changes direction because light waves slow down as they move from air into water.

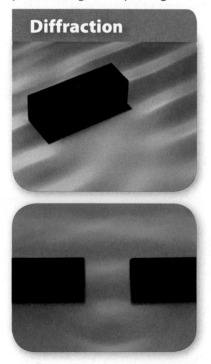

WORD ORIGIN

refraction
from Latin *refractus*, means "to break up"

Figure 16 Waves diffract as they pass by an object or pass through an opening.

Refraction

Sometimes waves change direction even if they are not reflected from a surface. The light beam in **Figure 15** changes direction as it travels from air into water. The speed of light waves in water is about three-fourths the speed of light waves in air. When light waves slow down, they change direction. **Refraction** *is the change in direction of a wave that occurs as the wave changes speed when moving from one medium to another.* The greater the change in speed, the more the wave changes direction.

Diffraction

Waves can also change direction as they travel by objects. Have you ever been walking down a hallway and heard people talking in a room before you got to the open door of the room? You heard some of the sound waves because they changed direction and spread out as they traveled through the doorway.

What is diffraction?

The change in direction of a wave when it travels by the edge of an object or through an opening is called **diffraction.** Examples of diffraction are shown in **Figure 16.** Diffraction causes the water waves to travel around the edges of the object and to spread out after they travel through the opening. More diffraction occurs as the size of the object or opening becomes similar in size to the wavelength of the wave.

Active Reading

7. Compare and Contrast Complete the Venn diagram to show how *refraction* and *diffraction* are similar and how they differ.

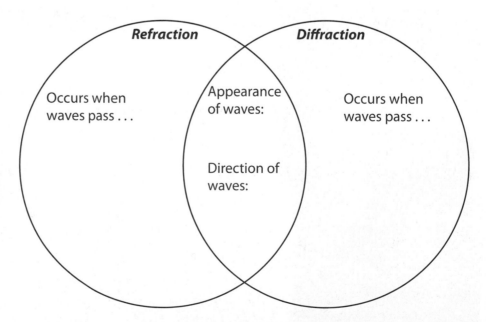

Refraction

Diffraction

Occurs when waves pass . . .

Appearance of waves:

Direction of waves:

Occurs when waves pass . . .

Diffraction of Sound Waves and Light Waves

The wavelengths of sound waves are similar in size to many common objects. Because of this size similarity, you often hear sound from sources that you can't see. For example, the wavelengths of sound waves are roughly the same size as the width of the doorway. Therefore, sound waves spread out as they travel through the doorway. The wavelengths of light waves are more than a million times smaller than the width of a doorway. As a result, light waves do not spread out as they travel through the doorway. Because the wavelengths of light waves are so much smaller than sound waves, you can't see into the room until you reach the doorway. However, you can hear the sounds much sooner.

Interference

Waves not only interact with matter. They also interact with each other. Suppose you throw two pebbles into a pond. Waves spread out from the impact of each pebble and move toward each other. When the waves meet, they overlap for a while as they travel through each other. **Interference** *occurs when waves that overlap combine, forming a new wave,* as shown in **Figure 17.** However, after the waves travel through each other, they keep moving without having been changed.

Active Reading
SC.7.P.10.2
LA.7.2.2.3
FOLDABLES®

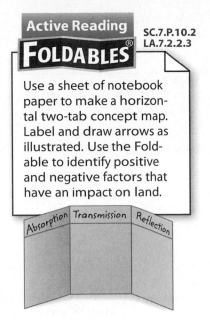

Use a sheet of notebook paper to make a horizontal two-tab concept map. Label and draw arrows as illustrated. Use the Foldable to identify positive and negative factors that have an impact on land.

Absorption | Transmission | Reflection

 8. NGSSS Check Restate Fill in the graphic organizer below to contrast reflection, refraction, and diffraction.

Interference	Explanation
Reflection	
Refraction	
Diffraction	

Wave Interference 🔑

1 Two waves approach each other from opposite directions.

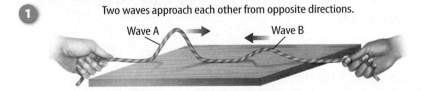

Wave A   Wave B

2 The waves interfere with each other and form a large amplitude wave.

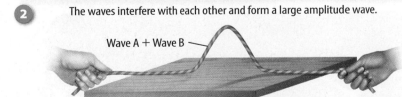

Wave A + Wave B

3 The waves keep traveling in opposite directions after they move through each other.

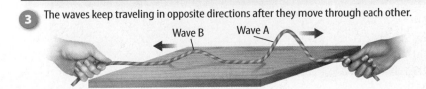

Wave B   Wave A

Figure 17 When waves interfere with each other, they create a new wave that has a different amplitude than either original wave.

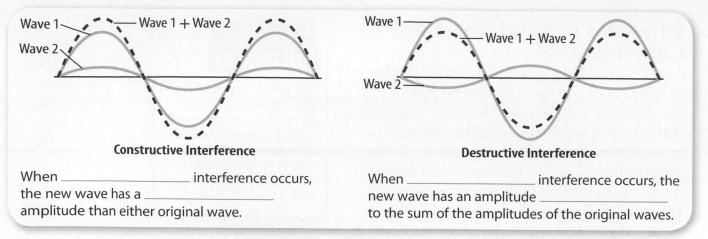

Constructive Interference

When _____ interference occurs, the new wave has a _____ amplitude than either original wave.

Destructive Interference

When _____ interference occurs, the new wave has an amplitude _____ to the sum of the amplitudes of the original waves.

Figure 18 Constructive and destructive interference occurs as waves travel through each other.

Active Reading

9. Illustrate Draw over the dashed waves in **Figure 18** to show the new waves formed. Then fill in the blanks for each sub-caption, summarizing the type of interference.

ACADEMIC VOCABULARY

constructive

(adjective) pertaining to building or putting parts together to make a whole

Constructive and Destructive Interference

As waves travel through each other, sometimes the crests of both waves overlap, as shown in **Figure 18**. A new wave forms with greater amplitude than either of the original waves. This type of interference is called **constructive** interference. It occurs when crests overlap with crests and troughs overlap with troughs.

Destructive interference occurs when a crest of one wave overlaps the trough of another wave. The new wave that forms has a smaller amplitude than the sum of the amplitudes of the original waves, also as shown in **Figure 18.** If the two waves have the same amplitude, they cancel each other when their crests and troughs overlap.

Standing Waves

Suppose you shake one end of a rope that has the other end attached to a wall. You create a wave that travels away from you and then reflects off the wall. As the wave you create and the reflected wave interact, interference occurs. For some values of the wavelength, the wave that forms from the combined waves seems to stand still. This wave is called a standing wave. An example is shown in **Figure 19.**

Figure 19 A standing wave can occur when two waves with the same wavelength travel in opposite directions and overlap. The wave that forms seems to be standing still.

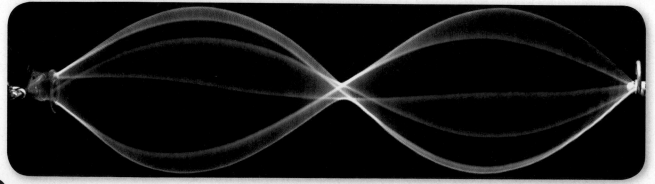

Visual Summary

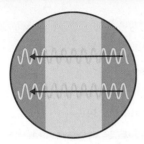

Transmission occurs when waves travel through a material.

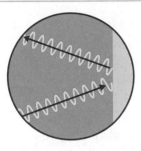

Reflection occurs when waves bounce off the surface of a material.

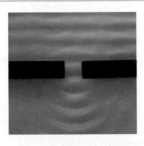

The change in direction of a wave when it travels through an opening is diffraction.

Use Vocabulary

1 Explain the law of reflection.

_____

_____

2 Distinguish between refraction and diffraction.

_____

_____

_____

Understand Key Concepts

3 Contrast the behavior of a water wave that travels by a stone barrier to a sound wave that travels through a door. **SC.7.P.10.3**

_____

_____

_____

_____

4 Which will NOT occur when a light ray interacts with a smooth pane of glass? **SC.7.P.10.2**

(A) absorption   (C) reflection

(B) diffraction   (D) transmission

Interpret Graphics

5 Organize Fill in the graphic organizer below. In each oval, name something that can happen to a wave as it interacts with matter. **LA.7.2.2.3**

Wave Interacting with Matter

Critical Thinking

6 Recommend An architect wants to design a conference room that reduces noise coming from outside the room. Suggest some design features that should be considered in this project.

_____

_____

_____

_____

Chapter 9 — Study Guide

 Think About It! Waves transfer energy but not matter as they travel. Waves, such as light waves and sound waves move at different speeds in different materials.

Key Concepts Summary

Vocabulary

LESSON 1 What are waves?

- Vibrations cause **waves.**
- **Transverse waves** make particles in a **medium** move at right angles to the direction that the wave travels. **Longitudinal waves** make particles in a medium move parallel to the direction that the wave travels.
- **Mechanical waves** cannot move through empty space, but **electromagnetic waves** can.

Direction wave moves

wave p. 341

mechanical wave p. 343

medium p. 343

transverse wave p. 343

crest p. 343

trough p. 343

longitudinal wave p. 344

compression p. 344

rarefaction p. 344

electromagnetic wave p. 347

LESSON 2 Wave Properties

- All waves have the properties of **amplitude, wavelength,** and **frequency.**
- Increasing the frequency of a wave decreases the wavelength, and decreasing the frequency increases the wavelength.
- The speed of a wave depends on the type of material in which it is moving and the temperature of the material.

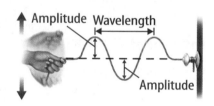

Amplitude Wavelength

Amplitude

amplitude p. 351

wavelength p. 353

frequency p. 354

LESSON 3 Wave Interactions

- When waves interact with matter, **absorption** and **transmission** can occur.
- Waves change direction as they interact with matter when **reflection, refraction,** or **diffraction** occurs.
- **Interference** occurs when waves that overlap combine to form a new wave.

absorption p. 360

transmission p. 360

reflection p. 360

law of reflection p. 361

refraction p. 362

diffraction p. 362

interference p. 363

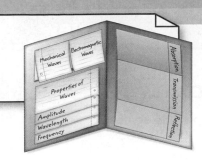

Active Reading
FOLDABLES® Chapter Project

Assemble your lesson Foldables as shown to make a Chapter Project. Use the project to review what you have learned in this chapter.

Use Vocabulary

1 A material though which a wave travels is a(n)

_____.

2 A(n) _____ is a region where matter is more closely spaced in a longitudinal wave.

3 The Sun gives off energy that travels through space in the form of _____.

4 The product of _____ and wavelength is the speed of the wave _____.

5 _____ is a property of waves that is measured in hertz.

6 The highest point on a transverse wave is a(n) _____.

7 _____ is when two waves pass through each other and keep going.

Link Vocabulary and Key Concepts

Use vocabulary terms from the previous page to complete the concept map.

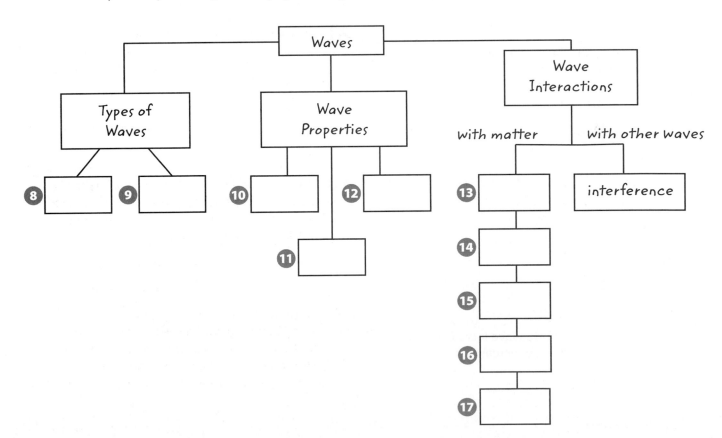

Chapter 9

Fill in the correct answer.

🔑 Understand Key Concepts

1 What is transferred by a radio wave? **SC.7.P.10.1**
- Ⓐ air
- Ⓑ energy
- Ⓒ matter
- Ⓓ space

2 In a longitudinal wave, where are the particles most spread out? **SC.7.P.10.3**
- Ⓐ compression
- Ⓑ crest
- Ⓒ rarefaction
- Ⓓ trough

3 Which would produce mechanical waves? **SC.7.P.10.3**
- Ⓐ burning a candle
- Ⓑ hitting a wall with a hammer
- Ⓒ turning on a flashlight
- Ⓓ tying a rope to a doorknob

4 Which is an electromagnetic wave? **SC.7.P.10.1**
- Ⓐ a flag waving in the wind
- Ⓑ a vibrating guitar string
- Ⓒ the changes in the air that result from blowing a horn
- Ⓓ the waves that heat a cup of water in a microwave oven

5 Identify the crest of the wave in the illustration below. **SC.7.P.10.3**

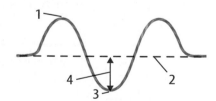

- Ⓐ 1
- Ⓑ 2
- Ⓒ 3
- Ⓓ 4

6 If the energy carried by a wave increases, which other wave property also increases. **SC.7.P.10.3**
- Ⓐ amplitude
- Ⓑ medium
- Ⓒ wavelength
- Ⓓ wave speed

Critical Thinking

7 **Assess** A student sets up a line of dominoes so that each is standing vertically next to another. He then pushes the first one and each falls down in succession. How does this demonstration represent a wave? How is it different? **SC.7.N.1.3**

_____

_____

_____

_____

8 **Infer** In the figure below, suppose wave 1 and wave 2 have the same amplitude. Describe the wave that forms when destructive interference occurs. **LA.7.2.2.3**

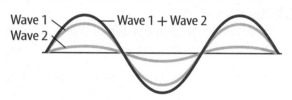

_____

_____

9 **Compare** A category 5 hurricane has more energy than a category 3 hurricane. Which hurricane will create water waves with greater amplitude? Why? **LA.7.2.2.3**

_____

_____

_____

10 **Infer** At a baseball game when you are far from the batter, you might see the batter hit the ball before you hear the sound of the bat hitting the ball. Explain why this happens. **SC.7.P.10.3**

_____

_____

11 **Evaluate** Geologists measure the amplitude of seismic waves using the Richter scale. If an earthquake of 7.3 has a greater amplitude than an earthquake of 4.4, which one carries more energy? Explain your answer. **SC.7.N.1.6**

_____

_____

_____

12 **Recommend** Some medicines lose their potency when exposed to ultraviolet light. Recommend the type of container in which these medicines should be stored. **SC.7.P.10.2**

_____

_____

13 **Explain** why the noise level rises in a room full of many talking people. **LA.7.2.2.3**

_____

_____

_____

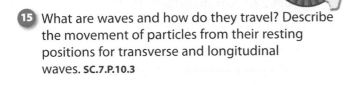

Writing in Science

14 **Write** a short essay, on a separate sheet of paper, explaining how an earthquake below the ocean floor can affect the seas near the earthquake area. **LA.7.2.2.3**

Big Idea Review

15 What are waves and how do they travel? Describe the movement of particles from their resting positions for transverse and longitudinal waves. **SC.7.P.10.3**

_____

_____

16 The chapter opener photo shows waves in the ocean. Describe the waves using vocabulary terms from the chapter. **LA.7.2.2.3**

_____

_____

_____

_____

Math Skills MA.6.A.3.6

Use Numbers

17 A hummingbird can flap its wings 200 times per second. If the hummingbird produces waves that travel at 340 m/s by flapping its wings, what is the wavelength of these waves?

18 A student did an experiment in which she collected the data shown in the table. What can you conclude about the wave speed and rope diameter in this experiment? What can you conclude about frequency and wavelength?

Wave Speed and Diameter			
Trial	Rope Diameter	Frequency	Wavelength
1	2.0 cm	2.0 Hz	8.0 m
2	2.0 cm	8.0 Hz	2.0 m
3	4.0 cm	2.0 Hz	10.0 m
4	4.0 cm	4.0 Hz	5.0 m

Record your answers on the answer sheet provided by your teacher or on a sheet of paper.

Multiple Choice

1 Through which medium would sound waves move most slowly? SC.7.P.10.3

Ⓐ air

Ⓑ aluminum

Ⓒ glass

Ⓓ water

Use the illustration below to answer question 2.

2 Which process enables the boy to see over the wall? SC.7.P.10.2

Ⓕ diffraction

Ⓖ interference

Ⓗ reflection

Ⓘ refraction

3 Which is an electromagnetic wave? SC.7.P.10.1

Ⓐ light

Ⓑ seismic

Ⓒ sound

Ⓓ water

4 Which statement best defines radiant energy? SC.7.P.10.1

Ⓕ Radiant energy is the energy that must travel through certain types of media.

Ⓖ Radiant energy is the energy carried by an electromagnetic wave.

Ⓗ Radiant energy is the energy carried only by magnetic waves.

Ⓘ Radiant energy travels at short wavelengths.

Use the table below to answer question 5.

Speed of Light Through Different Mediums			
Medium	air	water	glass
Speed of light (m/s × 10⁸)	3.0	2.2	1.5

5 According to the information in the table above, how does the speed of light in air compare to the speed of light through glass? SC.7.P.10.3

Ⓐ Light travels slower through air than through glass.

Ⓑ Light travels three times faster through air than through glass.

Ⓒ Light travels two times faster through air than through glass.

Ⓓ Light travels at the same speed through air and through glass.

6 How does solar energy travel to Earth? SC.7.P.10.1

Ⓕ through conduction

Ⓖ through convection

Ⓗ through emission

Ⓘ through radiation

7 Sound waves travel fastest through what type of material? SC.7.P.10.3

Ⓐ gases

Ⓑ liquids

Ⓒ solids

Ⓓ empty space

8 Which is an example of refraction? SC.7.P.10.2

(F) a flashlight beam hitting a mirror

(G) a shout crossing a crowded room

(H) a sunbeam striking a window

(I) a water wave bending around a rock

Use the table below to answer question 9.

Boulder Temperature (°C)	Speed of Sound (m/s)
−5	562
0	564
10	566
15	568
20	?

9 A student is studying how sound travels through a solid medium by timing the movement of sound through a boulder when the boulder is at different temperatures. Using the information above, how fast will sound move through the boulder when it reaches 20°C? SC.7.P.10.3

(A) 569 m/s

(B) 570 m/s

(C) 571 m/s

(D) 572 m/s

10 Given that sound waves are formed by vibrating molecules, which statement is true? SC.7.P.10.3

(F) Sound travels fastest through space.

(G) Sound travels faster through water than through air.

(H) Sound travels at the same rate through all mediums.

(I) Sound travels faster through air than through water.

Use the graph below to answer question 11.

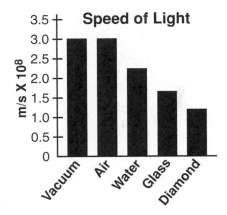

11 According to the graph, through which medium is the speed of light nearly half as fast as it is in a vacuum? SC.7.P.10.3

(A) air

(B) diamond

(C) glass

(D) water

NEED EXTRA HELP?

If You Missed Question...	1	2	3	4	5	6	7	8	9	10	11
Go to Lesson...	2	3	1	1	2	1	2	3	3	2	2

Multiple Choice *Bubble the correct answer.*

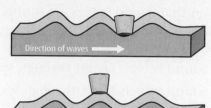

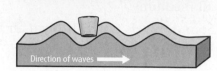

1. Identify the type of wave shown in the images above. **SC.7.P.10.1**

(A) asymmetric wave

(B) elliptical wave

(C) longitudinal wave

(D) transverse wave

2. Which type of electromagnetic waves do most organisms emit? **SC.7.P.10.1**

(F) infrared waves

(G) light waves

(H) radio waves

(I) ultraviolet waves

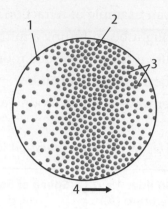

3. At which point in the diagram above are the particles of air closest together? **SC.7.P.10.1**

(A) Point 1

(B) Point 2

(C) Point 3

(D) Point 4

Copyright © Glencoe/McGraw-Hill, a division of The McGraw-Hill Companies, Inc.

Benchmark Mini-Assessment   Chapter 9 • Lesson 2

mini BAT

Multiple Choice *Bubble the correct answer.*

Use the images below to answer questions 1–3.

1

2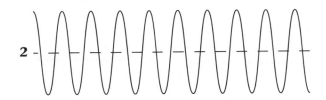

3

4

1. In which image is the wave amplitude the greatest? **SC.7.P.10.3**

- (A) 1
- (B) 2
- (C) 3
- (D) 4

2. In which image is the wavelength the greatest? **SC.7.P.10.1**

- (F) 1
- (G) 2
- (H) 3
- (I) 4

3. In which image is the energy the greatest? **SC.7.P.10.3**

- (A) 1
- (B) 2
- (C) 3
- (D) 4

4. In which medium will a sound wave move more quickly? **SC.7.P.10.3**

- (F) air
- (G) ethanol
- (H) rock
- (I) vacuum

Copyright © Glencoe/McGraw-Hill, a division of The McGraw-Hill Companies, Inc.

Benchmark Mini-Assessment **Chapter 9 • Lesson 3**

mini
BAT

Multiple Choice *Bubble the correct answer.*

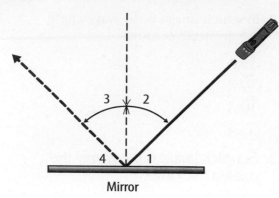

Mirror

1. The diagram above shows the angle of incidence from a light source. What will be the angle of reflection? **SC.7.P.10.1**

(A) Point 1

(B) Point 2

(C) Point 3

(D) Point 4

2. Transmission occurs when waves pass through **SC.7.P.10.2**

(F) aluminum foil.

(G) a mirror.

(H) plastic wrap.

(I) a steel beam.

3. Each diagram below shows two waves overlapping. Which diagram shows an example of a standing wave? **SC.7.P.10.2**

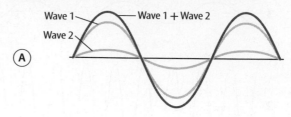

(A)

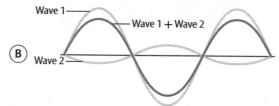

(B)

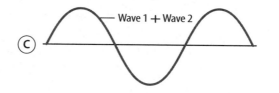

(C)

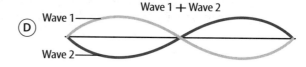

(D)

Copyright © Glencoe/McGraw-Hill, a division of The McGraw-Hill Companies, Inc.

Notes

Notes

Traveling Waves

Sound and light both travel by waves. In order to hear sounds or see light, waves must travel from one place to another and enter either your ears or eyes. How do you think these waves travel? Circle the response that best matches your thinking.

A: One needs to travel through matter.

B: Neither needs to travel through matter.

C: They both need to travel through matter.

Explain your thinking. Describe your ideas about how sound and light waves move.

Sound and LIGHT

FLORIDA BIG IDEAS

1   The Practice of Science

3   The Role of Theories, Laws, Hypotheses, and Models

10 Forms of Energy

Think About It!

How do sound and light waves travel and interact with matter?

The mirror's curved surface forms unusual images. You can see these images because of reflected light waves.

1 How do you think light waves are reflected from a surface?

_____

_____

2 How do you think your eyes see the reflection?

_____

_____

_____

Get Ready to Read

What do you think about sound and light?

Before you read, decide if you agree or disagree with each of these statements. As you read this chapter, see if you change your mind about any of the statements.

		AGREE	DISAGREE
1	Vibrating objects make sound waves.	☐	☐
2	Human ears are sensitive to more sound frequencies than any other animal's ears.	☐	☐
3	Unlike sound waves, light waves can travel through a vacuum.	☐	☐
4	Light waves always travel at the same speed.	☐	☐
5	All mirrors form images that appear identical to the object itself.	☐	☐
6	Lenses always magnify objects.	☐	☐

 **Connect ED**

There's More Online!
Video • Audio • Review • ⓘLab Station • WebQuest • Assessment • Concepts in Motion • Multilingual eGlossary   **379**

SOUND

 How are sound waves produced?

 Why does the speed of sound waves vary in different materials?

 How do your ears enable you to hear sounds?

Vocabulary

sound wave p. 381

pitch p. 385

echo p. 387

Florida NGSSS

LA.7.2.2.3 The student will organize information to show understanding (e.g., representing main ideas within text through charting, mapping, paraphrasing, summarizing, or comparing/contrasting);

MA.6.A.3.6 Construct and analyze tables, graphs, and equations to describe linear functions and other simple relations using both common language and algebraic notation.

SC.7.P.10.3 Recognize that light waves, sound waves, and other waves move at different speeds in different materials.

SC.7.N.1.3 Distinguish between an experiment (which must involve the identification and control of variables) and other forms of scientific investigation and explain that not all scientific knowledge is derived from experimentation.

SC.7.N.3.2 Identify the benefits and limitations of the use of scientific models.

 Launch Lab SC.7.N.1.3

15 minutes

How is sound produced?

When an object vibrates, it produces sound. How does the sound produced depend on how the object vibrates?

Procedure

1. Read and complete a lab safety form.

2. Place a **ruler** on a table so it extends over the table edge. Hold the ruler firmly on the table with one hand.

3. With the other hand, lightly bend the protruding end of the ruler down and then release it. Observe the ruler's motion and note the sound it produces.

4. Move the ruler back 2 cm so there is less of it extending over the edge of the table. Repeat step 3.

Think About This

1. How did the vibration rate and the sound change as the length of the ruler over the side of the table decreased?

2. **Key Concept** Were the sound and the ruler's vibration rate related? Explain.

1. The ears of this brown long-eared bat are nearly as long as its body. This bat finds its next meal by listening for the faint sounds that come from spiders and insects. How do large ears help a long-eared bat hear these sounds?

_____

_____

_____

_____

_____

REVIEW VOCABULARY

longitudinal wave
a wave in which particles in a material move along the same direction that the wave travels

Figure 1 Earbuds produce sound waves that travel into the listener's ears.

What is sound?

Have you ever walked down a busy city street and noticed all the sounds? You might hear many sounds every day, such as the music from an MP3 player, as shown in **Figure 1.** All sounds have one thing in common. The sounds travel from one place to another as sound waves. *A* **sound wave** *is a longitudinal wave that can travel only through matter.*

The sounds you might hear now are traveling through air—a mixture of solids and gases. You might have dived underwater and heard someone call to you. Then the sound waves traveled through a liquid. Sound waves travel through a solid when you knock on a door. As you will read, vibrating objects produce sound waves.

All sounds might have something else in common, too. Vibrating objects produce sound waves. For example, when you knock on a door, you produce sound waves by making the door vibrate. How do vibrating objects make sound waves?

Lesson 1 • EXPLAIN **381**

Figure 2 Vibrations of the drumhead produce sound waves.

Vibrations and Sound

Some objects, such as doors or drums, vibrate when you hit them. For example, when you hit a drum, the drumhead moves up and down, or vibrates, as shown in **Figure 2.** These vibrations produce sound waves by moving molecules in air.

Compressions and Rarefactions

As the drumhead moves up, it pushes the molecules in the air above it closer together. The region where molecules are closer together is a compression, as shown in **Figure 2.** When the drumhead moves down, it produces a rarefaction. This is a region where molecules are farther apart. As the drumhead vibrates down and up, it produces a series of rarefactions and compressions that travels away from the drumhead. This series of rarefactions and compressions is a sound wave.

The vibrating drumhead causes molecules in the air to move closer together and then farther apart. The molecules in air move back and forth in the same direction that the sound wave travels. As a result, a sound wave is a longitudinal wave.

Active Reading 2. **Explain** How do vibrating objects produce sound waves?

_____

_____

Wavelength and Frequency

A sound wave can be described by its wavelength and frequency. Wavelength is the distance between a point on a wave and the nearest point just like it, as shown in **Figure 3.** A sound wave's frequency is the number of wavelengths that pass a given point in one second. Recall that the SI unit of frequency is hertz (Hz). The faster an object vibrates, the higher the frequency of the sound wave produced.

Figure 3 Wavelength is the distance between one compression and the next compression or the distance between a rarefaction and the next rarefaction.

Active Reading 3. **Label** Which set of molecules is refraction and which is compression?

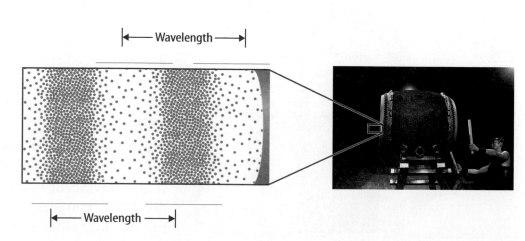

Table 1 The Speed of Sound Waves in Different Materials

Gases (0°C)		Liquids (25°C)		Solids	
Material	Speed (m/s)	Material	Speed (m/s)	Material	Speed (m/s)
Carbon dioxide	259	Ethanol	1,207	Brick	3,480
Dry Air	331	Mercury	1,450	Ice	3,850
Water vapor	405	Water	1,500	Aluminum	6,420
Helium	965	Glycerine	1,904	Diamond	17,500

Table 1 Sound waves travel at different speeds in different materials. Sound waves usually travel fastest in solids and slowest in gases.

Speeds of Sound Waves

Sound waves traveling through air cause the sounds you might hear every day. Like all types of waves, the speed of a sound wave depends on the material in which it travels.

Sound in Gases, Liquids, and Solids

Table 1 lists the speeds of sound waves in different materials. A sound wave's speed increases when the material's density increases. Solids and liquids are usually more dense than gases.

In addition, a sound wave's speed increases when the strengths of the forces between the particles—atoms or molecules—in the material increase. These forces are usually strongest in solids and weakest in gases. Overall, sound waves usually travel faster in solids than in liquids or gases.

Temperature and Sound Waves

The temperature of a material also affects the speed of a sound wave. The speed of a sound wave in a material increases as the temperature of the material increases. For example, the speed of a sound wave in dry air increases from 331 m/s to 343 m/s as the air temperature increases from 0°C to 20°C. A sound wave in air travels faster on a warm, summer day than on a cold, winter day.

Inquiry **LAB STATION** **Try It!**   SC.7.N.3.2

MiniLab *Can you model a sound wave?* at connectED.mcgraw-hill.com

Apply It!   After you complete the lab, answer these questions.

1. **Predict and Draw** How would the sound wave look if the spring toy were placed in a tank of water?

2. **Hypothesize** How would changing the room temperature affect the wave on the spring?

4. **Classify** Fill in the words *liquid*, *solid* and *gas* below. Describe how close the molecules are to each other in each phase.

← sound travels slowest   sound travels fastest →

		solid

molecules are ____   ____   ____

farthest apart ____   ____   ____

____

Use a Simple Equation
Speed (s) is equal to the distance (d) something travels divided by the time (t) it takes to cover that distance:

$$s = \frac{d}{t}$$

You can use this equation to calculate the speed of sound waves. For example, if a sound wave travels a distance of 662 meters in 2 seconds in air, its speed is:

$$s = \frac{d}{t} = \frac{662\ m}{2\ s} = 331\ m/s$$

Practice
4. How fast is a sound wave traveling if it travels 5,000 m in 5 s?

Figure 4 The human ear has three parts.

The Human Ear

When you think about your ears, you probably only think about the structure on each side of your head. However, there is more to your ears than those structures. The human ear has three parts—the outer ear, the middle ear, and the inner ear, as shown in **Figure 4.**

The Outer Ear

The outer ear collects sound waves. The structure on each side of your head is part of the outer ear. The ear canal is also part of the outer ear, as shown in **Figure 4.** The visible part of the outer ear funnels sound waves into the ear canal. The ear canal channels sound waves into the middle ear.

The Middle Ear

The middle ear amplifies sound waves. As shown in **Figure 4,** the middle ear includes the eardrum and three tiny bones. The eardrum is a thin membrane that stretches across the ear canal. The three tiny bones are called the hammer, the anvil, and the stirrup. A sound wave hitting the eardrum causes it to vibrate. The vibrations travel to the three bones, which amplify the sound.

The Inner Ear

The inner ear converts vibrations to nerve signals that travel to the brain. The inner ear consists of a small, fluid-filled chamber called the cochlea (KOH klee uh). Tiny hairlike cells line the inside of the cochlea. As a sound wave travels into the cochlea, it causes some hair cells to vibrate. The movements of these hair cells produce nerve signals that travel to the brain.

Structure of the Ear 🔑

Active Reading 5. **Label** Identify the three parts of the ear and explain the function of each.

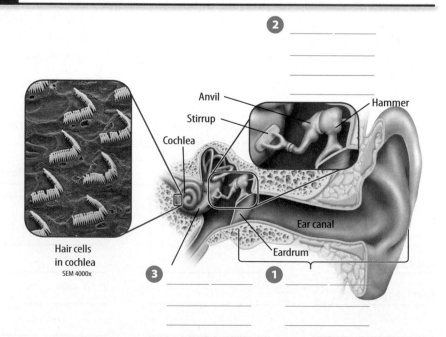

Anvil

Stirrup

Cochlea

Hammer

Ear canal

Eardrum

Hair cells in cochlea
SEM 4000x

2

3

1

Frequencies and the Human Ear

As you can see in **Table 2,** humans hear sounds with frequencies between about 20 and 20,000 Hz. Some mammals can hear sounds with frequencies greater than 100,000 Hz.

Sound and Pitch

Have you ever played a guitar? A guitar has strings with different thicknesses. If you pluck a thick string, you hear a low note. If you pluck a thin string, you hear a higher note. The sound a thick string makes has a lower pitch than the sound a thin string makes. *The* **pitch** *of a sound is the perception of how high or low a sound seems.* A sound wave with a higher frequency has a higher pitch. A sound wave with a lower frequency has a lower pitch.

Active Reading 7. **Classify** Use arrows to indicate how the pitch of a sound wave changes as the frequency changes.

↑	frequency		pitch
↓	frequency		pitch

You can produce sounds of different pitches by using your vocal cords. The vocal cords, shown in **Figure 5,** are two membranes in your neck above your windpipe, or trachea (TRAY kee uh). When you speak, you force air from your lungs through the space between the vocal cords. This causes the vocal cords to vibrate, creating sound waves you and other people hear as your voice.

Muscles connected to your vocal cords enable you to change the pitch of your voice. When these muscles contract, they pull on your vocal cords. This stretches the vocal cords and they become longer and thinner. The pitch of your voice is then higher, just as a thinner guitar string has a higher pitch than a thicker guitar string. When these muscles relax, the vocal cords become shorter and thicker, and the pitch of your voice becomes lower.

Table 2 Frequencies Different Mammals Can Hear

Creature	Frequency Range (Hz)
Human	20–20,000
Dog	67–45,000
Cat	45–64,000
Bat	2,000–110,000
Beluga whale	1,000–123,000
Porpoise	75–150,000

Active Reading 6. **Compare** Which mammals listed on the table can hear a sound with a frequency of 55 Hz?

Active Reading

FOLDABLES® LA.7.2.2.3

Make a two-tab concept map book. Label it as shown. Use it to organize information about pitch and loudness.

The Ear

| Pitch | Loudness |

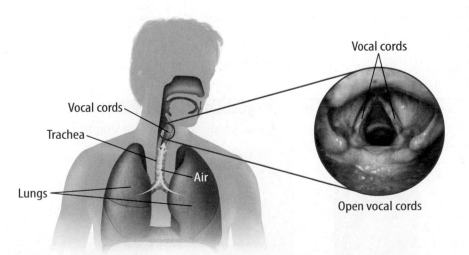

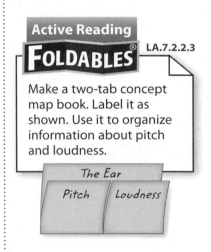

Vocal cords

Open vocal cords

Vocal cords

Trachea

Air

Lungs

Figure 5 The vocal cords vibrate by opening and closing when air is forced through them. These vibrations produce the sounds of the human voice.

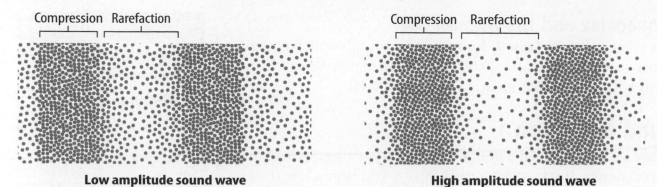

Low amplitude sound wave

High amplitude sound wave

Figure 6 The amplitude of a sound wave depends on how close together or far apart the particles are in the compressions and rarefactions.

Active Reading **8. Explain** How do distances between particles differ in high- and low-amplitude sound waves?

_____

_____

_____

Figure 7 The loudness of sounds can be compared on the decibel scale.

Sound and Loudness

Why is a shout louder than a whisper? Loudness is the human sensation of how much energy a sound wave carries. Sound waves produced by shouting carry more energy than sound waves produced by whispering. As a result, a shout sounds louder than a whisper.

Amplitude and Energy

The amplitude of a wave depends on the amount of energy that the wave carries. The more energy a wave has, the greater the amplitude. **Figure 6** shows the difference between a high-amplitude sound wave and a low-amplitude sound wave. High-amplitude sound waves have particles that are closer together in the compressions and farther apart in the rarefactions.

The Decibel Scale

The decibel scale, shown in **Figure 7,** is one way to compare the loudness of sounds. On this scale, the softest sound a person can hear is about 0 decibels 9dB, and a Space Shuttle launch at Cape Canaveral, FL is about 145 dB. Normal conversation is at about 50 dB. A sound wave that is 10 dB higher than another sound wave carries 10 times more energy. However, people hear the higher-energy sound wave as being only twice as loud.

The Decibel Scale

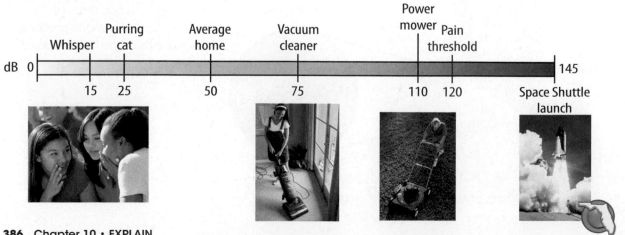

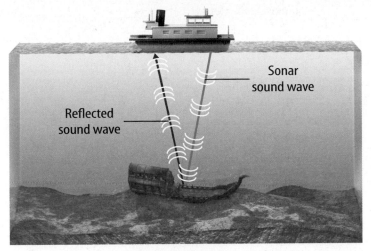
A Sonar System

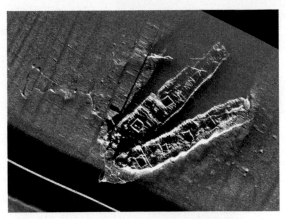
Image Made by Sonar System

Figure 8 A sonar system locates underwater objects by sending out sound waves and detecting the reflected sound waves. The photo on the right is a sonar image of two sunken ships.

Using Sound Waves

Have you ever yelled in a cave or a big, empty room? You might have heard the echo of your voice. *An* **echo** *is a reflected sound wave.* You may be able to hear echoes in a gymnasium or a cafeteria. You probably can't tell how far away a wall is by hearing an echo. However, sonar systems and some animals use reflected sound waves to determine how far away objects are.

Sonar and Echolocation

Sonar systems use reflected sound waves to locate objects under water, as shown in **Figure 8.** The sonar system emits a sound wave that reflects off an underwater object. The distance to the object can be calculated from the time difference between when the sound leaves the ship and when the sound returns to the ship. Sonar is used to map the ocean floor and to detect submarines, schools of fish, and other objects under water.

Active Reading **9. Explain** Highlight How do sonar systems use sound waves?

Some animals use a method called echolocation to navigate and hunt. Echolocation is a type of sonar. Bats and dolphins, for example, emit high-pitched sounds and interpret the echoes reflected from objects. Echolocation enables bats and dolphins to locate prey and detect objects.

Ultrasound

Ultrasound scanners, like the one shown in **Figure 9,** convert high-frequency sound waves to images of internal body parts. The sound waves reflect from structures within the body. The scanner analyzes the reflected waves and produces images, called sono-grams, of the body structures. These images can be used to help diagnose diseases or other medical conditions.

WORD ORIGIN

echo
from Greek *ekhe*, means "sound"

Figure 9 Ultrasound scanners produce images that doctors can use to diagnose diseases.

Lesson Review 1

Visual Summary

Compression

Drumhead moves up.

A sound wave is a longitudinal wave that can travel only through matter.

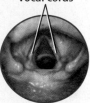
Vocal cords

The pitch is how high or low the frequency of a sound wave is. You create different pitches using your vocal cords.

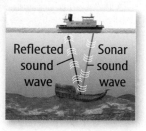
Reflected sound wave   Sonar sound wave

An echo is a reflected sound wave. Ships use sonar to find underwater objects.

Use Vocabulary

1. **Define** *echo* in your own words.

_____

_____

2. **Distinguish** between sound and a sound wave.

_____

_____

3. **Use** the word *pitch* in a sentence.

_____

_____

Understand Key Concepts

4. In which material do sound waves travel fastest? **SC.7.P.10.3**

   (A) aluminum   (C) ethanol
   (B) carbon dioxide (D) water

5. **Predict** Would a barking dog produce sound waves that travel faster during the day or at night? Explain your answer. **SC.7.P.10.3**

_____

_____

Interpret Graphics

6. **Sequence** Fill in the graphic organizer to show the path a sound wave travels from the air until it is interpreted by the brain. **LA.7.2.2.3**

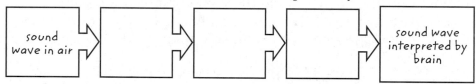

sound wave in air → ☐ → ☐ → ☐ → sound wave interpreted by brain

Critical Thinking

7. **Infer** A string vibrates with a frequency of 10 Hz. Why can't a person hear the sound waves produced by the vibrating string, no matter how large the amplitude of the waves?

_____

_____

Math Skills MA.6.A.3.6

8. If a sound wave travels 1,620 m in 3 s, what is its speed?

How do ultrasound machines work?

Using Ultrasound to Safely Monitor a Human Fetus

Ultrasound waves are sound waves with frequencies so high that humans cannot hear them. An ultrasound machine has a hand-held device, called a transducer, that emits and receives ultrasound waves. This enables medical professionals to see inside a human body.

1 A technician passes the transducer over the patient's skin, transmitting ultrasound waves into the patient's body.

2 When the ultrasound waves strike different surfaces, some of those waves reflect back to the transducer.

3 The transducer detects the reflected ultrasound waves and converts them into electronic signals.

4 A computer receives the signals from the transducer and produces an image.

It's Your Turn

RESEARCH other ways that medical professionals use ultrasound machines.

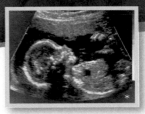

LIGHT

ESSENTIAL QUESTIONS

 How are light waves different from sound waves?

 How do waves in the electromagnetic spectrum differ?

 How do waves in the electromagnetic spectrum differ?

Vocabulary

light source p. 393

light ray p. 393

transparent p. 394

translucent p. 394

opaque p. 394

 Florida NGSSS

LA.7.2.2.3 The student will organize information to show understanding (e.g., representing main ideas within text through charting, mapping, paraphrasing, summarizing, or comparing/contrasting);

SC.7.P.10.1 Illustrate that the sun's energy arrives as radiation with a wide range of wavelengths, including infrared, visible, and ultraviolet, and that white light is made up of a spectrum of many different colors.

SC.7.P.10.2 Observe and explain that light can be reflected, refracted, and/or absorbed.

SC.7.P.10.3 Recognize that light waves, sound waves, and other waves move at different speeds in different materials.

SC.7.N.1.1 Define a problem from the seventh grade curriculum, use appropriate reference materials to support scientific understanding, plan and carry out scientific investigation of various types, such as systematic observations or experiments, identify variables, collect and organize data, interpret data in charts, tables, and graphics, analyze information, make predictions, and defend conclusions.

SC.7.N.3.2 Identify the benefits and limitations of the use of scientific models.

 **inquiry** Launch Lab   SC.7.P.10.3

15 minutes

What happens when light waves pass through water?

Do light waves always travel in a straight line? What happens to light waves when they travel through water?

Procedure

1. Read and complete a lab safety form.
2. Add **distilled water** to a **500-mL beaker** until it is two-thirds full.
3. Use **scissors** to cut a thin slit in a sheet of **paper. Tape** the paper over the lens of a **flashlight.**
4. Turn on the flashlight and tilt it slightly downward so the light beam is visible on the tabletop. Place the water-filled beaker in the light beam. Record your observations.

Data and Observations

Think About This

1. **Key Concept** Compare the direction of the light beam before it entered the water to after it left the water.

(Inquiry) **Are both men real?**

1. No, the man on the right is a hologram. A hologram is a type of image that seems to be three-dimensional. Laser light reflected from the person is used to create the life-like image. What are light waves and how do they interact with matter?

_____

_____

_____

_____

_____

_____

What is light?

As you read these words, you are probably looking at a page in a book. You might also see the desk on which the book is resting as well as the light from a lamp. What do your eyes detect when you see something?

Your eyes sense light waves. You see books and desks when light waves reflect off these objects and enter your eyes. Some objects, like a candle flame and a lightbulb that is lit, also emit light waves. You see a candle flame or a glowing lightbulb because the light waves they emit enter your eyes.

Light—An Electromagnetic Wave

Light is a type of wave called an electromagnetic wave. Like sound waves, electromagnetic waves can travel through matter. But they can also travel through a vacuum, where no matter is present. For example, light can travel through the space between Earth and the Sun.

Light travels through a vacuum at a speed of about 300,000 km/s. However, light waves slow down when they travel through matter. The speed of light in some different materials is listed in **Table 3.** Light waves travel much faster than sound waves. For example, in air the speed of light is about 900,000 times faster than the speed of sound.

Table 3 Light waves travel fastest in empty space. When light waves travel in matter, they move fastest in gases and slowest in solids.

Table 3 Speed of Light Waves in Some Materials	
Material	**Wave Speed (km/s)**
Vacuum	300,000
Air	299,920
Water	225,100
Glass	193,000

Active Reading

2. Compare Read each description. Decide if it applies to light, to sound, or to both. Put a check mark in the appropriate column.

Description	Light	Sound
Travels through matter		
Travels through a vacuum		
Travels at different speeds in different materials		

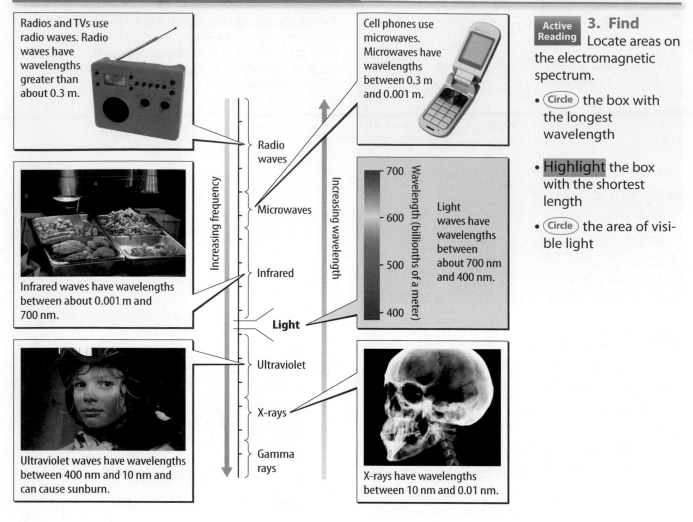

Radios and TVs use radio waves. Radio waves have wavelengths greater than about 0.3 m.

Cell phones use microwaves. Microwaves have wavelengths between 0.3 m and 0.001 m.

Infrared waves have wavelengths between about 0.001 m and 700 nm.

Light waves have wavelengths between about 700 nm and 400 nm.

Ultraviolet waves have wavelengths between 400 nm and 10 nm and can cause sunburn.

X-rays have wavelengths between 10 nm and 0.01 nm.

Increasing frequency

Increasing wavelength

Wavelength (billionths of a meter)
- 700
- 600
- 500
- 400

Radio waves
Microwaves
Infrared
Light
Ultraviolet
X-rays
Gamma rays

Active Reading 3. **Find** Locate areas on the electromagnetic spectrum.

- Ⓒircle the box with the longest wavelength

- Highlight the box with the shortest length

- Ⓒircle the area of visible light

Figure 10 Electromagnetic waves are classified according to their wavelength or frequency. Visible light waves are part of the electromagnetic spectrum.

Active Reading 4. **Differentiate** How are waves in the electromagnetic spectrum different?

_____
_____
_____
_____
_____

The Electromagnetic Spectrum

In addition to visible light waves, there are other types of electromagnetic waves, such as infrared and ultraviolet. Scientists classify electromagnetic waves into groups based on their wavelengths, as shown in **Figure 10.** The entire range of electromagnetic waves is called the electromagnetic spectrum.

Light waves are only a small part of the electromagnetic spectrum. The wavelengths of light waves are so short they are usually measured in nanometers (nm). One nanometer equals one-billionth of a meter. The wavelengths of light waves are from about 700 nm to about 400 nm. This is about one-hundredth the width of a human hair. You see different colors when different wavelengths of light waves enter your eyes.

Light-Emitting Objects

Think about walking into a dark room and turning on a light. The lightbulb produces light waves that travel away from the bulb in all directions, as shown in **Figure 11.** *A* **light source** *is something that emits light.* In order to emit light, the lightbulb transforms electric energy into light energy. Other examples of light sources are the Sun and burning candles. The Sun transforms nuclear energy into light energy. Burning candles transform chemical energy into light energy. In general, light sources convert other forms of energy into light energy.

Light Rays

As you just read, light waves spread out in all directions from a light source. You also can think of light in terms of light rays. *A* **light ray** *is a narrow beam of light that travels in a straight line.* The arrows in **Figure 11** represent some of the light rays moving away from the light source. Unless light rays come in contact with a surface or pass through a different material, they travel in straight lines.

Light Reflection

Suppose you are in a dark room. Do you see anything? Now you turn on a light. What do you see now? Light sources emit light. But other objects, like books, reflect light. In order to see an object that is not a light source, light waves must reflect from an object and enter your eyes.

Seeing Objects

When you see a light source, light rays travel directly from the light source into your eye. Light rays also reflect off objects, as shown in **Figure 12.** Light rays reflect from an object in many directions. Some of the light rays that reflect from an object enter your eye, enabling you to see the object.

| Active Reading | **5. Draw** Label the path of light from a light source to the eye. The complete the sequence diagram to explain what you have drawn. |

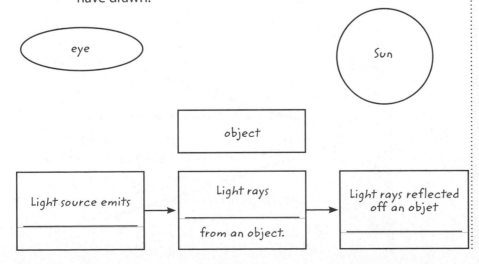

Figure 11 Light travels in all directions away from its source.

Active Reading SC.7.P.10.2
FOLDABLES® LA.7.2.2.3

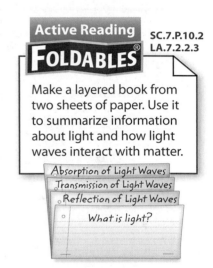

Make a layered book from two sheets of paper. Use it to summarize information about light and how light waves interact with matter.

Absorption of Light Waves
Transmission of Light Waves
Reflection of Light Waves
What is light?

Figure 12 Some light waves from a light source reach the page and reflect off it. The girl sees the page when some of the reflected light enters her eyes.

Figure 13 The butter-fly's wing, the frosted glass, and the curtains interact with light waves differently.

Transparent

Translucent

Opaque

WORD ORIGIN

opaque
from Latin *opacus*, means "shady, dark"

The Interaction of Light and Matter

Like all waves, when light waves interact with matter they can be transmitted, absorbed, or reflected.

- Reflection occurs when light waves come in contact with the surface of a material and bounce off.

- Transmission occurs when light waves travel through a material.

- Absorption occurs when interactions with a material convert light energy into other forms such as thermal energy.

In many materials, reflection, transmission, and absorption occur at the same time. For example, the tinted glass of an office building reflects some light, transmits some light, and absorbs some light.

 6. NGSSS Check Identify <u>Underline</u> the three interactions that light can have with matter. **SC.7.P.10**

Depending on how they interact with light, materials can be classified as transparent, translucent, or opaque, as shown in **Figure 13.** *A material is* **transparent** *if it allows almost all light that strikes it to pass through and forms a clear image. A material is* **translucent** *if it allows most of the light that strikes it to pass but forms a blurry image. A material is* **opaque** *if light does not pass through it.* An opaque material, such as light-blocking cloth curtains, does not transmit light.

The Reflection of Light Waves

Figure 14 shows what happens when a surface reflects light waves. All waves, including light waves, obey the law of reflection. According to the law of reflection, the angle of incidence always equals the angle of reflection. In **Figure 14,** the line perpendicular to a surface is called the normal. The angle between the normal and the incoming light ray is the angle of incidence. The angle between the reflected light ray and the normal is the angle of reflection.

Figure 14 When a surface reflects a light ray, the angle of incidence equals the angle of reflection.

Active Reading **7. Draw** Using arrows, draw the angle of reflection changes if the angle of incidence increases.

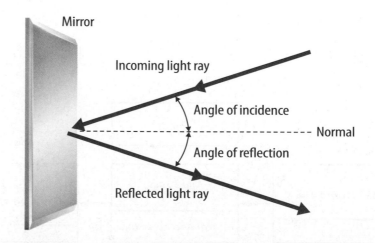

Mirror

Incoming light ray

Angle of incidence

Normal

Angle of reflection

Reflected light ray

Figure 15 Particles of dust floating in the air scatter light rays in a sunbeam. When light rays strike these particles, light rays reflect in many different directions.

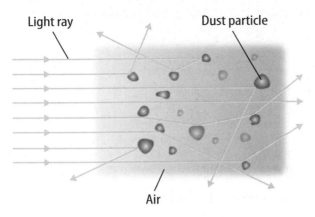

Light ray

Dust particle

Air

Scattering

When a beam of sunlight shines through a window, you might notice tiny particles of dust. You see the dust particles because they reflect light waves. As **Figure 15** shows, dust particles reflect light waves in many different directions because they have different shapes. This is an example of scattering. Scattering occurs when light waves traveling in one direction are made to travel in many directions. The dust particles scatter the light waves in the sunbeam.

The Refraction of Light Waves

Like all types of waves, light waves can change direction when they travel from one material to another. The light beam in **Figure 16** changes direction as it goes from the air into the glass and from the glass into the air. A wave that changes direction as it travels from one material into another is refracting.

Recall that light waves travel at different speeds in different materials. Refraction occurs when a wave changes speed. The greater the change in speed, the more the light wave refracts or changes direction.

Active Reading 8. **Explain** When does refraction occur?

_____

_____

Inquiry | **Try It!** SC.7.N.1.1 SC.7.P.10.2

LAB STATION

MiniLab *Can you see a light beam in water?* at connectED.mcgraw-hill.com

Apply It! After you complete the lab, answer these questions.

1. **Predict** What would happen to the beam of light if different types of milk were used, such as skim milk, buttermilk, or evaporated milk?

2. **Explain** Based on what you discovered in the lab, what would happen to car headlights on a foggy day?

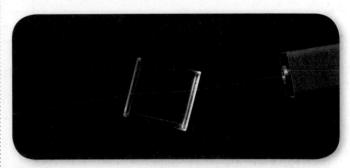

▲ **Figure 16** The red beam of light slows down as it enters the glass rectangle. It speeds up as it leaves the rectangle and enters the air.

Visual Summary

An object is seen when light waves emitted by the object or reflected by the object enter the eye

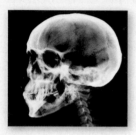

The electromagnetic spectrum includes electromagnetic waves of different wavelengths, such as X-rays.

When light waves interact with matter, they can be absorbed, reflected, or transmitted.

Inquiry SC.7.N.1.1, SC.7.N.3.2, SC.7.P.10.2

iLAB STATION **Try It!**

Skill Lab *How are light rays reflected from a plane mirror?*

Use Vocabulary

1 **Explain** the difference between transparent and translucent.

_____

_____

2 **Define** *light source* using your own words.

_____

_____

Understand Key Concepts

3 **Apply** Do light waves refract more when they travel from air to water or air to glass? **SC.7.P.10.2**

_____

_____

4 **Which** electromagnetic wave has the shortest wavelength? **SC.7.P.10.1**

(A) gamma   (C) a rope

(B) a rarefaction   (D) a vibration

Interpret Graphics

5 **Evaluate** If a light wave has an angle of incidence of 30° as shown below, what is its angle of reflection? **SC.7.P.10.2**

_____

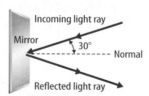

Incoming light ray
Mirror
30°
Normal
Reflected light ray

6 **Organize Information** Fill in the table below. **LA.7.2.2.3**

Interaction	Description
Absorption	
	Light wave bounces off a surface.
Transmission	

Critical Thinking

7 **Describe** how the speed of light waves changes when they travel from air into a water-filled aquarium and back into air. **SC.7.P.10.3**

_____

_____

Interactions of Light and Matter

Summarize each term below in your own words. Use your classroom to find an example of each and circle how light interacts with matter to produce the effect.

Translucent

③ Definition

Example

Reflected
Transmitted
Absorbed

Opaque

② Definition

Example

Reflected
Transmitted
Absorbed

Transparent

① Definition

Example

Reflected
Transmitted
Absorbed

Mirrors, Lenses and THE EYE

ESSENTIAL QUESTIONS

What is the difference between regular and diffuse reflection?

What types of images are formed by mirrors and lenses?

How does the human eye enable a person to see?

Vocabulary

mirror p. 400

lens p. 401

cornea p. 402

iris p. 403

pupil p. 403

retina p. 404

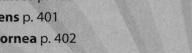

Florida NGSSS

LA.7.2.2.3 The student will organize information to show understanding (e.g., representing main ideas within text through charting, mapping, paraphrasing, summarizing, or comparing/contrasting);

MA.6.A.3.6 Construct and analyze tables, graphs, and equations to describe linear functions and other simple relations using both common language and algebraic notation.

SC.7.P.10.1 Illustrate that the sun's energy arrives as radiation with a wide range of wavelengths, including infrared, visible, and ultraviolet, and that white light is made up of a spectrum of many different colors.

SC.7.P.10.2 Observe and explain that light can be reflected, refracted, and/or absorbed.

SC.7.N.1.1 Define a problem from the seventh grade curriculum, use appropriate reference materials to support scientific understanding, plan and carry out scientific investigation of various types, such as systematic observations or experiments, identify variables, collect and organize data, interpret data in charts, tables, and graphics, analyze information, make predictions, and defend conclusions.

 **Launch Lab** SC.7.N.1.1

15 minutes

Are there different types of reflections?

Some surfaces are like mirrors. Other surfaces do not form reflected images you can see. What is the difference between a surface that forms a sharp reflected image and one that does not?

Procedure

1 Read and complete a lab safety form.

2 Place a **black bowl** on a **paper towel.** Look straight down at the bottom of the bowl. Record your observations.

3 Carefully add about 3 cm of water to the bowl. Look at the bottom of the bowl. Record your observations.

4 Tap the side of the bowl gently and look again. Record your observations.

Data and Observations

Think About This

1. Compare your observations in steps 2, 3, and 4.

2. **Key Concept** What do you think caused the differences in the images you observed?

1. These Cypress trees, shown here from the Apalachicola National Forest in Florida, are reflected on the lake. Have you ever seen an image on the surface of a lake? If so, you have observed light waves reflecting from a mirrorlike surface. Are all reflected images the same? How does the shape of a mirror's surface affect the image that you see?

_____

_____

_____

_____

Why are some surfaces mirrors?

When you look at a smooth pond, you can see a sharp image of yourself reflected off the water surface. When you look at a lake on a windy day, you do not see a sharp image. Why are these images different? A smooth surface reflects light rays traveling in the same direction at the same angle. This is called regular reflection, as shown in **Figure 17.** Because the light rays travel the same way relative to each other before and after reflection, the reflected light rays form a sharp image.

When a surface is not smooth, light rays still follow the law of reflection. However, light rays traveling in the same direction hit the rough surface at different angles. The reflected light rays travel in many different directions, as shown in **Figure 17.** This is called diffuse reflection. You do not see a clear image when diffuse reflection occurs.

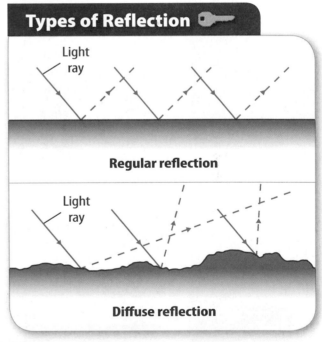

Types of Reflection 🔑

Light ray

Regular reflection

Light ray

Diffuse reflection

Figure 17 Light waves always obey the law of reflection, whether the surface is smooth or rough.

2. **NGSSS Check** **Contrast** Provide two examples of regular and diffuse reflection. SC.7.P.10.2

_____

Table 4 Images and Mirrors 🔑

Concave Mirror

Focal length

Optical axis

Focal point

Concave mirror

The focal length is the distance from the center of the mirror to the focal point.

Concave mirror

Focal point

Focal length

The image in a concave mirror is _____ when an object is more than _____ focal length from the mirror.

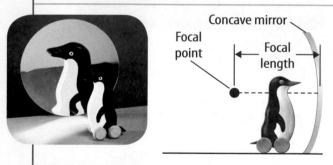

Concave mirror

Focal point

Focal length

The image in a concave mirror is _____ when an object is _____ than one focal length from the mirror.

Convex Mirror

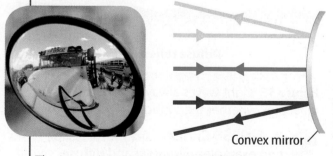

Convex mirror

The image in a convex mirror is always _____ and _____ than the object.

Types of Mirrors

When you look at a wall mirror, the image you see is about the same size that you are and right-side up. *A **mirror** is any reflecting surface that forms an image by regular reflection.* The image formed by a mirror depends on the shape of the mirror's surface.

Plane Mirrors

A plane mirror is a mirror that has a flat reflecting surface. The image formed by the mirror looks like a photograph of the object except that the image is reversed left to right. The size of the image in the mirror depends on how far the object is from the mirror. The image gets smaller as the object gets farther from the mirror.

Concave Mirrors

Reflecting surfaces that are curved inward are concave mirrors, as shown in **Table 4.** Light rays that are parallel to a line called the optical axis, which is shown in **Table 4,** are reflected through one point—the focal point. The distance from the mirror to the focal point is called the focal length.

The type of image formed depends on where the object is, as shown in **Table 4.** If an object is more than one focal length from a concave mirror, the image will be upside down. If the object is closer than one focal length, the image will be right-side up. If an object is placed exactly at the focal point, no image forms.

Convex Mirrors

A convex mirror has a reflecting surface that is curved outward, as shown in **Table 4.** The image is always right-side up and smaller than the object. Store security mirrors and passenger-side car mirrors are usually convex mirrors.

Active Reading

3. Identify Fill in the blanks in **Table 4** to explain how the images formed by concave mirrors and convex mirrors depend on the distance of an object from the mirror.

Types of Lenses

Have you ever used a magnifying lens or binoculars? Or you might wear glasses that help you see more clearly. All of these items use lenses to change the way an image of an object forms. *A **lens** is a transparent object with at least one curved side that causes light to change direction.* The more curved the sides of a lens, the more the light changes direction as it passes through the lens.

Convex Lenses

A convex lens is curved outward on at least one side so it is thicker in the middle than at its edges. Just like a concave mirror, a convex lens has a focal point and a focal length, as shown in **Table 5.** The more curved the lens is, the shorter the focal length.

The image formed by a convex lens depends on where the object is just like it does for a concave mirror. When an object is farther than one focal length from a convex lens, the image is upside down, as shown to the right.

When an object is less than one focal length from a convex lens, the image is larger and right side up. For example, the dollar bill in **Table 5** is less than one focal length from the lens. As a result, the image of the dollar bill is larger than the actual dollar bill and is right-side up. Both a magnifying lens and a camera lens are convex lenses.

Active Reading 4. **Identify** Fill in the blanks in **Table 5** to explain how the images formed by concave lenses and convex lenses depend on the distance of an object from the mirror.

Concave Lenses

A concave lens is curved inward on at least one side and thicker at its edges. The image formed by a concave lens is upright and smaller than the object, as shown in **Table 5.** Concave lenses are usually used in combinations with other lenses in instruments such as telescopes and microscopes.

Table 5 Images and Lenses 🔑

Convex Lens

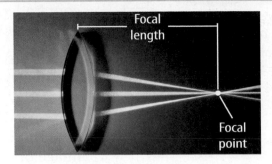

A convex lens is thicker in its middle than its edges. The focal length is the distance from the center of the lens to the focal point.

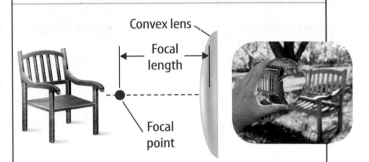

The image formed by a convex lens is _____ when an object is _____ than one focal length from the lens.

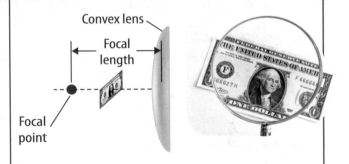

The image formed by a convex lens is _____ when an object is less than _____ focal length from the lens.

Concave Lens

A concave lens is thinner in its middle than its edge. The image formed by a concave lens is always _____ and _____ than the object.

Active Reading

FOLDABLES® LA.7.2.2.3

Make a half book from a sheet of paper. Use it to identify the parts of the eye and their function.

ACADEMIC VOCABULARY

convert

(verb) to change from one form into another

Light and the Human Eye

Microscopes, binoculars, and telescopes are instruments that contain lenses that form images of objects. Human eyes also contain lenses, as well as other parts, that can enable a person to see.

The structure of a human eye is shown in **Figure 18.** To see an object, light waves from an object travel through two convex lenses in the eye. The first of these lenses is called the cornea, and the second is simply called the lens. These lenses form an image of the object on a thin layer of tissue at the back of the eye. Special cells in this layer **convert** the image into electrical signals. Nerves carry these signals to the brain.

Active Reading
5. **Label** Underline the function of the lenses in the eye.

Cornea

Light waves first travel through the cornea (KOR nee uh), as shown in **Figure 18.** *The* **cornea** *is a convex lens made of transparent tissue located on the outside of the eye.* Most of the change in direction of light rays occurs at the cornea. Some vision problems are corrected by changing the cornea's shape.

The Structure of the Human Eye 🔑

Figure 18 The eye is made of a number of parts that have different functions.

Active Reading
6. **Draw** If this eye is looking at a tree, draw the tree in the part of the eye where the image is formed.

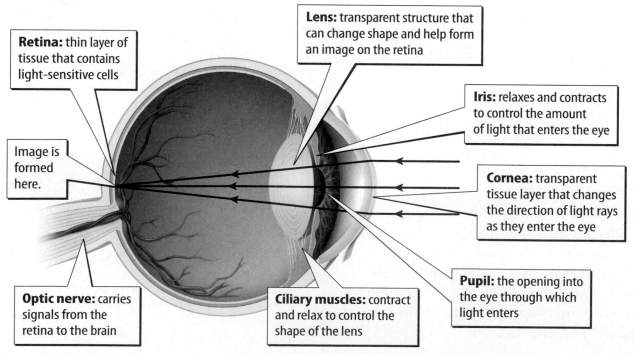

Retina: thin layer of tissue that contains light-sensitive cells

Lens: transparent structure that can change shape and help form an image on the retina

Iris: relaxes and contracts to control the amount of light that enters the eye

Image is formed here.

Cornea: transparent tissue layer that changes the direction of light rays as they enter the eye

Optic nerve: carries signals from the retina to the brain

Ciliary muscles: contract and relax to control the shape of the lens

Pupil: the opening into the eye through which light enters

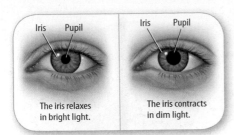

Figure 19 The iris controls the amount of light that enters the eye.

The iris relaxes in bright light.

The iris contracts in dim light.

Iris   Pupil

Iris   Pupil

Iris and Pupil

The **iris** is the colored part of the eye. The **pupil** is an opening into the interior of the eye at the center of the iris. When the size of the pupil changes, the amount of light that enters the eye changes. As shown in **Figure 19,** in bright light, the iris relaxes and the pupil becomes smaller. Then less light enters the eye. In dim light, the iris contracts and the pupil becomes larger. Then more light enters the eye.

Lens

Behind the iris is the lens, as shown in **Figure 18.** It is made of flexible, transparent tissue. The lens enables the eye to form a sharp image of nearby and distant objects. The muscles surrounding the lens change the lens's shape. To focus on nearby objects, these muscles relax and the lens becomes more curved, as shown in **Figure 20.** To focus on distant objects, these muscles pull on the lens and make it flatter.

Active Reading **7. Compare and Contrast** List similarities and differences between the cornea and the lens.

Similar	Different

Figure 20 The lens in the eye changes shape, enabling the formation of sharp images of objects that are either nearby or far away.

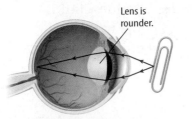

Lens is rounder.

Lens becomes rounder and a sharp image forms of a nearby object.

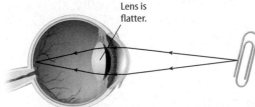

Lens is flatter.

Lens becomes flatter and a sharp image forms of a distant object.

Inquiry LAB STATION **Try It!** MA.6.A.3.6, MA.6.S.6.2

MiniLab *How does the size of an image change?* at connectED.mcgraw-hill.com

Apply It! After you complete the lab, answer these questions.

1. **Apply** Think about what you know about the eye's lens and what you learned in the MiniLab. Explain why you hold an object closer to your eyes to be able to see it better.

2. **Analyze** Look at the graph you created in the MiniLab. Is there a limit to how much a convex lens can magnify an object? Justify your answer.

Active Reading **8. Explain** How do the muscles in the eye change the lens to help it focus on objects that are nearby or far away?

9. Identify
Indicate which parts of the eye form a sharp image of an object and the parts that convert an image into electrical signals.

Lens:

_____

_____

Iris:

_____

_____

Cornea:

_____

_____

Pupil:

_____

_____

Retina:

_____

_____

Contains 2 light-sensitive cells:

a. rod cells:

_____

_____

b. cone cells:

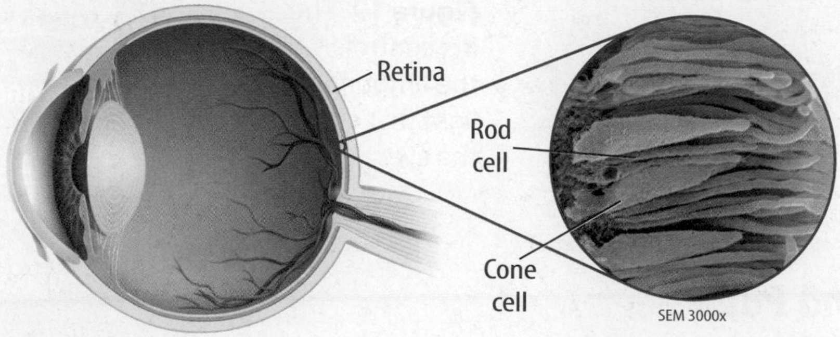

_____

_____

Retina ————

Rod cell

Cone cell

SEM 3000x

Figure 21 Rod cells in the retina respond to dim light. Cone cells in the retina enable you to see colors.

Retina

The **retina** *is a layer of special light-sensitive cells in the back of the eye,* as shown in **Figure 21.** After light travels through the lens, an image forms on your retina. There, chemical reactions produce nerve signals that the optic nerve sends to your brain. There are two types of light-sensitive cells in your retina—rod cells and cone cells.

Rod Cells There are more than 100 million rod cells in a human retina. Rod cells are sensitive to low-light levels. They enable people to see objects in dim light. However, the signals rod cells send to your brain do not enable you to see colors.

Cone Cells A retina contains over 6 million cone cells. Cone cells enable a person to see colors. However, cone cells need brighter light than rod cells to function. In very dim light, only rod cells function. That is why objects seem to have no color in very dim light.

How do cone cells enable you to see colors? The responses of cone cells to light waves with different wavelengths enable you to see different colors.

The retina has three types of cone cells. Each type of cone cell responds to a different range of wavelengths. This means that different wavelengths of light cause each type of cone cell to send different signals to the brain. Your brain interprets the different combinations of signals from the three types of cone cells as different colors. However, in some people not all three types of cone cells function properly. These people cannot detect certain colors. This condition is commonly known as color blindness but is more appropriately called color deficiency. People with some kinds of color deficiency cannot see the number 74 in **Figure 22.**

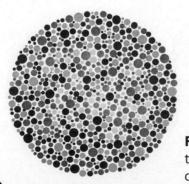

Figure 22 An image like this is used to test for color deficiency.

The Colors of Objects

The objects you see around you are different colors. A banana is mostly yellow, but a rose might be red. Why is a banana a different color from a rose? Bananas and roses do not give off, or emit, light. Instead, they reflect light. The colors of an object depend on the wavelengths of the light waves it reflects.

Reflection of Light and Color

When light waves of different wavelengths interact with an object, the object absorbs some light waves and reflects others. The wavelengths of light waves absorbed and reflected depend on the materials from which the object is made.

For example, **Figure 23** shows that the rose is red because the petals of the rose reflect light waves with certain wavelengths. When these light waves enter your eye, they cause the cone cells in your retina to send certain nerve signals to your brain. These signals cause you to see the rose as red.

A banana reflects different wavelengths of light than a rose. These different wavelengths cause cone cells in the retina to send different signals to your brain. These signals cause you to see the banana as yellow instead of red.

You might think that light waves have colors. Color, however, is a sensation produced by your brain when light waves enter your eyes. Light waves have no color as they travel from an object to your eyes.

Active Reading

10. Explain Why do you experience the sensation of color?

_____

_____

_____

The Color of Objects that Emit Light

Some objects such as the Sun, lightbulbs, and neon lights emit light. The color of an object that emits light depends on the wavelengths of the light waves it emits. For example, a red neon light emits light waves with wavelengths that you see as red.

Reflected light waves

The rose reflects light waves with wavelengths that you see as red. It absorbs all other wavelengths of light.

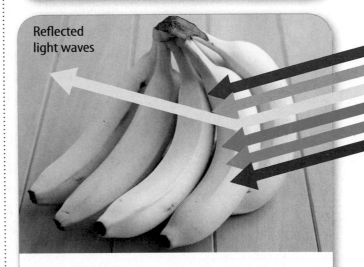

Reflected light waves

The banana reflects light waves with wavelengths that you see as yellow. It absorbs all other wavelengths of light.

Figure 23 The color of an object depends on the wavelengths of the light waves the object reflects.

Figure 24 Different wavelengths of light change direction by different amounts when they move into and out of a prism. This causes the waves to spread out.

White Light A Combination of Light Waves

You might have noticed at a concert that the colors of objects on stage depend on the colors of the spotlights. A shirt that appears blue when a white spotlight shines on it might appear black when a red spotlight shines on it. That same shirt will appear blue when a blue spotlight shines on it. How is light that is white different from light that is red or blue?

Light that you see as white is actually a combination of light waves of many different wavelengths. **Figure 24** shows what happens when white light travels through a prism. Light waves with different wavelengths spread out after passing through the prism and form a color spectrum.

Changing Colors

Figure 25 shows why the color of a blue shirt appears different when different spotlights shine on it. The shirt reflects only those wavelengths that are seen as blue. It absorbs all other wavelengths of light. When white light or blue light hits the shirt, it reflects the wavelengths you see as blue. However, when red light strikes the shirt, almost no light is reflected. This causes the shirt to appear black. An object appears black when it absorbs almost all light waves that strike it.

11. Describe How would a rainbow look if viewed through an indigo filter?

_____

_____

_____

_____

_____

_____

Figure 25 The appearance of an object changes under different colors of light.

When white light strikes the shirt, only the wavelengths that you see as blue are reflected. The shirt appears blue under white light.

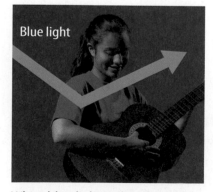

When blue light strikes the shirt, the blue light is reflected. This makes the shirt appear blue under blue light.

When red light strikes the shirt, the light is absorbed and no light is reflected. This makes the shirt appear black under red light.

Visual Summary

A mirror is a surface that causes a regular reflection. The shape of the reflecting surface and the position of the object determine what the image looks like.

A lens is a transparent object with at least one curved side that causes light waves to change direction. The shape of the lens and the position of the object determine how the image appears

The eye has different parts with different functions. The iris is the colored part of your eye. The iris opens and closes, controlling the amount of light that enters the eye.

Inquiry MA.6.A.3.6, MA.6.S.6.2

LAB STATION Try It!

Skill Lab *The Images Formed by a Lens* at connectED.mcgraw-hill.com

Use Vocabulary

1. The layer of tissue in the eye that contains cells sensitive to light is the _____.

2. **Define** *cornea* in your own words.

3. **Distinguish** between a lens and a mirror.

Understand Key Concepts

4. **Draw** a picture of two light rays that reflect off a smooth surface. **SC.7.P.10.2**

5. **Compare** the function of the cornea and the lens of the eye.

6. Which statement describes the image formed by a convex mirror?
   - (A) It will be caused by refraction.
   - (B) It will be smaller than the object.
   - (C) It will be upside down.
   - (D) It will produce a beam of light.

Interpret Graphics

7. **Sequence** Fill in the graphic organizer below to trace the path of light through the different parts of the human eye. **LA.7.2.2.3**

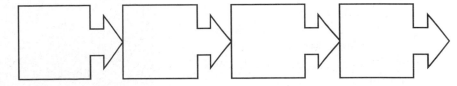

Critical Thinking

8. **Infer** A person cannot see well in dim light. What part of his or her eye is damaged?

Think About It! Sound waves must travel through matter, while light waves can also travel in a vacuum. Waves interact with matter through absorption, transmission, and reflection.

Key Concepts Summary

Vocabulary

LESSON 1 Sound

- Vibrating objects produce **sound waves.**
- Sound waves travel at different speeds in different materials. Sound waves usually travel fastest in solids and slowest in gases.
- The outer ear collects sound waves. The middle ear amplifies sound waves. The inner ear converts sound waves to nerve signals.

sound wave p. 381

pitch p. 385

echo p. 387

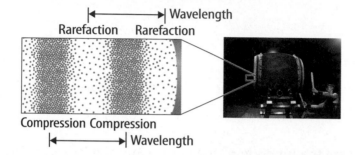

LESSON 2 Light

- Light waves are electromagnetic waves that can travel in matter and through a vacuum.
- Electromagnetic waves have different wavelengths and frequencies.
- When light waves interact with matter, they are reflected, transmitted, or absorbed.

light source p. 393

light ray p. 393

transparent p. 394

translucent p. 394

opaque p. 394

LESSON 3 Mirrors, Lenses, and the Eye

- When regular reflection occurs from a surface, a clear image forms and the surface is a **mirror.** When diffuse reflection occurs from a surface, a clear image does not form.
- The shape of a mirror or a **lens** and the distance of an object from the mirror or lens determine how the image appears.
- When light rays enter the eye through the **cornea** and pass through the **pupil,** an image forms on the **retina.** Rod and cone cells convert the image to nerve signals that travel to the brain.

mirror p. 400

lens p. 401

cornea p. 402

iris p. 403

pupil p. 403

retina p. 404

Assemble your lesson Foldables as shown to make a Chapter Project. Use the project to review what you have learned in this chapter.

Use Vocabulary

1 A vibrating object produces _____.

2 A sonar system detects the _____ of a sound wave.

3 A(n) _____ travels in a straight path until it is refracted, reflected, or absorbed.

4 A window is a(n) _____ object because you can see objects clearly through it.

5 The Sun is a(n) _____ because it emits light waves.

6 The _____ of the eye controls how much light enters it.

7 A(n) _____ is a surface that produces a regular reflection.

Link Vocabulary and Key Concepts

Use vocabulary terms from the previous page to complete the concept map.

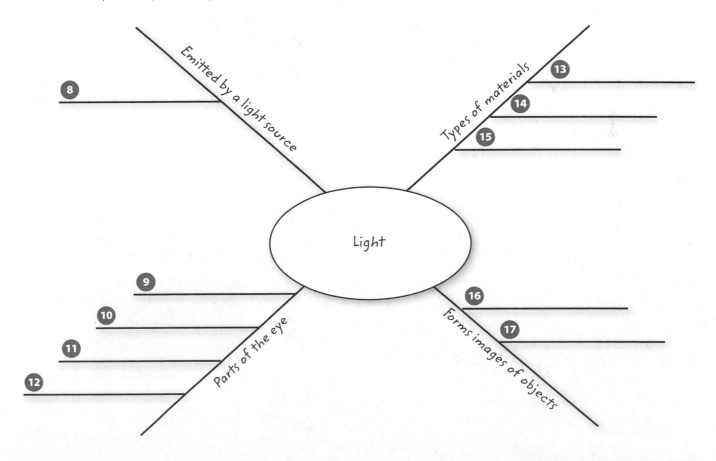

Fill in the correct answer answer.

🔑 Understand Key Concepts

1 Which part of the ear acts as the amplifier? **SC.7.P.10.3**

- Ⓐ cochlea
- Ⓑ nerves
- Ⓒ inner ear
- Ⓓ middle ear

2 Identify the part of the ear that serves as the sound-wave collector. **SC.7.P.10.3**

- Ⓐ I
- Ⓑ II
- Ⓒ III
- Ⓓ IV

3 Which produces sound that has the highest pitch? **SC.7.P.10.3**

- Ⓐ a tuba
- Ⓑ emergency siren
- Ⓒ lion's roar
- Ⓓ thunder

4 The speed of light is slowest in which medium? **SC.7.P.10.3**

- Ⓐ cold air
- Ⓑ outer space
- Ⓒ pond water
- Ⓓ window glass

5 A shirt appears red under white light. What color would it appear under blue light? **SC.7.P.10.2**

- Ⓐ black
- Ⓑ blue
- Ⓒ red
- Ⓓ white

6 Which enables a person to see color? **SC.7.P.10.2**

- Ⓐ cone cells
- Ⓑ cornea
- Ⓒ iris
- Ⓓ rod cells

Critical Thinking

7 **Compare** an echo to light that hits a mirror. **SC.7.P.10.2**

_____

_____

_____

8 **Summarize** Listen to the sounds around you. Choose one sound and describe its path from its source to the point where the nerve signal is sent to your brain. **SC.7.P.10.2**

_____

_____

_____

_____

_____

9 **Judge** Frosted glass is made of glass with a scratched surface. Decide whether frosted glass is opaque, translucent, or transparent and explain your reasoning. **SC.7.P.10.2**

_____

_____

_____

10 **Infer** On a hot summer day, black pavement feels much hotter than light-gray concrete. Explain why this is so. **SC.7.P.10.2**

_____

_____

_____

11 **Evaluate** Stores are often equipped with mirrors like the one shown below. What type of mirror is this and why is it useful? **SC.7.P.10.2**

_____

_____

12 **Compare** When you enter a room filled with many hard surfaces and start talking, your voice echoes all around you. How is this similar to what happens when light rays scatter? **SC.7.P.10.2**

_____

_____

_____

13 **Infer** A film camera forms an upside-down image of a tree on the film. What is the location of the tree relative to the focal length of the lens? **SC.7.P.10.2**

_____

_____

_____

Writing in Science

14 **Write** a 500–700-word essay on a separate piece of paper about an object such as a telescope, a microscope, or a periscope that uses mirrors or lenses to create an image. Describe how and why the mirrors and lenses are used in the object. **LA.7.2.2.3**

Big Idea Review

15 What are three possible results when light waves strike matter? **SC.7.P.10.2**

_____

_____

_____

16 In many movies, you can both see and hear explosions that happen in outer space. Explain how this is inaccurate. **LA.7.2.2.3**

_____

_____

17 The photo below shows the Chicago bean display. What happens to light waves when they strike the surface of the bean? What type of surface is this? **LA.7.2.2.3**

_____

_____

_____

Math Skills MA.6.A.3.6

Use Equations

18 What is the speed of a sound wave if it travels 2,500 m in 2 s?

19 A sound wave travels through air at 331 m/s. How far would it travel at 0°C in 5 s?

20 A sound wave travels through water at 1,500 m/s. How far would it travel in 5 s?

Fill in the correct answer.

Multiple Choice

1 In which material is the speed of sound waves the fastest? SC.7.P.10.3

Ⓐ solids

Ⓑ liquids

Ⓒ gases

Ⓓ air

Use the diagram below to answer question 2.

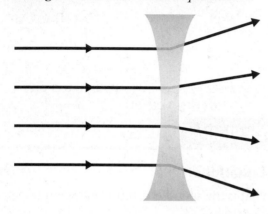

2 Which describes the image formed by the lens shown above? SC.7.P.10.2

Ⓕ right-side up, larger than the object

Ⓖ right-side up, smaller than the object

Ⓗ upside down, larger than the object

Ⓘ upside down, smaller than the object

3 A shirt that appears red under red light will also appear red under which color light?
SC.7.P.10.2

Ⓐ blue

Ⓑ green

Ⓒ yellow

Ⓓ white

Use the table below to answer question 4.

Wave Type	Wavelength
Radio waves	More than 0.001 m
Microwaves	0.3 m to 0.001 m
Visible light	700 nm to 400 nm
Ultraviolet light	400 nm to 10 nm
X-rays	10 nm to 0.01 nm

4 Which electromagnetic waves can have wavelengths of 500 nm? SC.7.P.10.1

Ⓕ microwaves

Ⓖ ultraviolet light

Ⓗ visible light

Ⓘ X-rays

5 When does the refraction of light waves occur? SC.7.P.10.2

Ⓐ when they bounce off a reflecting surface

Ⓑ when they are absorbed

Ⓒ when they move far from their source

Ⓓ when they pass from one material to another and change speed

6 On which property of the particles in a material does the speed of sound depend? SC.7.P.10.3

Ⓕ the dimensions of the particles

Ⓖ the forces between the particles

Ⓗ the shape of the particles

Ⓘ the number of particles

7 How does the speed of sound in air change as air becomes warmer? SC.7.P.10.3

Ⓐ It increases.

Ⓑ It decreases.

Ⓒ It does not change.

Ⓓ It increases, then decreases.

Use the diagram below to answer question 8.

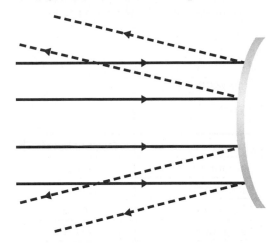

8 Which describes the image formed by the mirror shown above? **SC.7.P.10.2**

    Ⓕ larger than the object, right-side up

    Ⓖ larger than the object, upside down

    Ⓗ smaller than the object, right-side up

    Ⓘ smaller than the object, upside down

9 Which surface produces a diffuse reflection of light? **SC.7.P.10.2**

    Ⓐ a bathroom mirror

    Ⓑ the surface of a calm lake

    Ⓒ the hood of a newly waxed car

    Ⓓ a white painted wall

10 Which type of wave requires a medium through which to travel? **SC.7.P.10.1**

    Ⓕ visible light

    Ⓖ ultraviolet light

    Ⓗ sound waves

    Ⓘ microwaves

Use the illustration below to answer question 11.

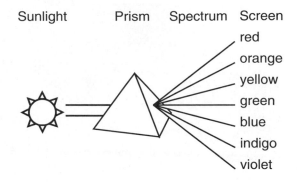

11 The figure above shows sunlight passing through a prism. How does a prism separate sunlight into the colors of the spectrum? **SC.7.P.10.1**

    Ⓐ absorbing certain wavelengths and transmitting others

    Ⓑ reflecting each wavelength a different amount

    Ⓒ reflecting each wavelength in a different direction

    Ⓓ refracting each wavelength a different amount

12 What causes a stop sign to appear red? **SC.7.P.10.2**

    Ⓕ It absorbs red light.

    Ⓖ It reflects red light.

    Ⓗ It transmits red light.

    Ⓘ It reflects all colors except red.

NEED EXTRA HELP?

If You Missed Question...	1	2	3	4	5	6	7	8	9	10	11	12
Go to Lesson...	1	3	3	2	2	1	1	3	3	2	3	3

Multiple Choice *Bubble the correct answer.*

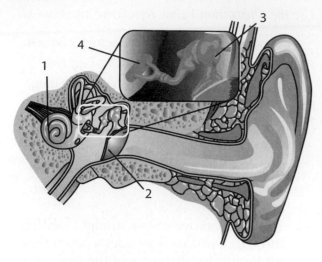

1. Which structure of the ear in the diagram above converts sound waves to nerve signals that the brain can interpret? **SC.7.P.10.3**

   (A) 1

   (B) 2

   (C) 3

   (D) 4

2. Isabel plays a scale on her clarinet. As she plays from low notes to high notes, what happens to the sound waves that the clarinet creates? **SC.7.P.10.3**

   (F) The amplitude of the sound waves decreases.

   (G) The frequency of the sound waves increases.

   (H) The speed of the sound waves will decrease.

   (I) The wavelengths of the sound waves increase.

3. Through which material would a sound wave move most slowly? **SC.7.P.10.3**

   (A) gelatin

   (B) milk

   (C) oxygen

   (D) steel

	Decibel Level
Tornado siren	140 dB
Jackhammer	130 dB
Chain saw	100 dB
Lawn mower	90 dB
Vacuum cleaner	75 dB
Dishwasher	60 dB

4. The chart above shows the decibels produced by a number of items. According to the chart, which sound has 1,000 times more energy than the sound of a dishwasher? **SC.7.P.10.3**

   (F) the sound of a tornado siren

   (G) the sound of a chain saw

   (H) the sound of a jackhammer

   (I) the sound of a lawn mower

Copyright © Glencoe/McGraw-Hill, a division of The McGraw-Hill Companies, Inc.

Multiple Choice *Bubble the correct answer.*

Use the following diagram to answer questions 1 and 2.

The Electromagnetic Spectrum

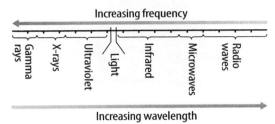

1. You would expect electromagnetic waves with a wavelength of 550 nm to **SC.7.P.10.1**

   (A) cause a sunburn.

   (B) reflect off skin.

   (C) heat water in a microwave.

   (D) transmit through wood.

2. Which type of electromagnetic waves is used with cellular phones? **SC.7.P.10.1**

   (F) infrared

   (G) microwaves

   (H) radio

   (I) ultraviolet

3. Which substance is translucent? **SC.7.P.10.2**

   (A) clear glass

   (B) cream cheese

   (C) tissue paper

   (D) orange juice

4. Through which material does the light travel the fastest? **SC.7.P.10.3**

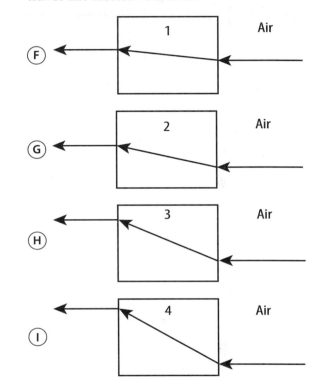

Copyright © Glencoe/McGraw-Hill, a division of The McGraw-Hill Companies, Inc.

Multiple Choice *Bubble the correct answer.*

1. If Rebecca stands exactly one focal length away from a concave mirror, what will her image look like? **SC.7.P.10.2**

   (A) Her image will be right-side up.

   (B) Her image will be reversed.

   (C) Her image will be upside down.

   (D) Her image will not appear in the mirror.

2. The fur of a black cat **SC.7.P.10.2**

   (F) absorbs red and blue light only.

   (G) reflects red and blue light only.

   (H) absorbs all wavelengths of visible light.

   (I) reflects all wavelengths of visible light.

Use the image below to answer questions 3 and 4.

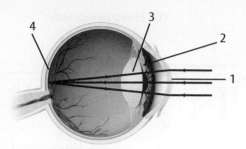

3. Which part of the eye can change shape and help form an image? **SC.7.P.10.1**

   (A) 1

   (B) 2

   (C) 3

   (D) 4

4. Which part of the eye bends light rays as the rays enter the eye? **SC.7.P.10.1**

   (F) 1

   (G) 2

   (H) 3

   (I) 4

Copyright © Glencoe/McGraw-Hill, a division of The McGraw-Hill Companies, Inc.

Notes

Unit 3

Heredity & Reproduction

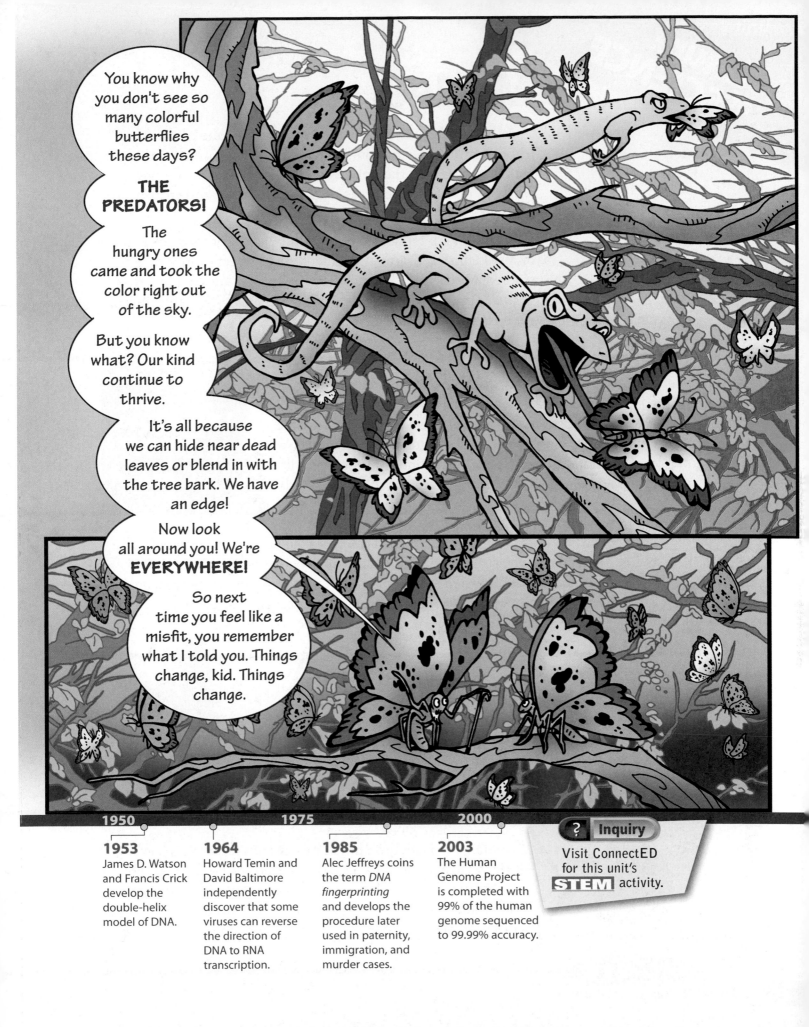

Visit ConnectED for this unit's STEM activity.

Nature of SCIENCE

Models

What would you do without your heart—one of the most important muscles in your body? Worldwide, people on donor lists wait for heart transplants because their hearts are not working properly. Today, doctors can diagnose and treat heart problems with the help of models.

A **model** is a representation of an object, a process, an event, or a system that is similar to the physical object or idea being studied. Models are used to study things that are too big or too small, happen too quickly or too slowly, or are too dangerous or expensive to study directly.

Active Reading
1. Determine
Underline the term *model* and its definition.

A magnetic resonance image (MRI) is a type of model created by using a strong magnetic field and radio waves. MRI machines produce high-resolution images from a series of images of layers. An MRI model of the heart helps cardiologists diagnose heart disease or damage. To obtain a clear MRI, the patient must be still. Even the beating of the heart can limit the ability of an MRI to capture clear images.

A computer tomography (CT) scan combines multiple X-ray images into a detailed 3-D visual model of structures in the body. Cardiologists use this model to diagnose a malfunctioning heart or blocked arteries. A limitation of a CT scan is that some coronary artery diseases, especially if they do not involve a buildup of calcium, may not be detected by the scan.

An artificial heart is a physical model of a heart that can pump blood throughout the body. For a patient with heart failure, a doctor might temporarily replace the heart with an artificial model while they wait for a transplant. Because of its size, the artificial heart is suitable for about only 50 percent of the male population. It is stable for about 2 years before it wears out.

A cardiologist might use a physical model of a heart to explain a diagnosis to a patient. The parts of the heart can be touched and manipulated to explain how a heart works and the location of any complications. However, this model does not function like a real heart, and it cannot be used to diagnose disease.

Maps as Models

One way to think of a computer model, such as an MRI or a CT scan, is as a map. A map is a model that shows how locations are arranged in space. A map can be a model of a small area, such as your street. Or, maps can be models of very large areas, such as a state, a country, or the world.

Biologists study maps to understand where different animal species live, how they interact, and how they migrate. Most animals travel in search of food, water, specific weather, or a place to mate. By placing small electronic tracking devices on migrating animals biologists can create maps of their movements, such as the map of elephant movement in **Figure 1.** These maps are models that help determine how animals survive, repeat the patterns of their life cycle, and respond to environmental changes.

Limitations of Models

It is impossible to include all the details about an object or an idea in one model. A map of elephant migration does not tell you whether the elephant is eating, sleeping, or playing with other elephants. Scientists must consider the limitations of the models they use when drawing conclusions about animal behavior.

All models have limitations. When making decisions about a patient's diagnosis and treatment, a cardiologist must be aware of the information each type of model does and does not provide. Scientists and doctors consider the purpose and limitations of the models they use to ensure that they draw the most accurate conclusions possible.

Active Reading 2. **Revisit** Review all the information about models and their uses. (Circle) limitations and restrictions to the use of models in research, diagnosis, and general use.

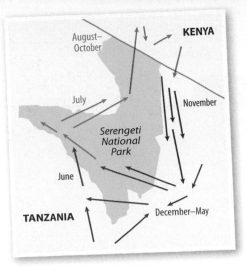

Figure 1 This map is a model of elephants' movements. The colored lines show the paths of three elephants that were equipped with tracking devices for a year.

Inquiry **iLAB STATION** **Try It!** SC.7.N.3.2

MiniLab *How can you model an elephant enclosure?* at connectED.mcgraw-hill.com

Apply It!
After you complete the lab, answer these questions.

1. **Propose** How might you use a computer program to help model the design of a wildlife enclosure?

2. **Recall** List several types of models you have constructed and how you used them.

Notes

Reproduction

Some organisms reproduce by sexual reproduction. Put an X next to each organism you think uses sexual reproduction to produce offspring.

_____ frogs   _____ chickens   _____ humans

_____ fish   _____ trees   _____ snakes

_____ cats   _____ worms   _____ carrots

_____ whales   _____ bean plants   _____ penguins

_____ dogs   _____ horses   _____ weeds

Explain your thinking. Describe your ideas about the types of organisms that reproduce sexually.

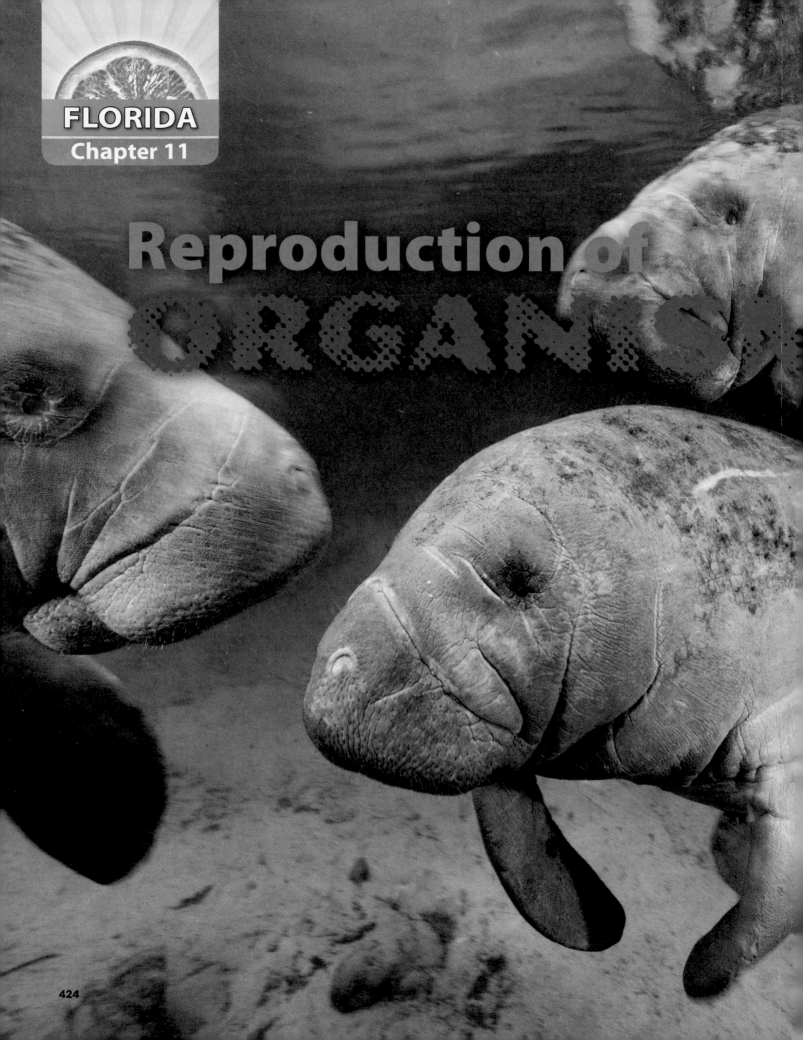

Reproduction of ORGANISMS

The Big Idea

Think About It!

Why do living things reproduce?

Have you ever seen a family of animals, such as the one of manatees shown here? Notice the baby manatee beside its parents. Like all living things, manatees reproduce.

1 Do you think all living things have two parents?

_____
_____
_____

2 What might happen if the manatees did not reproduce?

_____
_____
_____

3 Why do living things reproduce?

_____
_____
_____

Get Ready to Read

What do you think about reproduction?

Before you read, decide if you agree or disagree with each of these statements. As you read this chapter, see if you change your mind about any of the statements.

	AGREE	DISAGREE
1 Humans produce two types of cells: body cells and sex cells.	☐	☐
2 Environmental factors can cause variation among individuals.	☐	☐
3 Two parents always produce the best offspring.	☐	☐
4 Cloning produces identical individuals from one cell.	☐	☐
5 All organisms have two parents.	☐	☐
6 Asexual reproduction occurs only in microorganisms.	☐	☐

Sexual Reproduction and MEIOSIS

 What is sexual reproduction, and why is it beneficial?

 What is the order of the phases of meiosis, and what happens during each phase?

 Why is meiosis important?

Vocabulary

sexual reproduction p. 427

egg p. 427

sperm p. 427

fertilization p. 427

zygote p. 427

diploid p. 428

homologous chromosomes p. 428

haploid p. 429

meiosis p. 429

 Florida NGSSS

LA.7.2.2.3 The student will organize information to show understanding (e.g., representing main ideas within text through charting, mapping, paraphrasing, summarizing, or comparing/contrasting);

MA.6.A.3.6 Construct and analyze tables, graphs, and equations to describe linear functions and other simple relations using both common language and algebraic notation.

SC.7.L.16.3 Compare and contrast the general processes of sexual reproduction requiring meiosis and asexual reproduction requiring mitosis.

SC.7.L.16.4 Recognize and explore the impact of biotechnology (cloning, genetic engineering, artificial selection) on the individual, society and the environment.

SC.7.N.1.1 Define a problem from the seventh grade curriculum, use appropriate reference materials to support scientific understanding, plan and carry out scientific investigation of various types, such as systematic observations or experiments, identify variables, collect and organize data, interpret data in charts, tables, and graphics, analyze information, make predictions, and defend conclusions.

SC.7.N.1.3 Distinguish between an experiment (which must involve the identification and control of variables) and other forms of scientific investigation and explain that not all scientific knowledge is derived from experimentation.

 Launch Lab

SC.7.N.1.3

15 minutes

Why do offspring look different?

Unless you're an identical twin, you probably don't look exactly like any siblings you might have. You might have differences in physical characteristics, such as eye color, hair color, ear shape, or height. Why are there differences in the offspring from the same parents?

Procedure

1. Read and complete a lab safety form.

2. Open the **paper bag** labeled *Male Parent,* and, without looking, remove three **beads.** Record the bead colors and replace the beads.

3. Open the **paper bag** labeled *Female Parent,* and remove three **beads.** Record the bead colors, and replace the beads

4. Repeat steps 2 and 3 for each member of the group.

5. After each member has recorded his or her bead colors, study the results. Each combination of male and female beads represents an offspring.

Think About This

1. Compare your group's offspring to another group's offspring. What similarities or differences do you observe?

2. What caused any differences you observed? Explain.

3. **Key Concept** Why might this type of reproduction be beneficial to an organism?

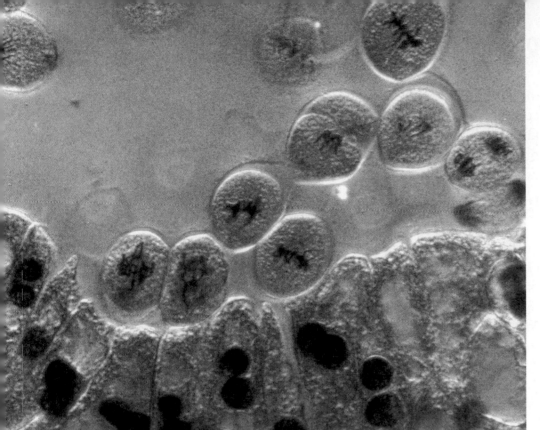

1. This photo looks like a piece of modern art. Look closely at the image. The cells are dividing by a process that occurs during the production of sex cells. From the photo, are you able to tell if these are animal cells or plant cells?

_____

_____

_____

_____

What is sexual reproduction?

Have you ever seen a litter of kittens? One kitten might have orange fur like its mother. A second kitten might have gray fur like its father. Still another kitten might look like a combination of both parents. How is this possible?

The kittens look different because of sexual reproduction. **Sexual reproduction** *is a type of reproduction in which the genetic materials from two different cells combine, producing an offspring.* The cells that combine are called sex cells. Sex cells form in reproductive organs. *The female sex cell, an* **egg,** *forms in an ovary. The male sex cell, a* **sperm,** *forms in a testis. During a process called* **fertilization** (fur tuh luh ZAY shun), *an egg cell and a sperm cell join together.* This produces a new cell. *The new cell that forms from fertilization is called a* **zygote.** As shown in **Figure 1,** the zygote develops into a new organism.

| **Active Reading** | **3. Identify** (Circle) the name of the female sex cell in Figure 1. Place a [Box] around the name of the male sex cell in Figure 1. |

| **Active Reading** | **2. Label** Fill in the blanks in Figure 1 using these terms:
• egg
• sperm
• zygote |

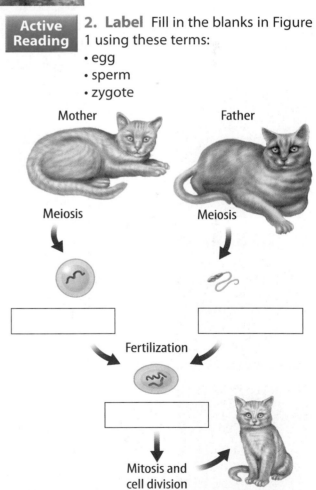

Figure 1 The zygote that forms during fertilization can become a multicellular organism.

Diploid Cells

Following fertilization, a zygote goes through mitosis and cell division. These processes produce nearly all the cells in a multicellular organism. Organisms that reproduce sexually form two kinds of cells—body cells and sex cells. In body cells of most organisms, similar chromosomes occur in pairs. **Diploid** *cells are cells that have pairs of chromosomes.*

Chromosomes

Pairs of chromosomes that have genes for the same traits arranged in the same order are called **homologous** (huh MAH luh gus) **chromosomes.** Because one chromosome is inherited from each parent, the chromosomes are not identical. For example, the kittens mentioned earlier in this lesson inherited a gene for orange fur color from their mother. They also inherited a gene for gray fur color from their father. So, some kittens might be orange, and some might be gray. Both genes for fur color are at the same place on homologous chromosomes, but they code for different colors.

Different organisms have different numbers of chromosomes. Recall that diploid cells have pairs of chromosomes. Notice in **Table 1** that human diploid cells have 23 pairs of chromosomes for a total of 46 chromosomes. A fruit fly diploid cell has 4 pairs of chromosomes, and a rice diploid cell has 12 pairs.

Table 1 An organism's chromosomes can be matched as pairs of chromosomes that have genes for the same traits.

Active Reading **4. Find** Complete the table with the correct information.

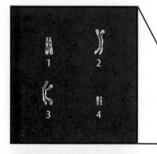
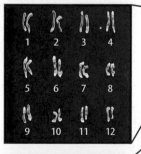

Table 1 Chromosomes of Selected Organisms		
Organism	Number of Chromosomes	Number of Homologous Pairs
Fruit fly	8	
Rice	24	
Yeast	32	16
Cat	38	19
Human		23
Dog	78	39
Fern	1,260	630

Having the correct number of chromosomes is very important. If a zygote has too many or too few chromosomes, it will not develop properly. For example, a genetic condition called Down syndrome occurs when a person has an extra copy of chromosome 21. A person with Down syndrome can have short stature, heart defects, or mental disabilities.

Haploid Cells

Organisms that reproduce sexually also form egg and sperm cells, or sex cells. Sex cells have only one chromosome from each pair of chromosomes. **Haploid** *cells are cells that have only one chromosome from each pair.* Organisms produce sex cells using a special type of cell division called meiosis. *In* **meiosis,** *one diploid cell divides and makes four haploid sex cells.* Meiosis occurs only during the formation of sex cells.

Active Reading

5. **Contrast** How do diploid cells differ from haploid cells?

_____

_____

Active Reading

6. **Detail** Discuss the relationship between diploid cells and homologous chromosomes.

_____

_____

Active Reading

7. **Explain** Define haploid cells, and explain how they are produced.

_____

_____

_____

WORD ORIGIN

haploid
from Greek *haploeides*, means "single"

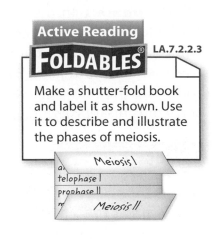

Active Reading

FOLDABLES®   LA.7.2.2.3

Make a shutter-fold book and label it as shown. Use it to describe and illustrate the phases of meiosis.

Meiosis I
telophase I
prophase II
Meiosis II

The Phases of Meiosis

Next, you will read about the phases of meiosis. Many of the phases might seem familiar to you because they also occur during mitosis. Recall that mitosis and cytokinesis involve one division of the nucleus and the cytoplasm. Meiosis involves two divisions of the nucleus and the cytoplasm called meiosis I and meiosis II. They result in four haploid cells with half the number of chromosomes as the original cell. When the number of chromosomes is reduced during cell division, it is called a reduction division.

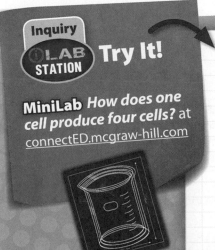

Inquiry

LAB STATION Try It!

MiniLab *How does one cell produce four cells?* at connectED.mcgraw-hill.com

SC.7.N.1.1, SC.7.L.16.3

Apply It! After you complete the lab, answer the question.

1. What is the purpose of meiosis?

Phases of Meiosis I

A reproductive cell goes through interphase before beginning meiosis I, which is shown in **Figure 2**. During interphase, the reproductive cell grows and copies, or duplicates, its chromosomes. Each duplicated chromosome consists of two sister chromatids joined together by a centromere.

1 Prophase I In the first phase of meiosis I, duplicated chromosomes condense and thicken. Homologous chromosomes come together and form pairs. The membrane surrounding the nucleus breaks apart, and the nucleolus disappears.

2 Metaphase I Homologous chromosome pairs line up along the middle of the cell. A spindle fiber attaches to each chromosome.

3 Anaphase I Chromosome pairs separate and are pulled toward the opposite ends of the cell. Notice that the sister chromatids stay together.

4 Telophase I A nuclear membrane forms around each group of duplicated chromosomes. The cytoplasm divides through cytokinesis and two daughter cells form. Sister chromatids remain together.

Meiosis 🔑

8. Apply Complete the graphic of meiosis I by labeling each phase.

Meiosis I

Active Reading

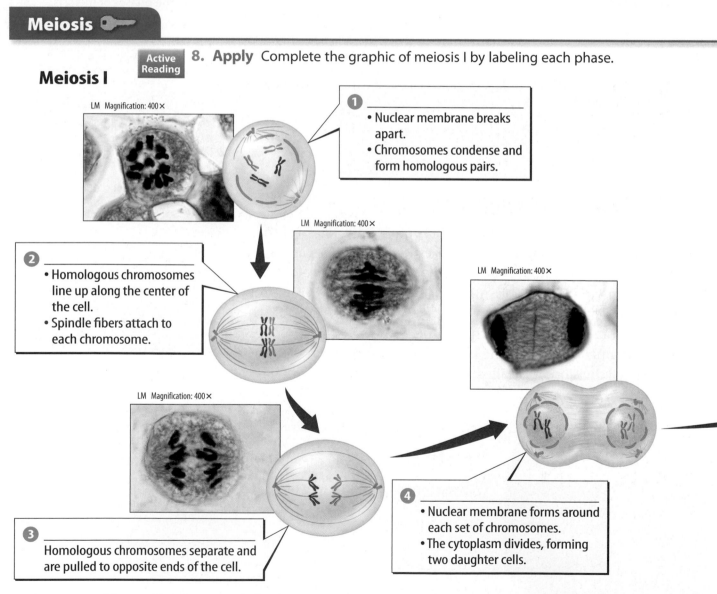

LM Magnification: 400×

1 _____
• Nuclear membrane breaks apart.
• Chromosomes condense and form homologous pairs.

LM Magnification: 400×

2 _____
• Homologous chromosomes line up along the center of the cell.
• Spindle fibers attach to each chromosome.

LM Magnification: 400×

LM Magnification: 400×

3 _____
Homologous chromosomes separate and are pulled to opposite ends of the cell.

4 _____
• Nuclear membrane forms around each set of chromosomes.
• The cytoplasm divides, forming two daughter cells.

Figure 2 Unlike mitosis, meiosis involves two divisions of the nucleus and the cytoplasm.

Phases of Meiosis II

During meiosis II, the two cells formed previously go through a second division of the nucleus and the cytoplasm, as shown in **Figure 2.** This reduction division results in a haploid gamete or spore.

5 Prophase II Chromosomes are not copied again before prophase II. They remain as condensed, thickened sister chromatids. The nuclear membrane breaks apart, and the nucleolus disappears in each cell.

6 Metaphase II The pairs of sister chromatids line up along the middle of the cell in single file.

7 Anaphase II The sister chromatids of each duplicated chromosome are pulled away from each other and move toward opposite ends of the cells.

8 Telophase II During the final phase of meiosis—telophase II—a nuclear membrane forms around each set of chromatids, which are again called chromosomes. The cytoplasm divides through cytokinesis, and four haploid cells form.

Active Reading 9. **Identify** List the phases of meiosis in order.

_____

_____

_____

_____

Active Reading 10. **Apply** Complete the graphic of meiosis II by labeling each phase.

Meiosis II

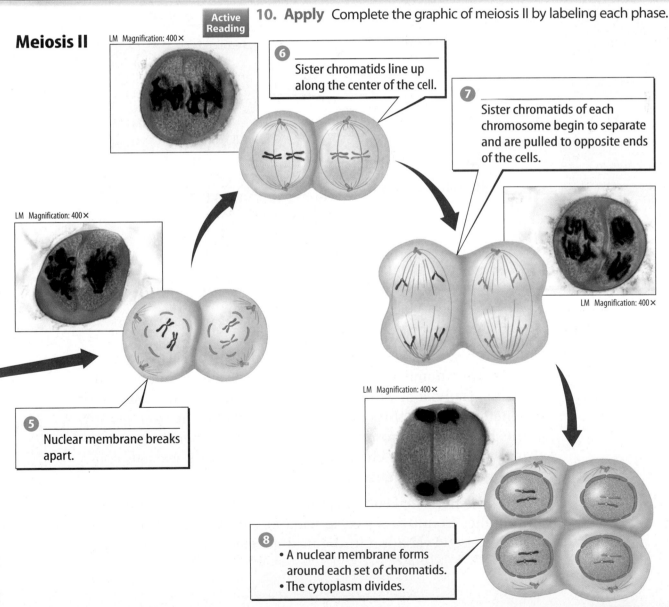

LM Magnification: 400×

6 _____
Sister chromatids line up along the center of the cell.

7 _____
Sister chromatids of each chromosome begin to separate and are pulled to opposite ends of the cells.

LM Magnification: 400×

LM Magnification: 400×

5 _____
Nuclear membrane breaks apart.

LM Magnification: 400×

8 _____
• A nuclear membrane forms around each set of chromatids.
• The cytoplasm divides.

Why is meiosis important?

Meiosis forms sex cells with the correct haploid number of chromosomes. This maintains the correct diploid number of chromosomes in organisms when sex cells join. Meiosis also creates genetic variation by producing haploid cells.

Maintaining Diploid Cells

Recall that diploid cells have pairs of chromosomes. Meiosis helps to maintain diploid cells in offspring by making haploid sex cells. When haploid sex cells join together during fertilization, they make a diploid zygote, or fertilized egg. The zygote then divides by mitosis and cell division and creates a diploid organism. **Figure 3** illustrates how the diploid number is maintained in ducks.

Figure 3 Meiosis ensures that the chromosome number of a species stays the same from generation to generation.

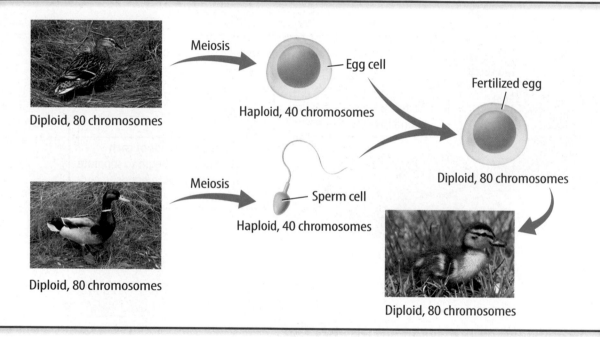

Diploid, 80 chromosomes

Meiosis

Egg cell

Haploid, 40 chromosomes

Fertilized egg

Diploid, 80 chromosomes

Meiosis

Sperm cell

Haploid, 40 chromosomes

Diploid, 80 chromosomes

Diploid, 80 chromosomes

Active Reading

11. Explain Why is meiosis important?

Creating Haploid Cells

The result of meiosis is haploid sex cells. This helps maintain the correct number of chromosomes in each generation of offspring. Haploid cell formation is important because it allows for genetic variation through a process called crossing over. During this process, chromosomal segments are exchanged between a pair of homologous chromosomes. The resulting sex cells contain new combinations of genes.

The genetic makeup of offspring is a combination of chromosomes from two sex cells. Variation in the sex cells results in more genetic variation in the next generation.

How do mitosis and meiosis differ?

Sometimes, it's hard to remember the differences between mitosis and meiosis. Use **Table 2** to review these processes.

During mitosis and cell division, a body cell and its nucleus divide once and produce two identical cells. These processes are important for growth and repair or replacement of damaged tissue. Some organisms reproduce by these processes. The two daughter cells produced by mitosis and cell division have the same genetic information.

During meiosis, a reproductive cell and its nucleus divide twice and produce four cells—two pairs of identical haploid cells. Each cell has half the number of chromosomes as the original cell. Meiosis happens in the reproductive organs of multicellular organisms. Meiosis forms sex cells used for sexual reproduction.

Active Reading 12. **Compare and contrast** Use Table 2 to complete the Venn diagram below.

Mitosis has
_1_ division of nucleus
_____ daughter cells produced

Both have
1 diploid parent cell

Meiosis has
_____ divisions of nucleus
_____ daughter cells produced

Table 2 Comparison of Types of Cell Division		
Characteristic	**Meiosis**	**Mitosis and Cell Division**
Number of chromosomes in parent cell	diploid	diploid
Type of parent cell	reproductive	body
Number of divisions of nucleus	2	1
Number of daughter cells produced	4	2
Chromosome number in daughter cells	haploid	diploid
Function	forms sperm and egg cells	growth, cell repair, some types of reproduction

Active Reading 13. **Compare** How many cells are produced during mitosis? During meiosis?

_____

_____

Math Skills   MA.6.A.3.6

Use Proportions

An equation that shows that two ratios are equivalent is a proportion. The ratios $\frac{1}{2}$ and $\frac{3}{6}$ are equivalent, so they can be written as $\frac{1}{2} = \frac{3}{6}$.

You can use proportions to figure out how many daughter cells will be produced during mitosis. If you know that one cell produces two daughter cells at the end of mitosis, you can use proportions to calculate how many daughter cells will be produced by eight cells undergoing mitosis.

Set up an equation of the two ratios. $\frac{1}{2} = \frac{8}{y}$

Cross-multiply. $1 \times y = 8 \times 2$

$1y = 16$

Divide each side by 1. $y = 16$

Practice

14. You know that one cell produces four daughter cells at the end of meiosis. How many daughter cells would be produced if eight sex cells undergo meiosis?

Advantages of Sexual Reproduction

Did you ever wonder why a brother and a sister might not look alike? The answer is sexual reproduction. The main advantage of sexual reproduction is that offspring inherit half their **DNA** from each parent. Offspring are not likely to inherit the same DNA from the same parents. Different DNA means that each offspring has a different set of traits. This results in genetic variation among the offspring.

Active Reading **15. Identify** Why is sexual reproduction beneficial?

Genetic Variation

As you just read, genetic variation exists among humans. You can look at your friends to see genetic variation. Genetic variation occurs in all organisms that reproduce sexually. Consider the plants shown in **Figure 4.** The plants are members of the same species, but they have different traits, such as the ability to resist disease.

The inheritance of one trait does not influence the inheritance of another trait. This trend is called independent assortment. Independent assortment means genetic variation can vary widely. These differences might be an advantage if the environment changes. This helps some individuals survive unusually harsh conditions, such as drought or severe cold.

REVIEW VOCABULARY

DNA
the genetic information in a cell

16. Visual Check
Describe How does cassava mosaic disease affect cassava leaves?

Disease-resistant cassava leaves

Genetic Variation 🔑

Cassava leaves with cassava mosaic disease

Figure 4 These plants belong to the same species. However, one is more disease-resistant than the other.

Selective Breeding

Did you know that broccoli, kohlrabi, kale, and cabbage all descended from one type of mustard plant? It's true. More than 2,000 years ago farmers noticed that some mustard plants had different traits, such as larger leaves or bigger flower buds. The farmers started to choose which traits they wanted by selecting certain plants to reproduce and grow. For example, some farmers chose only the plants with the biggest flowers and stems and planted their seeds. Over time, the offspring of these plants became what we know today as broccoli, shown in **Figure 5.** This process is called selective breeding or artificial selection. Artificial selection is used to develop many types of plants and animals with desirable traits. It is another example of the benefits of sexual reproduction.

Active Reading **17. Explain** Write why genetic variation and selective breeding are advantages of sexual reproduction.

Advantage	Explanation
Genetic variation	
Selective breeding	

Selective Breeding 🔑

Broccoli

Cabbage

Kale

Wild mustard

Kohlrabi

Figure 5 The wild mustard is the common ancestor to all these plants.

Disadvantages of Sexual Reproduction

Although sexual reproduction produces more genetic variation, it does have some disadvantages. Sexual reproduction takes time and energy. Organisms have to grow and develop until they are mature enough to produce sex cells. Then the organisms have to form sex cells—either eggs or sperm. Before they can reproduce, organisms usually have to find mates. Searching for a mate can take a long time and requires energy. The search for a mate might also expose individuals to predators, diseases, or harsh environmental conditions. In addition, sexual reproduction is limited by certain factors. For example, fertilization cannot take place during pregnancy, which can last as long as two years in some mammals.

Active Reading **18. Locate** Underline two disadvantages of sexual reproduction.

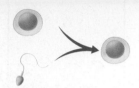

Fertilization occurs when an egg cell and a sperm cell join together.

Organisms produce sex cells through meiosis.

Sexual reproduction results in genetic variation among individuals.

Use Vocabulary

1 **Extend** Use the terms *egg*, *sperm*, and *zygote* in a sentence.

_____

2 **Distinguish** between haploid and diploid. SC.7.L.16.3

_____

_____

Understand Key Concepts 🔑

3 **Define** sexual reproduction. SC.7.L.16.3

_____

_____

4 Homologous chromosomes separate during which phase of meiosis?

(A) anaphase I   (C) metaphase I

(B) anaphase II   (D) metaphase II

Interpret Graphics

5 **Organize** Fill in the graphic organizer below to sequence the phases of meiosis I and meiosis II.

Meiosis I ☐ → ☐ → ☐ → ☐

Meiosis II ☐ → ☐ → ☐ → ☐

Critical Thinking

6 **Analyze** Why is the result of this stage of meiosis an advantage for organisms that reproduce sexually?

_____

_____

Math Skills

7 If 15 cells undergo meiosis, how many daughter cells would be produced?

The Spider
Mating Dance

Meet Norman Platnick, a scientist studying spiders.

Norman Platnick is fascinated by all spider species—from the dwarf tarantula-like spiders of Panama to the blind spiders of New Zealand. These are just two of the over 1,400 species he's discovered worldwide.

How does Platnick identify new species? One way is the pedipalps. Every spider has two pedipalps, but they vary in shape and size among the over 40,000 species. Pedipalps look like legs but function more like antennae and mouthparts. Male spiders use their pedipalps to aid in reproduction.

Getting Ready When a male spider is ready to mate, he places a drop of sperm onto a sheet of silk he constructs. Then he dips his pedipalps into the drop to draw up the sperm.

Finding a Mate The male finds a female of the same species by touch or by sensing certain chemicals she releases.

Courting and Mating Males of some species court a female with a special dance. For other species, a male might present a female with a gift, such as a fly wrapped in silk. During mating, the male uses his pedipalps to transfer sperm to the female.

What happens to the male after mating? That depends on the species. Some are eaten by the female, while others move on to find new mates.

▲ Spiders reproduce sexually, so each offspring has a unique combination of genes from its parents. Over many generations, this genetic variation has led to the incredible diversity of spiders in the world today.

◄ Norman Platnick is an arachnologist (uh rak NAH luh just) at the American Museum of Natural History. Arachnologists are scientists who study spiders.

It's Your Turn

RESEARCH Select a species of spider and research its mating rituals. What does a male do to court a female? What is the role of the female? What happens to the spiderlings after they hatch? Use images to illustrate a report on your research.

Asexual REPRODUCTION

 What is asexual reproduction, and why is it beneficial?

How do the types of asexual reproduction differ?

Vocabulary

asexual reproduction p. 439

fission p. 440

budding p. 441

regeneration p. 442

vegetative reproduction p. 443

cloning p. 444

 Florida NGSSS

LA.7.2.2.3 The student will organize information to show understanding (e.g., representing main ideas within text through charting, mapping, paraphrasing, summarizing, or comparing/contrasting);

SC.7.L.16.1 Understand and explain that every organism requires a set of instructions that specifies its traits, that this hereditary information (DNA) contains genes located in the chromosomes of each cell, and that heredity is the passage of these instructions from one generation to another.

SC.7.L.16.3 Compare and contrast the general processes of sexual reproduction requiring meiosis and asexual reproduction requiring mitosis.

SC.7.L.16.4 Recognize and explore the impact of biotechnology (cloning, genetic engineering, artificial selection) on the individual, society and the environment.

SC.7.N.1.1 Define a problem from the seventh grade curriculum, use appropriate reference materials to support scientific understanding, plan and carry out scientific investigation of various types, such as systematic observations or experiments, identify variables, collect and organize data, interpret data in charts, tables, and graphics, analyze information, make predictions, and defend conclusions.

Inquiry Launch Lab

SC.7.N.1.1

20 minutes

How do yeast reproduce?

Some organisms can produce offspring without meiosis or fertilization. You can observe this process when you add sugar and warm water to dried yeast.

Procedure

1. Read and complete a lab safety form.

2. Pour 125 mL of water into a **beaker.** The water should be at a temperature of 34°C.

3. Add 5 g of **sugar** and 5 g of **yeast** to the water. Stir slightly. Record your observations after 5 minutes.

4. Using a **dropper,** put a drop of the yeast solution on a **microscope slide.** Place a **coverslip** over the drop.

5. View the yeast solution under a **microscope.** Draw what you see.

Data and Observations

Think About This

1. What evidence did you observe that yeast reproduce?

2. **Key Concept** How do you think this process differs from sexual reproduction?

1. Look closely at the edges of the plant's leaves. Tiny plants are growing there. How do you think this type of plant can reproduce without meiosis and fertilization?

_____

_____

_____

Active Reading LA.7.2.2.3, SC.7.L.16.3

FOLDABLES®

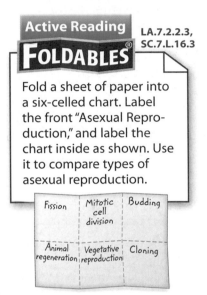

Fold a sheet of paper into a six-celled chart. Label the front "Asexual Reproduction," and label the chart inside as shown. Use it to compare types of asexual reproduction.

Fission	Mitotic cell division	Budding
Animal regeneration	Vegetative reproduction	Cloning

What is asexual reproduction?

Lunch is over and you are in a rush to get to class. You wrap up your half-eaten sandwich and toss it into your locker. A week goes by before you spot the sandwich in the corner of your locker. The surface of the bread is now covered with fuzzy mold—not very appetizing. How did that happen?

The mold on the sandwich is a type of fungus (FUN gus). A fungus releases enzymes that break down organic matter, such as food. It has structures that penetrate and anchor to food, much like roots anchor plants to soil. A fungus can multiply quickly in part because generally a fungus can reproduce either sexually or asexually. Recall that sexual reproduction involves two parent organisms and the processes of meiosis and fertilization. Offspring inherit half their DNA from each parent, resulting in genetic variation among the offspring.

_In_ **asexual reproduction,** _one parent organism produces offspring without meiosis and fertilization._ Because the offspring inherit all their DNA from one parent, they are genetically identical to each other and to their parent.

Active Reading **2. Restate** Describe asexual reproduction in your own words.

_____

_____

_____

_____

_____

_____

Types of Asexual Reproduction

There are many different types of organisms that reproduce by asexual reproduction. In addition to fungi, bacteria, protists, plants, and animals can reproduce asexually. In this lesson, you will learn how organisms reproduce asexually.

Fission

Recall that prokaryotes have a simpler cell structure than eukaryotes. A prokaryote's DNA is not contained in a nucleus. For this reason, mitosis does not occur and cell division in a prokaryote is a simpler process than in a eukaryote. *Cell division in prokaryotes that forms two genetically identical cells is known as* **fission.**

Fission begins when a prokaryote's DNA molecule is copied. Each copy attaches to the cell membrane. Then the cell begins to grow longer, pulling the two copies of DNA apart. At the same time, the cell membrane begins to pinch inward along the middle of the cell. Finally the cell splits and forms two new identical offspring. The original cell no longer exists.

As shown in **Figure 6,** *E. coli,* a common bacterium, divides through fission. Some bacteria can divide every 20 minutes. At that rate, 512 bacteria can be produced from one original bacterium in about three hours.

WORD ORIGIN

fission
from Latin *fissionem,* means "a breaking up, cleaving"

Active Reading **3. Explain** What advantage might asexual reproduction by fission have over sexual reproduction?

Fission 🔑

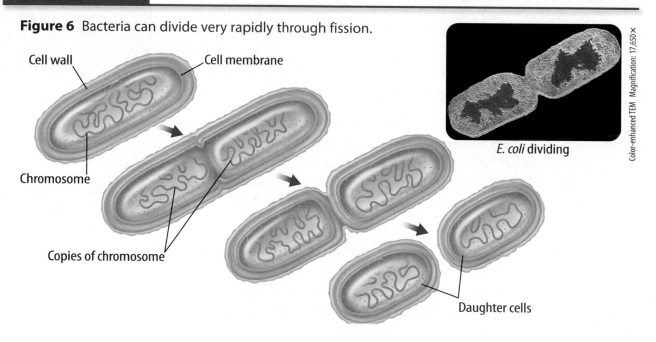

Figure 6 Bacteria can divide very rapidly through fission.

Cell wall

Cell membrane

Chromosome

Copies of chromosome

Daughter cells

E. coli dividing

Color-enhanced TEM Magnification: 17,650×

4. Visual Check **Analyze** What happens to the original cell's chromosones during fission?

Mitotic Cell Division

Many unicellular eukaryotes reproduce by mitotic cell division. In this type of asexual reproduction, an organism forms two offspring through mitosis and cell division. In **Figure 7,** an amoeba's nucleus has divided by mitosis. Next, the cytoplasm and its contents divide through cytokinesis and two new amoebas form.

Budding

In **budding**, *a new organism grows by mitosis and cell division on the body of its parent.* The bud, or offspring, is genetically identical to its parent. When the bud becomes large enough, it can break from the parent and live on its own. In some cases, an offspring remains attached to its parent and starts to form a colony. **Figure 8** shows a hydra in the process of budding. The hydra is an example of a multicellular organism that can reproduce asexually. Unicellular eukaryotes, such as yeast, can also reproduce through budding, as you saw in the Launch Lab.

LM Magnification: 50·

▲ **Figure 7** During mitotic cell division, an amoeba divides its chromosomes and cell contents evenly between the daughter cells.

Active Reading 5. **Identify** What type of reproduction occurs by mitotic cell division?

_____

Budding 🔑

Figure 8 The hydra bud has the same genetic makeup as its parent.

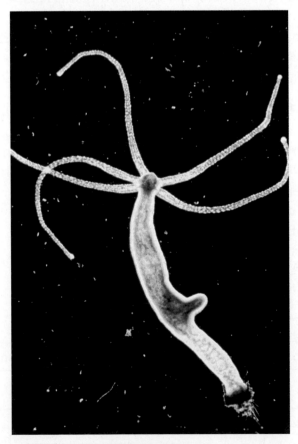

Bud forms.

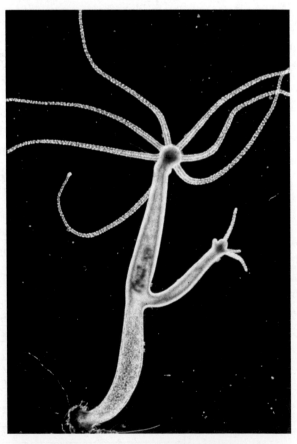

Bud develops a mouth and tentacles.

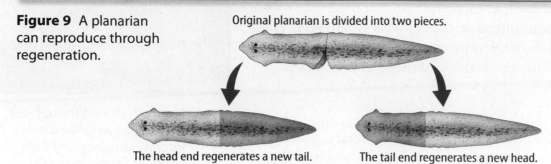

Figure 9 A planarian can reproduce through regeneration.

Original planarian is divided into two pieces.

The head end regenerates a new tail.

The tail end regenerates a new head.

Animal Regeneration

Another type of asexual reproduction, **regeneration,** *occurs when an offspring grows from a piece of its parent.* The ability to regenerate a new organism varies greatly among animals.

Producing New Organisms Some sea stars have five arms. If separated from the parent sea star, each arm has the **potential** to grow into a new organism. To regenerate a new sea star, the arm must contain a part of the central disk of the parent. If conditions are right, one five-armed sea star can produce as many as five new organisms.

Sea urchins, sea cucumbers, sponges, and planarians, such as the one shown in **Figure 9,** can also reproduce through regeneration. Notice that each piece of the original planarian becomes a new organism. As with all types of asexual reproduction, the offspring is genetically identical to the parent.

Producing New Parts When you hear the term *regeneration,* you might think about a salamander regrowing a lost tail or leg. Regeneration of damaged or lost body parts is common in many animals. Newts, tadpoles, crabs, hydra, and zebra fish are all able to regenerate body parts. Even humans are able to regenerate some damaged body parts, such as the skin and the liver. This type of regeneration, however, is not considered asexual reproduction. It does not produce a new organism.

ACADEMIC VOCABULARY
potential
(noun) possibility

Active Reading **6. Compare** What is true of all cases of asexual reproduction?

_____

_____

Active Reading **7. Explain** Complete the graphic below to explain how animal regeneration can produce two results.

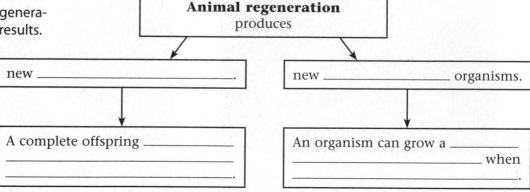

Animal regeneration
produces

new _____ .

A complete offspring _____ _____ _____ .

new _____ organisms.

An organism can grow a _____ _____ when _____ .

Vegetative Reproduction

Plants can also reproduce asexually in a process similar to regeneration. **Vegetative reproduction** *is a form of asexual reproduction in which offspring grow from a part of a parent plant.* For example, the strawberry plants shown in **Figure 10** send out long horizontal stems called stolons. Wherever a stolon touches the ground, it can produce roots. Once the stolons have grown roots, a new plant can grow—even if the stolons have broken off the parent plant. Each new plant grown from a stolon is genetically identical to the parent plant.

Vegetative reproduction usually involves structures such as the roots, the stems, and the leaves of plants. In addition to strawberries, many other plants can reproduce by this method, including raspberries, potatoes, and geraniums.

Active Reading **9. Identify** Write the correct terms to identify the structures of plants usually involved with vegetative reproduction.

_____	_____	_____

Figure 10 The smaller plants were grown from stolons produced by the parent plant.

8. Visual Check Identify Which plants in Figure 10 are the parent plants?

_____

_____

Inquiry SC.7.N.1.1, SC.7.L.16.3

LAB STATION **Try It!**

MiniLab *What parts of plants can grow?* at connectED.mcgraw-hill.com

Apply It! After you complete the lab, answer these questions.

1. Infer How is the process of vegetative reproduction similar to animal regeneration?

Cloning

Fission, budding, and regeneration are all types of asexual reproduction that can produce genetically identical offspring in nature. In the past, the term *cloning* described any process that produced genetically identical offspring. Today, however, the word usually refers to a technique developed by scientists and performed in laboratories. **Cloning** *is a type of asexual reproduction performed in a laboratory that produces identical individuals from a cell or from a cluster of cells taken from a multicellular organism.* Farmers and scientists often use cloning to make copies of organisms or cells that have desirable traits, such as large flowers.

Plant Cloning Some plants can be cloned using a method called tissue **culture,** as shown in **Figure 11.** Tissue culture enables plant growers and scientists to make many copies of a plant with desirable traits, such as sweet fruit. Also, a greater number of plants can be produced more quickly than by vegetative reproduction.

Tissue culture also enables plant growers to reproduce plants that might have become infected with a disease. To clone such a plant, a scientist can use cells from a part of a plant where they are rapidly undergoing mitosis and cell division. This part of a plant is called a meristem. Cells in meristems are disease-free. Therefore, if a plant becomes infected with a disease, it can be cloned using meristem cells.

Active Reading

10. Identify <u>Underline</u> three advantages of using tissue cultures.

SCIENCE USE v. COMMON USE

culture

Science Use the process of growing living tissue in a laboratory

Common Use the social customs of a group of people

Figure 11 New carrot plants can be produced from cells of a carrot root using tissue culture techniques.

Plant Cloning

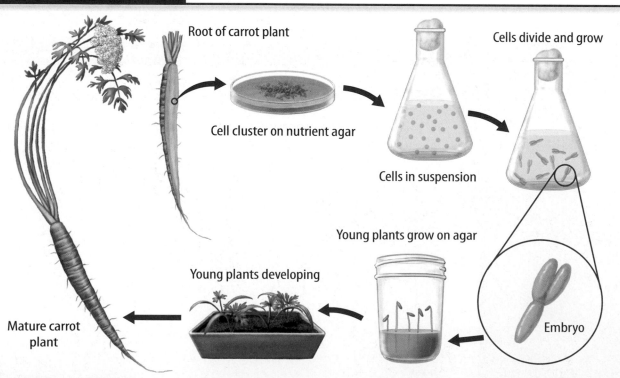

Root of carrot plant

Cell cluster on nutrient agar

Cells in suspension

Cells divide and grow

Embryo

Young plants grow on agar

Young plants developing

Mature carrot plant

Animal Cloning In addition to cloning plants, scientists have been able to clone many animals. Because all of a clone's chromosomes come from one parent (the donor of the nucleus), the clone is a genetic copy of its parent. The first mammal cloned was a sheep named Dolly. **Figure 12** illustrates how this was done.

Scientists are currently working to save some endangered species from extinction by cloning. Although cloning is an exciting advancement in science, some people are concerned about the high cost and the ethics of this technique. Ethical issues include the possibility of human cloning. You might be asked to consider issues like this during your lifetime.

 11. NGSSS Check Compare and contrast Discuss the different types of asexual reproduction. SC.7.L.16.3

_____

_____

Animal Cloning 🔑

Figure 12 Scientists used two different sheep to produce the cloned sheep known as Dolly.

Active Reading 12. Identify (Circle) the two sheep that are genetically identical.

Sheep X

Sheep Z

Remove cell from sheep X.

Remove unfertilized egg cell from sheep Z. Remove DNA from egg cell.

Fuse cells.

New cell contains only DNA from sheep X.

Cell develops into embryo in the laboratory.

Sheep Z

Embryo is implanted in sheep Z.

Dolly

Clone of sheep X

Dolly   Sheep Z

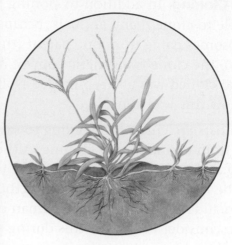

Active Reading 13. **Explain** How can crabgrass spread so quickly?

_____

_____

_____

_____

Active Reading 15. **Define** Write a question about each vocabulary term in this lesson on a separate piece of paper. Exchange questions with another student. Together, discuss the answers to the questions.

Advantages of Asexual Reproduction

What are the advantages to organisms of reproducing asexually? Asexual reproduction enables organisms to reproduce without a mate. Recall that searching for a mate takes time and energy. Asexual reproduction also enables some organisms to rapidly produce a large number of offspring. For example, the crabgrass shown in **Figure 13** reproduces asexually by underground stems called stolons. This enables one plant to spread and colonize an area in a short period of time.

Active Reading 14. **Locate** <u>Underline</u> one way in which asexual reproduction is beneficial.

Disadvantages of Asexual Reproduction

Although asexual reproduction usually enables organisms to reproduce quickly, it does have some disadvantages. Asexual reproduction produces offspring that are genetically identical to their parent. This results in little genetic variation within a population. Why is genetic variation important? Recall from Lesson 1 that genetic variation can give organisms a better chance of surviving if the environment changes. Think of the crabgrass. Imagine that all the crabgrass plants in a lawn are genetically identical to their parent plant. If a certain weed killer can kill the parent plant, then it can kill all the crabgrass plants in the lawn. This might be good for your lawn, but it is a disadvantage for the crabgrass.

Another disadvantage of asexual reproduction involves genetic changes, called mutations, that can occur. If an organism has a harmful mutation in its cells, the mutation will be passed to asexually reproduced offspring. This could affect the offspring's ability to survive.

Visual Summary

In asexual reproduction, offspring are produced without meiosis and fertilization.

Cloning is one type of asexual reproduction.

Asexual reproduction enables organisms to reproduce quickly.

SC.7.N.1.1,
SC.7.L.16.3,
LA.7.2.2.3

Inquiry

LAB STATION **Try It!**

Inquiry *Mitosis and Meiosis* at connectED.mcgraw-hill.com

Use Vocabulary

1 In _____, only one parent organism produces offspring.

2 **Define** *cloning* in your own words. **SC.7.L.16.4**

_____

_____

3 **Use** the term *regeneration* in a sentence.

_____

Understand Key Concepts

4 **State** two reasons why asexual reproduction is beneficial.

_____

_____

5 Which is an example of asexual reproduction by regeneration?

   (A) cloning sheep
   (B) lizard regrowing a tail
   (C) sea star arm producing a new organism
   (D) strawberry plant producing stolons

Interpret Graphics

6 **Organize** Fill in the graphic organizer below to list the different types of asexual reproduction that occur in multicellular organisms. **SC.7.L.16.3**

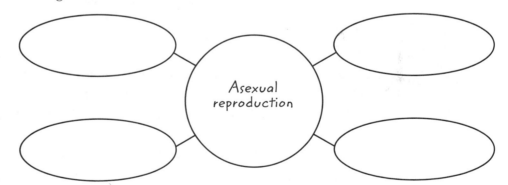

Critical Thinking

7 **Justify** the use of cloning to save endangered animals. **SC.7.L.16.4**

_____

_____

_____

Bunnies

Natalie's white rabbit had a litter of six bunnies. Four bunnies were black and two bunnies were white. The father of the bunnies is black. Natalie wondered why there were more black bunnies than white bunnies in the litter. Circle the response you think is true about the litter of bunnies.

A. All of the black bunnies are male and the white bunnies are female.

B. The fur color has nothing to do with the parents. It depends on the environment.

C. The bunnies got more traits for fur color from their father than from their mother.

D. There are fewer white bunnies because they do not survive as well as black bunnies.

E. There are more black bunnies because father rabbits have stronger traits for fur color than mother rabbits.

F. Each parent contributed the same amount of information about fur color during reproduction.

G. There must have been something wrong because black rabbits and white rabbits should produce grey rabbits.

Explain your thinking. Describe your ideas about how traits such as fur color are determined.

Genetics

FLORIDA BIG IDEAS
1 **The Practice of Science**
2 **The Characteristics of Scientific Knowledge**
3 **The Role of Theories, Laws, Hypotheses, and Models**
16 **Heredity and Reproduction**

Think About It!

How are traits passed from parents to offspring?

The color of this fawn is caused by a genetic trait called albinism. Albinism is the absence of body pigment. Notice that the fawn's mother has brown fur, the normal fur color of an adult whitetail deer.

1 Why do you think the fawn looks so different from its mother?

_____

_____

_____

2 What do you think determines the color of the offspring?

_____

_____

_____

_____

3 How do you think traits are passed from generation to generation?

_____

_____

_____

Get Ready to Read

What do you think about genetics?

Before you read, decide if you agree or disagree with each of these statements. As you read this chapter, see if you change your mind about any of the statements.

		AGREE	DISAGREE
1	Like mixing paints, parents' traits always blend in their offspring.	☐	☐
2	If you look more like your mother than you look like your father, then you received more traits from your mother.	☐	☐
3	All inherited traits follow Mendel's patterns of inheritance.	☐	☐
4	Scientists have tools to predict the form of a trait an offspring might inherit.	☐	☐
5	Any condition present at birth is genetic.	☐	☐
6	A change in the sequence of an organism's DNA always changes the organism's traits.	☐	☐

 **Connect ED**

There's More Online!
Video • Audio • Review • ⓘLab Station • WebQuest • Assessment • Concepts in Motion • Multilingual eGlossary   **459**

Mendel and His PEAS

Vocabulary

heredity p. 461

genetics p. 461

dominant trait p. 467

recessive trait p. 467

Florida NGSSS

LA.7.2.2.3 The student will organize information to show understanding (e.g., representing main ideas within text through charting, mapping, paraphrasing, summarizing, or comparing/contrasting);

MA.6.A.3.6 Construct and analyze tables, graphs, and equations to describe linear functions and other simple relations using both common language and algebraic notation.

SC.7.L.16.1 Understand and explain that every organism requires a set of instructions that specifies its traits, that this hereditary information (DNA) contains genes located in the chromosomes of each cell, and that heredity is the passage of these instructions from one generation to another.

SC.7.N.1.1 Define a problem from the seventh grade curriculum, use appropriate reference materials to support scientific understanding, plan and carry out scientific investigation of various types, such as systematic observations or experiments, identify variables, collect and organize data, interpret data in charts, tables, and graphics, analyze information, make predictions, and defend conclusions.

SC.7.N.2.1 Identify an instance from the history of science in which scientific knowledge has changed when new evidence or new interpretations are encountered.

Inquiry Launch Lab   SC.7.L.16.1

10 minutes

What makes you unique?

Traits such as eye color have many different types, but some traits have only two types. By a show of hands, determine how many students in your class have each type of trait below.

Student Traits		
Trait	Type 1	Type 2
Earlobes	Unattached	Attached
Thumbs	Curved	Straight
Interlacing fingers	Left thumb over right thumb	Right thumb over left thumb

Think About This

1. Why might some students have types of traits that others do not have?

2. If a person has dimples, do you think his or her offspring will have dimples? Explain.

3. **Key Concept** What do you think determines the types of traits you inherit?

Inquiry **Same Species?**

1. Have you ever seen a black ladybug? It is less common than the orange variety you might know, but both are the same species of beetle. So why do you think they look different?

_____

_____

_____

_____

_____

Early Ideas About Heredity

Have you ever mixed two paint colors to make a new color? Long ago, people thought an organism's characteristics, or traits, mixed like colors of paint because offspring resembled both parents. This is known as blending inheritance.

Today, scientists know that **heredity** (huh REH duh tee)—*the passing of traits from parents to offspring*—is more complex. Every new generation of organisms requires a set of instructions. These instructions determine what an organism will look like. More than 150 years ago, Gregor Mendel, an Austrian monk, performed experiments that helped answer these questions and disprove the idea of blending inheritance. Because of his research, Mendel is known as the father of **genetics** (juh NEH tihks)—*the study of how traits are passed from parents to offspring.*

WORD ORIGIN

genetics
from Greek *genesis*, means "origin"

Active Reading **2. Summarize** Describe genetics, and explain why Gregor Mendel is known as the father of genetics.

_____

_____

_____

_____

_____

3. Identify
<u>Underline</u> three characterisics that make pea plants ideal for genetic studies.

Mendel's Experimental Methods

During the 1850s, Mendel studied genetics by doing controlled breeding experiments with pea plants. Pea plants were ideal for genetic studies because

- they reproduce quickly. This enabled Mendel to grow many plants and collect a lot of data.

- they have easily observed traits, such as flower color and pea shape. This enabled Mendel to observe whether a trait was passed from one generation to the next.

- Mendel could control which pairs of plants reproduced. This enabled him to determine which traits came from which plant pairs.

Pollination in Pea Plants

To observe how a trait was inherited, Mendel controlled which plants pollinated other plants. Pollination occurs when pollen lands on the pistil of a flower. **Sperm** cells from the pollen then can fertilize **egg** cells in the pistil. Pollination in pea plants can occur in two ways. Self-pollination occurs when pollen from one plant lands on the pistil of a flower on the same plant, as shown in **Figure 1.** Cross-pollination occurs when pollen from one plant reaches the pistil of a flower on a different plant. Cross-pollination occurs naturally when wind, water, or animals such as bees carry pollen from one flower to another. Mendel allowed one group of flowers to self-pollinate. With another group, he cross-pollinated the plants himself.

REVIEW VOCABULARY

sperm
a haploid sex cell formed in the male reproductive organs

egg
a haploid sex cell formed in the female reproductive organs

Self-Pollination

Figure 1 Self-pollination occurs when pollen from a stamen lands on a pistil of the same flower or on another flower on the same plant.

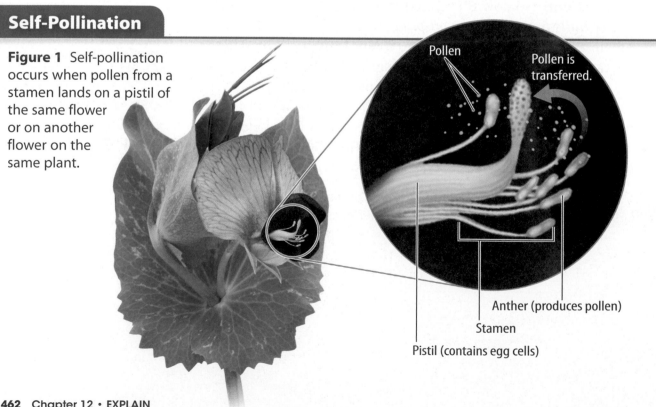

Pollen

Pollen is transferred.

Anther (produces pollen)

Stamen

Pistil (contains egg cells)

True-Breeding Plants

Mendel began his experiments with plants that were true-breeding for the trait he would test. When a true-breeding plant self-pollinates, it always produces offspring with traits that match the parent. For example, when a true-breeding pea plant with wrinkled seeds self-pollinates, it produces only plants with wrinkled seeds. In fact, plants with wrinkled seeds appear generation after generation.

Mendel's Cross-Pollination

By cross-pollinating plants himself, Mendel was able to select which plants pollinated other plants. **Figure 2** shows an example of a manual cross between a plant with white flowers and one with purple flowers.

Figure 2 To control pollination, Mendel removed the stamens of one flower and pollinated that flower with pollen from a flower of a different plant.

Active Reading 4. **Identify** (Circle) the part of **Figure 2** that shows Mendel cross-pollinating two plants.

Cross-Pollination 🔑

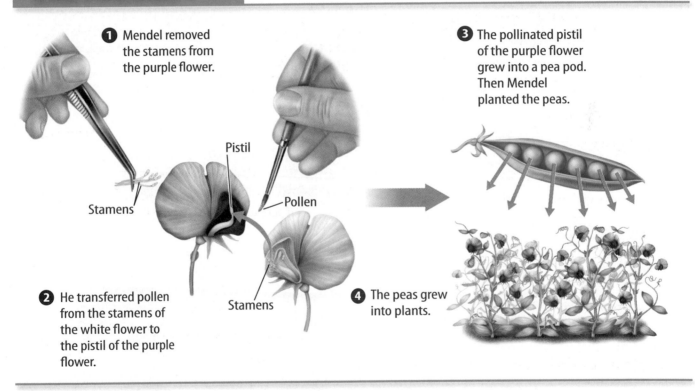

❶ Mendel removed the stamens from the purple flower.

❷ He transferred pollen from the stamens of the white flower to the pistil of the purple flower.

Pistil

Stamens

Pollen

Stamens

❸ The pollinated pistil of the purple flower grew into a pea pod. Then Mendel planted the peas.

❹ The peas grew into plants.

Mendel cross-pollinated hundreds of plants for each set of traits, such as flower color—purple or white; seed color—green or yellow; and seed shape—round or wrinkled. With each cross-pollination, Mendel recorded the traits that appeared in the off-spring. By testing such a large number of plants, Mendel was able to predict which crosses would produce which traits.

 5. **NGSSS Check** **Explain** Why did Mendel perform cross-pollination experiments? SC.7.L.16.1

_____

_____

_____

Mendel's Results

Once Mendel had enough true-breeding plants for a trait that he wanted to test, he cross-pollinated selected plants. His results are shown in **Figure 3.**

First-Generation Crosses

A cross between true-breeding plants with purple flowers produced plants with only purple flowers. A cross between true-breeding plants with white flowers produced plants with only white flowers. But something unexpected happened when Mendel crossed true-breeding plants with purple flowers and true-breeding plants with white flowers— all the offspring had purple flowers.

New Questions Raised

The results of the crosses between true-breeding plants with purple flowers and true-breeding plants with white flowers led to more questions for Mendel. Why did all the offspring always have purple flowers? Why were there no white flowers? Why didn't the cross produce offspring with pink flowers—a combination of the white and purple flower colors? Mendel carried out more experiments with pea plants to answer these questions.

Active Reading **6. Analyze** Use **Figure 3** to predict the offspring of a cross between two true-breeding pea plants with smooth seeds.

_____

_____

First-Generation Crosses

Figure 3 Mendel crossed three combinations of true-breeding plants and recorded the flower colors of the offspring.

Purple × Purple

All purple flowers (true-breeding)

White × White

All white flowers (true-breeding)

Purple (true-breeding) × White (true-breeding)

All purple flowers (hybrids)

✓**Visual Check 7. Apply** Suppose you cross hundreds of true-breeding plants with purple flowers with hundreds of true-breeding plants with white flowers. Based on the results of this cross in the figure above, would any offspring produce white flowers? Explain.

_____

_____

_____

Second-Generation (Hybrid) Crosses

The first-generation purple-flowering plants are called **hybrid** plants. This means they came from true-breeding parent plants with different forms of the same trait. Mendel wondered what would happen if he cross-pollinated two purple-flowering hybrid plants.

As shown in **Figure 4,** some of the offspring had white flowers, even though both parents had purple flowers. The results were similar each time Mendel cross-pollinated two hybrid plants. The trait that had disappeared in the first generation always reappeared in the second generation.

The same result happened when Mendel cross-pollinated pea plants for other traits. For example, he found that cross-pollinating a true-breeding yellow-seeded pea plant with a true-breeding green-seeded pea plant always produced yellow-seeded hybrids. A second-generation cross of two yellow-seeded hybrids always yielded plants with yellow seeds and plants with green seeds.

Active Reading
8. **Explain** What is a hybrid plant?

_____

_____

_____

SCIENCE USE v. COMMON USE

hybrid

Science Use the offspring of two animals or plants with different forms of the same trait

Common Use having two types of components that perform the same function, such as a vehicle powered by both a gas engine and an electric motor

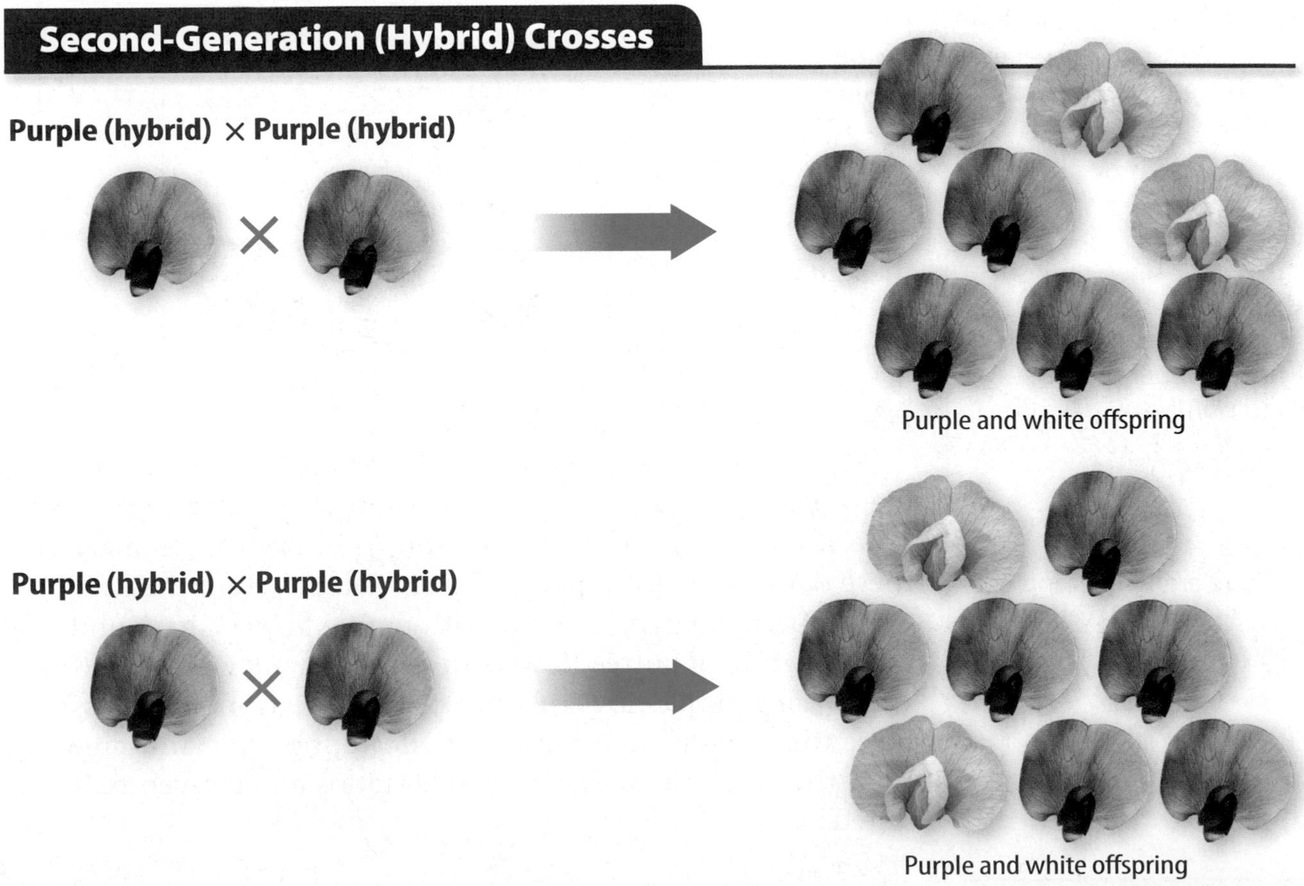

Figure 4 When Mendel cross-pollinated first-generation hybrid offspring, the trait that had disappeared from the first generation reappeared in the second generation.

Second-Generation (Hybrid) Crosses

Purple (hybrid) × Purple (hybrid)

Purple and white offspring

Purple (hybrid) × Purple (hybrid)

Purple and white offspring

Table 1 When Mendel crossed two hybrids for a given trait, the trait that had disappeared then reappeared in a ratio of about 3:1.

Active Reading 9. **Identify** In **Table 1**, (circle) the trait that appeared most often for each characteristic.

Table 1 Results of Hybrid Crosses

Characteristic	Trait and Number of Offspring		Trait and Number of Offspring		Ratio
Flower color	Purple 705		White 224		3.15:1
Flower position	Axial (Side of stem) 651		Terminal (End of stem) 207		3.14:1
Seed color	Yellow 6,022		Green 2,001		3.01:1
Seed shape	Round 5,474		Wrinkled 1,850		2.96:1
Pod shape	Inflated (Smooth) 882		Constricted (Bumpy) 299		2.95:1
Pod color	Green 428		Yellow 152		2.82:1
Stem length	Long 787		Short 277		2.84:1

Math Skills MA.6.A.3.6

Use Ratios

A ratio is a comparison of two numbers or quantities by division. For example, the ratio comparing 6,022 yellow seeds to 2,001 green seeds can be written as follows:

6,022 to 2,001 or

6,022 : 2,001 or

$\frac{6,022}{2,001}$

To simplify the ratio, divide the first number by the second number.

$\frac{6,022}{2,001} = \frac{3}{1}$ or 3:1

Practice

10. There are 14 girls and 7 boys in a science class. Simplify the ratio.

More Hybrid Crosses

Mendel counted and recorded the traits of offspring from many experiments in which he cross-pollinated hybrid plants. Data from these experiments are shown in **Table 1.** He analyzed these data and noticed patterns. For example, from the data of crosses between hybrid plants with purple flowers, he found that the ratio of purple flowers to white flowers was about 3:1. This means purple-flowering pea plants grew from this cross three times more often than white-flowering pea plants grew from the cross. He calculated similar ratios for all seven traits he tested.

Mendel's Conclusions

After analyzing the results of his experiments, Mendel concluded that two genetic factors control each inherited trait. He also proposed that when organisms reproduce, each reproductive cell—sperm or egg—contributes one factor for each trait.

Active Reading **11. Recall** <u>Underline</u> what Mendel concluded about inherited traits.

Dominant and Recessive Traits

Recall that when Mendel cross-pollinated a true-breeding plant with purple flowers and a true-breeding plant with white flowers, the hybrid offspring had only purple flowers. Mendel hypothesized that the hybrid offspring had one genetic factor for purple flowers and one genetic factor for white flowers.

Mendel also hypothesized that the purple factor is the only factor expressed because it blocks the white factor. *A genetic factor that blocks another genetic factor is called a* **dominant** (DAH muh nunt) **trait.** A dominant trait is observed when offspring have either one or two dominant factors. *A genetic factor that is blocked by the presence of a dominant factor is called a* **recessive** (rih SE sihv) **trait.** A recessive trait is observed only when two recessive genetic factors are present in offspring.

From Parents to Second Generation

For the second generation, Mendel cross-pollinated two hybrids with purple flowers. About 75 percent of the second-generation plants had purple flowers. These plants had at least one dominant factor. Twenty-five percent of the second-generation plants had white flowers. These plants had the same two recessive factors.

Active Reading **12. Describe** How do dominant and recessive factors interact?

_____

_____

_____

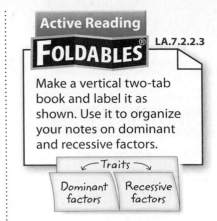

Active Reading

FOLDABLES® LA.7.2.2.3

Make a vertical two-tab book and label it as shown. Use it to organize your notes on dominant and recessive factors.

Traits

Dominant factors | Recessive factors

Inquiry SC.7.N.1.1, SC.7.L.16.1

LAB STATION **Try It!**

MiniLab *Which is the dominant trait?* at connectED.mcgraw-hill.com

Apply It! After you complete the lab, answer this question.

1. How does the data from the first and second generations help you determine that the terminal flower position is a recessive trait?

Genetics is the study of how traits are passed from parents to offspring.

Mendel studied genetics by doing cross-breeding experiments with pea plants.

Purple
705

White
224

Mendel's experiments with pea plants showed that some traits are dominant and others are recessive.

Use Vocabulary

1 Distinguish between heredity and genetics. SC.7.L.16.1

_____

_____

2 Define the terms *dominant* and *recessive*.

_____

_____

Understand Key Concepts

3 A recessive trait is observed when an organism has _____ recessive genetic factor(s).

   (A) 0   (C) 2

   (B) 1   (D) 3

4 **Summarize** Mendel's conclusions about how traits pass from parents to offspring. SC.7.L.16.1

_____

_____

_____

Interpret Graphics

5 **Suppose** the two true-breeding plants shown at right were crossed. What color would the flowers of the offspring be? Explain.

_____

_____

Critical Thinking

6 **Examine** how Mendel's conclusions disprove blending inheritance.

_____

_____

_____

Math Skills MA.6.A.3.6

7 A cross between two pink camellia plants produced the following offspring: 7 plants with red flowers, 7 with white flowers, and 14 with pink flowers. What is the ratio of red to white to pink?

Pioneering
the Science of Genetics

One man's curiosity leads to a branch of science.

Gregor Mendel—monk, scientist, gardener, and beekeeper—was a keen observer of the world around him. Curious about how traits pass from one generation to the next, he grew and tested almost 30,000 pea plants. Today, Mendel is called the father of genetics. After Mendel published his findings, however, his "laws of heredity" were overlooked for several decades.

In 1900, three European scientists, working independently of one another, rediscovered Mendel's work and replicated his results. Then, other biologists quickly began to recognize the importance of Mendel's work.

Gregor Mendel ▶

1902: American physician Walter Sutton demonstrates that Mendel's laws of inheritance can be applied to chromosomes. He concludes that chromosomes contain a cell's hereditary material on genes.

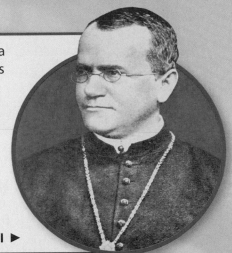

1906: William Bateson, a United Kingdom scientist, coins the term *genetics.* He uses it to describe the study of inheritance and the science of biological inheritance.

1952: American geneticists Martha Chase and Alfred Hershey prove that DNA transmits inherited traits from one generation to the next.

1953: Francis Crick and James Watson determine the structure of the DNA molecule. Their work begins the field of molecular biology and leads to important scientific and medical research in genetics.

2003: The National Human Genome Research Institute (NHGRI) completes mapping and sequencing human DNA. Researchers and scientists are now trying to discover the genetic basis for human health and disease.

It's Your Turn

RESEARCH What are some genetic diseases? Report on how genome-based research might help cure these diseases in the future.

Understanding INHERITANCE

ESSENTIAL QUESTIONS

🔑 What determines the expression of traits?

🔑 How can inheritance be modeled?

🔑 How do some patterns of inheritance differ from Mendel's model?

Vocabulary

gene p. 472

allele p. 472

phenotype p. 472

genotype p. 472

homozygous p. 473

heterozygous p. 473

Punnett square p. 474

incomplete dominance p. 476

codominance p. 476

polygenic inheritance p. 477

 Florida NGSSS

LA.7.2.2.3 The student will organize information to show understanding (e.g., representing main ideas within text through charting, mapping, paraphrasing, summarizing, or comparing/contrasting);

SC.7.L.16.1 Understand and explain that every organism requires a set of instructions that specifies its traits, that this hereditary information (DNA) contains genes located in the chromosomes of each cell, and that heredity is the passage of these instructions from one generation to another.

SC.7.L.16.2 Determine the probabilities for genotype and phenotype combinations using Punnett Squares and pedigrees.

SC.7.N.1.1 Define a problem from the seventh grade curriculum, use appropriate reference materials to support scientific understanding, plan and carry out scientific investigation of various types, such as systematic observations or experiments, identify variables, collect and organize data, interpret data in charts, tables, and graphics, analyze information, make predictions, and defend conclusions.

SC.7.N.2.1 Identify an instance from the history of science in which scientific knowledge has changed when new evidence or new interpretations are encountered.

Inquiry Launch Lab   SC.7.N.1.1
15 minutes

What is the span of your hand?

Mendel discovered some traits have a simple pattern of inheritance—dominant or recessive. However, some traits, such as eye color, have more variation. Is human hand span a Mendelian trait?

Procedure 🔒

1 Read and complete a lab safety form.

2 Use a **metric ruler** to measure the distance (in cm) between the tips of your thumb and little finger with your hand stretched out.

3 As a class, record everyone's name and hand span in a data table.

Data and Observations

Think About This

1. What range of hand span measurements did you observe?

2. 🔑 **Key Concept** Do you think hand span is a simple Mendelian trait like pea plant flower color?

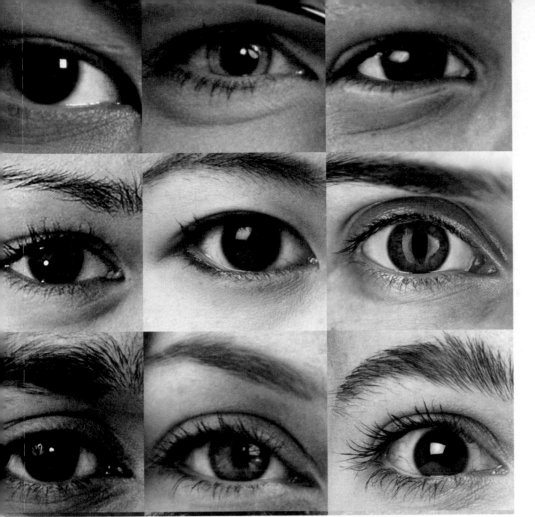

Inquiry Make the Connection

1. Physical traits, such as those shown in these eyes, can vary widely from person to person. Take a closer look at the eyes on this page. What traits can you identify among them? How do they differ?

_____

_____

_____

_____

_____

_____

_____

What controls traits?

Mendel concluded that two factors—one from each parent—control each trait. Mendel hypothesized that one factor came from the egg cell and one factor came from the sperm cell. What are these factors? How are they passed from parents to offspring?

Chromosomes

When other scientists studied the parts of a cell and combined Mendel's work with their work, these factors were more clearly understood. Scientists discovered that inside each cell is a nucleus that contains threadlike structures called chromosomes. Over time, scientists learned that chromosomes contain genetic information that controls traits. We now know that Mendel's "factors" are part of chromosomes and that each cell in offspring contains chromosomes from both parents. As shown in **Figure 5,** these chromosomes exist as pairs—one chromosome from each parent.

Figure 5 Humans have 23 pairs of chromosomes. Each pair has one chromosome from the father and one chromosome from the mother.

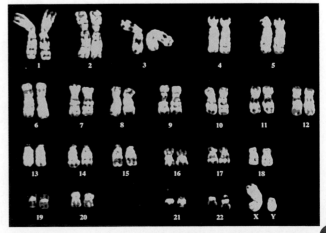

Genes and Alleles

Scientists have discovered that each chromosome can have information about hundreds or even thousands of traits. *A* **gene** (JEEN) *is a section on a chromosome that has genetic information for one trait.* For example, a gene of a pea plant might have information about flower color. Recall that an offspring inherits two genes (factors) for each trait—one from each parent. The genes can be the same or different, such as purple or white for pea flower color. *The different forms of a gene are called* **alleles** (uh LEELs). Pea plants can have two purple alleles, two white alleles, or one of each allele. In **Figure 6,** the chromosome pair has information about three traits—flower position, pod shape, and stem length.

Active Reading 2. **Determine** How many alleles controlled flower color in Mendel's experiments?

Genotype and Phenotype

Look again at the photo at the beginning of this lesson. What human trait can you observe? You might observe that eye color can be shades of blue or brown. *Geneticists call how a trait appears, or is expressed, the trait's* **phenotype** (FEE nuh tipe). What other phenotypes can you observe in the photo?

Mendel concluded that two alleles control the expression or phenotype of each trait. *The two alleles that control the phenotype of a trait are called the trait's* **genotype** (JEE nuh tipe). Although you cannot see an organism's genotype, you can make inferences about a genotype based on its phenotype. For example, you have already learned that a pea plant with white flowers has two recessive alleles for that trait. These two alleles are its genotype. The white flower is its phenotype.

WORD ORIGIN

phenotype
from Greek *phainein*, means "to show"

Figure 6 This chromosome pair has information about flower position, pod shape, and stem length.

3. **Visual Check** **Identify** Which alleles are the same?

Which alleles are different?

Chromosome Pair

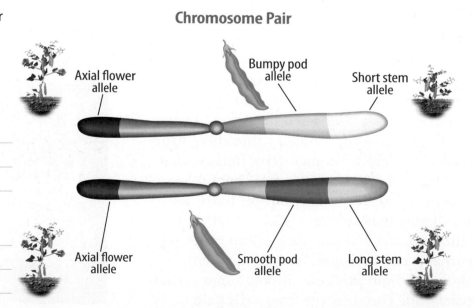

Axial flower allele

Bumpy pod allele

Short stem allele

Axial flower allele

Smooth pod allele

Long stem allele

Symbols for Genotypes Scientists use symbols to represent the alleles in a genotype. In genetics, uppercase letters represent dominant alleles and lowercase letters represent recessive alleles. **Table 2** shows the possible genotypes for both round and wrinkled seed phenotypes. Notice that the dominant allele, if present, is written first.

Table 2 Phenotype and Genotype

Phenotypes (observed traits)	Genotypes (alleles of a gene)
 **Round**	Homozygous dominant (*RR*)
	Heterozygous (*Rr*)
Wrinkled	Homozygous recessive (*rr*)

A round seed can have two genotypes—*RR* and *Rr*. Both genotypes have a round phenotype. Why does *Rr* result in round seeds? This is because the round allele (*R*) is dominant to the wrinkled allele (*r*).

A wrinkled seed has the recessive genotype, *rr*. The wrinkled-seed phenotype is possible only when the same two recessive alleles (*rr*) are present in the genotype.

Homozygous and Heterozygous *When the two alleles of a gene are the same, its genotype is* **homozygous** (hoh muh ZI gus). *Both RR and rr are homozygous genotypes, as shown in* **Table 2.**

If the two alleles of a gene are different, its genotype is **heterozygous** (he tuh roh ZI gus). *Rr is a heterozygous genotype.*

 4. NGSSS Check Explain How do alleles determine the expression of traits? **SC.7.L.16.1**

_____

_____

_____

Inquiry **SC.7.N.1.1, SC.7.L.16.1**

LAB STATION **Try It!**

MiniLab *Can you infer genotype?* at connectED.mcgraw-hill.com

Apply It! After you complete the lab, answer these questions.

1. What are the possible genotypes for a green dragon with four legs and short wings?

2. **Apply** In a pea plant, an allele for a tall stem is dominant to an allele for a short stem. You see a pea plant with a short stem. What can you conclude about the genotype of this plant?

Making Proteins

Recall that proteins are important for every cellular process. The DNA of each cell carries a complete set of genes that provides instructions for making all the proteins a cell requires. Most genes contain instructions for making proteins. Some genes contain instructions for when and how quickly proteins are made.

Junk DNA

As you have learned, all genes are segments of DNA on a chromosome. However, you might be surprised to learn that most of your DNA is not part of any gene. For example, about 97 percent of the DNA on human chromosomes does not form genes. Segments of DNA that are not parts of genes are often called junk DNA. It is not yet known whether junk DNA segments have functions that are important to cells.

The Role of RNA in Making Proteins

How does a cell use the instructions in a gene to make proteins? Proteins are made with the help of ribonucleic acid (**RNA**)— *a type of nucleic acid that carries the code for making proteins from the nucleus to the cytoplasm.* RNA also carries amino acids around inside a cell and forms a part of ribosomes.

RNA, like DNA, is made of nucleotides. However, there are key differences between DNA and RNA. DNA is double-stranded, but RNA is single-stranded. RNA has the nitrogen base uracil (U) instead of thymine (T) and the sugar ribose instead of deoxyribose.

The first step in making a protein is to make mRNA from DNA. *The process of making mRNA from DNA is called* **transcription. Figure 15** shows how mRNA is transcribed from DNA.

Active Reading

7. Explain What is the role of RNA in protein production?

_____

_____

_____

_____

Transcription 🔑

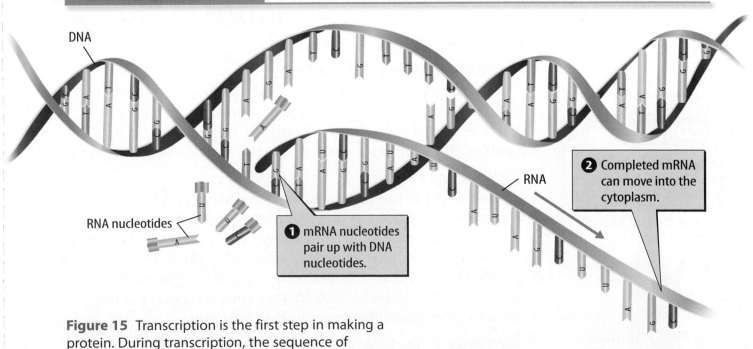

DNA

RNA nucleotides

❶ mRNA nucleotides pair up with DNA nucleotides.

RNA

❷ Completed mRNA can move into the cytoplasm.

Figure 15 Transcription is the first step in making a protein. During transcription, the sequence of nitrogen bases on a gene determines the sequence of bases on mRNA.

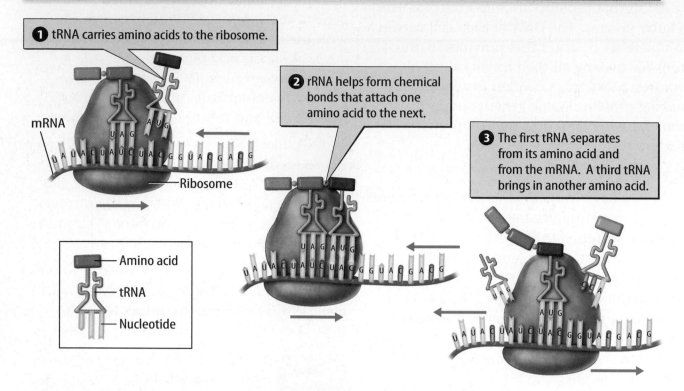

① tRNA carries amino acids to the ribosome.

② rRNA helps form chemical bonds that attach one amino acid to the next.

③ The first tRNA separates from its amino acid and from the mRNA. A third tRNA brings in another amino acid.

mRNA

Ribosome

Amino acid

tRNA

Nucleotide

Figure 16 A protein forms as mRNA moves through a ribosome. Different amino acid sequences make different proteins. A complete protein is a folded chain of amino acids.

Active Reading

FOLDABLES® LA.7.2.2.3

Make a vertical three-tab book and label it as shown. Use your book to record information about the three types of RNA and their functions.

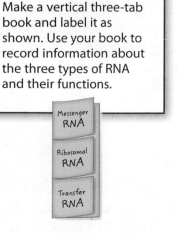

Messenger RNA

Ribosomal RNA

Transfer RNA

Three Types of RNA

On the previous page, you read about messenger RNA (mRNA). There are two other types of RNA, transfer RNA (tRNA) and ribosomal RNA (rRNA). **Figure 16** illustrates how the three work together to make proteins. *The process of making a protein from RNA is called* **translation.** Translation occurs in ribosomes. Recall that ribosomes are cell organelles that are attached to the rough endoplasmic reticulum (rough ER). Ribosomes are also in a cell's cytoplasm.

Translating the RNA Code

Making a protein from mRNA is like using a secret code. Proteins are made of amino acids. The order of the nitrogen bases in mRNA determines the order of the amino acids in a protein. Three nitrogen bases on mRNA form the code for one amino acid.

Each series of three nitrogen bases on mRNA is called a codon. There are 64 codons, but only 20 amino acids. Some of the codons code for the same amino acid. One of the codons codes for an amino acid that is the beginning of a protein. This codon signals that translation should start. Three of the codons do not code for any amino acid. Instead, they code for the end of the protein. They signal that translation should stop.

Active Reading

8. Identify Underline the definition of a codon.

Mutations

You have read that the sequence of nitrogen bases in DNA determines the sequence of nitrogen bases in mRNA, and that the mRNA sequence determines the sequence of amino acids in a protein. You might think these sequences always stay the same, but they can change. *A change in the nucleotide sequence of a gene is called a* **mutation.**

The 46 human chromosomes contain between 20,000 and 25,000 genes that are copied during DNA replication. Sometimes, mistakes can happen during replication. Most mistakes are corrected before replication is completed. A mistake that is not corrected can result in a mutation. Mutations can be triggered by exposure to X-rays, ultraviolet light, radioactive materials, and some kinds of chemicals.

Types of Mutations

There are several types of DNA mutations. Three types are shown in **Figure 17.** In a deletion mutation, one or more nitrogen bases are left out of the DNA sequence. In an insertion mutation, one or more nitrogen bases are added to the DNA. In a substitution mutation, one nitrogen base is replaced by a different nitrogen base.

Each type of mutation changes the sequence of nitrogen base pairs. This can cause a mutated gene to code for a different protein than a normal gene. Some mutated genes do not code for any protein. For example, a cell might lose the ability to make one of the proteins it needs.

9. Identify
Underline the causes of mutations.

WORD ORIGIN

mutation
from Latin *mutare,* means "to change"

Figure 17 Three types of mutations are substitution, insertion, and deletion.

10. Visual Check
Determine Which base pairs were omitted during replication in the deletion mutation?

Mutations 🔑

Original DNA sequence

Substitution
The C-G base pair has been replaced with a T-A pair.

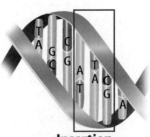

Insertion
Three base pairs have been added.

Deletion
Three base pairs have been removed. Other base pairs will move in to take their place.

Results of a Mutation

The effects of a mutation depend on where in the DNA sequence the mutation happens and the type of mutation. Proteins express traits. Because mutations can change proteins, they can cause traits to change. Some mutations in human DNA cause genetic disorders, such as those described in **Table 4.**

However, not all mutations have negative effects. Some mutations don't cause changes in proteins, so they don't affect traits. Other mutations might cause a trait to change in a way that benefits the organism.

Active Reading

11. Explain How do changes in the sequence of DNA affect traits?

_____

_____

_____

Scientists still have much to learn about genes and how they determine an organism's traits. Scientists are researching and experimenting to identify all genes that cause specific traits. With this knowledge, we might be one step closer to finding cures and treatments for genetic disorders.

Table 4 Genetic Disorders

Defective Gene or Chromosome	Disorder	Description
Chromosome 12, PAH gene	Phenylketonuria (PKU)	People with defective PAH genes cannot break down the amino acid phenylalanine. If phenylalanine builds up in the blood, it poisons nerve cells.
Chromosome 7, CFTR gene	Cystic fibrosis	In people with defective CFTR genes, salt cannot move in and out of cells normally. Mucus builds up outside cells. The mucus can block airways in lungs and affect digestion.
Chromosome 7, elastin gene	Williams syndrome	People with Williams syndrome are missing part of chromosome 7, including the elastin gene. The protein made from the elastin gene makes blood vessels strong and stretchy.
Chromosome 17, BRCA 1; Chromosome 13, BRCA 2	Breast cancer and ovarian cancer	A defect in BRCA1 and/or BRCA2 does not mean the person will have breast cancer or ovarian cancer. People with defective BRCA1 or BRCA2 genes have an increased risk of developing breast cancer and ovarian cancer.

Lesson Review 3

Visual Summary

DNA is a complex molecule that contains the code for an organism's genetic information.

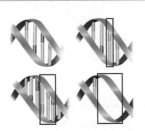

RNA carries the codes for making proteins.

An organism's nucleotide sequence can change through the deletion, insertion, or substitution of nitrogen bases.

Inquiry SC.7.N.1.1, SC.7.L.16.1, SC.7.L.16.2

LAB STATION **Try It!**

Inquiry *Gummy Bear Genetics* at connectED.mcgraw-hill.com

Use Vocabulary

1. **Use the terms** *DNA* and *nucleotide* in a sentence.

_____

_____

2. A change in the sequence of nitrogen bases in a gene is called a(n)

_____.

Understand Key Concepts

3. Where does the process of transcription occur?
   - (A) cytoplasm
   - (C) cell nucleus
   - (B) ribosomes
   - (D) outside the cell

4. **Illustrate** On a separate sheet of paper, make a drawing that illustrates the process of translation.

5. **Distinguish** between the sides of the DNA double helix and the teeth of the DNA double helix.

_____

_____

Interpret Graphics

6. **Sequence** Fill in the graphic organizer below about important steps in making a protein, beginning with DNA and ending with protein.

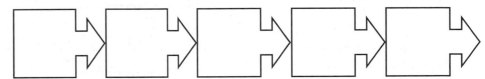

Critical Thinking

7. **Hypothesize** What would happen if a cell were unable to make mRNA?

_____

_____

8. **Assess** What is the importance of DNA replication occurring without any mistakes?

_____

_____

_____

Think About It! Genetic information is passed from generation to generation by DNA; DNA controls the traits of an organism.

🔑 Key Concepts Summary

Vocabulary

LESSON 1 Mendel and His Peas

- Mendel performed cross-pollination experiments to track which traits were produced by specific parental crosses.
- Mendel found that two genetic factors—one from a sperm cell and one from an egg cell—control each trait.
- **Dominant** traits block the expression of **recessive** traits. Recessive traits are expressed only when two recessive factors are present.

heredity p. 461

genetics p. 461

dominant trait p. 467

recessive trait p. 467

LESSON 2 Understanding Inheritance

- **Phenotype** describes how a trait appears.
- **Genotype** describes alleles that control a trait.
- **Punnett squares** and pedigrees are tools to model patterns of inheritance.
- Many patterns of inheritance, such as **codominance** and **polygenic inheritance,** are more complex than Mendel described.

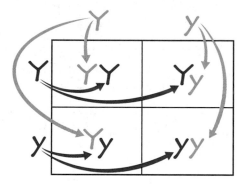

gene p. 472

allele p. 472

phenotype p. 472

genotype p. 472

homozygous p. 473

heterozygous p. 473

Punnett square p. 474

incomplete dominance p. 476

codominance p. 476

polygenic inheritance p. 477

LESSON 3 DNA and Genetics

- **DNA** contains an organism's genetic information.
- **RNA** carries the codes for making proteins from the nucleus to the cytoplasm. RNA also forms part of ribosomes.
- A change in the sequence of DNA, called a **mutation,** can change the traits of an organism.

DNA p. 482

nucleotide p. 483

replication p. 484

RNA p. 485

transcription p. 485

translation p. 486

mutation p. 487

FOLDABLES **Chapter Project**

Assemble your lesson Foldables as shown to make a Chapter Project. Use the project to review what you have learned in this chapter.

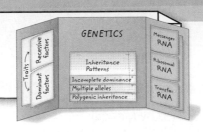

Use Vocabulary

1. The study of how traits are passed from parents to offspring is called _____.

2. The passing of traits from parents to offspring is _____.

3. Human height, weight, and skin color are examples of characteristics determined by _____ _____.

4. A helpful device for predicting the ratios of possible genotypes is a(n) _____.

5. The code for a protein is called a(n) _____.

6. An error made during the copying of DNA is called a(n) _____.

Link Vocabulary and Key Concepts

Use vocabulary terms from the previous page to complete the concept map.

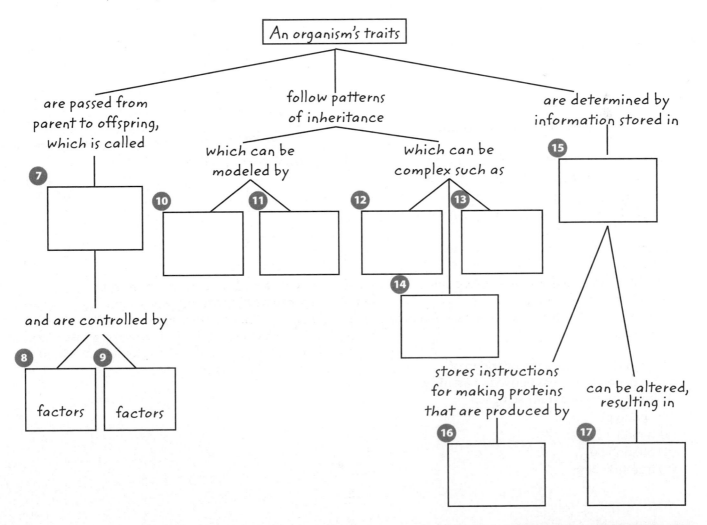

Fill in the correct answer choice.

🔑 Understand Key Concepts

1 The process shown below was used by Mendel during his experiments.

What is the process called? **SC.7.N.2.1**

Ⓐ cross-pollination
Ⓑ segregation
Ⓒ asexual reproduction
Ⓓ blending inheritance

2 Which statement best describes Mendel's experiments? **SC.7.N.2.1**

Ⓐ He began with hybrid plants.
Ⓑ He controlled pollination.
Ⓒ He observed only one generation.
Ⓓ He used plants that reproduce slowly.

3 Before Mendel's discoveries, which statement describes how people believed traits were inherited? **SC.7.N.2.1**

Ⓐ Parental traits blend like colors of paint to produce offspring.
Ⓑ Parental traits have no effect on their offspring.
Ⓒ Traits from only the female parent are inherited by offspring.
Ⓓ Traits from only the male parent are inherited by offspring.

4 Which term describes the offspring of a first-generation cross between parents with different forms of a trait? **SC.7.L.16.1**

Ⓐ genotype
Ⓑ hybrid
Ⓒ phenotype
Ⓓ true-breeding

Critical Thinking

5 **Compare** heterozygous genotype and homozygous genotype. **LA.7.2.2.3**

6 **Distinguish** between multiple alleles and polygenic inheritance. **LA.7.2.2.3**

7 **Give an example** of how the environment can affect an organism's phenotype.

8 **Predict** In pea plants, the allele for smooth pods is dominant to the allele for bumpy pods. Predict the genotype of a plant with bumpy pods. Can you predict the genotype of a plant with smooth pods? Explain. **SC.7.L.16.1**

9 **Compare and contrast** characteristics of replication, transcription, translation, and mutation. Which of these processes takes place only in the nucleus of a cell? Which can take place in both the nucleus and the cytoplasm? How do you know? **LA.7.2.2.3**

10 **Interpret Graphics** In tomato plants, red fruit (R) is dominant to yellow fruit (r). Interpret the Punnett square below, which shows a cross between a heterozygous red plant and a yellow plant. Include the possible genotypes and corresponding phenotypes. SC.7.L.16.2

	R	r
r	Rr	rr
r	Rr	rr

Writing in Science

11 **Write** a paragraph contrasting the blending theory of inheritance with the current theory of inheritance. Include a main idea, supporting details, and a concluding sentence. SC.7.N.2.1

Big Idea Review

12 How are traits passed from generation to generation? Explain how dominant and recessive alleles interact to determine the expression of traits. SC.7.L.16.1

13 The photo at the beginning of the chapter shows an albino offspring from a non-albino mother. If albinism is a recessive trait, what are the possible genotypes of the mother, the father, and the offspring? SC.7.L.16.2

Math Skills   MA.6.A.3.6

Use Ratios

14 A cross between two heterozygous pea plants with yellow seeds produced 1,719 yellow seeds and 573 green seeds. What is the ratio of yellow to green seeds?

15 A cross between two heterozygous pea plants with smooth green pea pods produced 87 bumpy yellow pea pods, 261 smooth yellow pea pods, 261 bumpy green pea pods, and 783 smooth green pea pods. What is the ratio of bumpy yellow to smooth yellow to bumpy green to smooth green pea pods?

16 A jar contains three red, five green, two blue, and six yellow marbles. What is the ratio of red to green to blue to yellow marbles?

Fill in the correct answer choice.

Multiple Choice

Use the diagram below to answer questions 1 and 2.

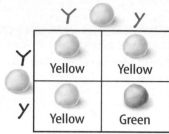

1 Which genotype belongs in the lower right square? SC.7.L.16.2

Ⓐ YY

Ⓑ Yy

Ⓒ yY

Ⓓ yy

2 What percentage of plants from this cross will produce yellow seeds? SC.7.L.16.2

Ⓕ 25 percent

Ⓖ 50 percent

Ⓗ 75 percent

Ⓘ 100 percent

3 What is heredity? SC.7.L.16.1

Ⓐ the study of how traits are passed from parents to offspring

Ⓑ the study of how DNA replicates

Ⓒ the process of chromosomes mutating

Ⓓ the passing of traits from parents to offspring

4 Which can be determined by using a Punnett Square or pedigree? SC.7.L.16.2

Ⓕ phenotypes of polygenic traits

Ⓖ phenotypes of dominant and recessive traits

Ⓗ phenotypes of codominant traits

Ⓘ genotype mutations

Use the chart below to answer questions 5 and 6.

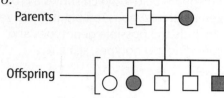

Phenotypes

◯ Female, dominant   ● Female, recessive
□ Male, dominant   ■ Male, recessive

5 Based on the pedigree above, how many off-spring from this cross had the recessive phenotype? SC.7.L.16.2

Ⓐ 1

Ⓑ 2

Ⓒ 3

Ⓓ 5

6 Based on the pedigree above, how many off-spring from this cross are homozygous dominant? SC.7.L.16.2

Ⓕ 0

Ⓖ 1

Ⓗ 3

Ⓘ 5

7 Which is a section on a chromosome that has genetic information for one trait? SC.7.L.16.1

Ⓐ allele

Ⓑ genotype

Ⓒ phenotype

Ⓓ gene

8 Which is a model used to predict possible genotypes and phenotypes of offspring? SC.7.L.16.2

Ⓕ polygenic inheritance

Ⓖ ratio

Ⓗ Punnett Square

Ⓘ incomplete dominance

Use the diagrams below to answer questions 9 and 10.

	R	r
R	RR	Rr
r	Rr	rr

Genotype	Phenotype
RR	Red
Rr	Pink
rr	White

9 According to the information in the diagrams above, what is the ratio of the offspring? SC.7.L.16.2

Ⓐ 0 red: 4 pink: 0 white

Ⓑ 1 red: 2 pink: 1 white

Ⓒ 3 red: 0 pink: 1 white

Ⓓ 4 red: 0 pink: 0 white

10 According to the information in the diagrams above, what is the phenotype of homozygous dominant offspring? SC.7.L.16.2

Ⓕ red

Ⓖ pink

Ⓗ white

Ⓘ yellow

11 Mendel crossed a true-breeding plant with round seeds and a true-breeding plant with wrinkled seeds. Which was true of every offspring of this cross? SC.7.L.16.2

Ⓐ They had the recessive phenotype.

Ⓑ They showed a combination of traits.

Ⓒ They were homozygous.

Ⓓ They were hybrid plants.

Use the diagram below to answer question 12.

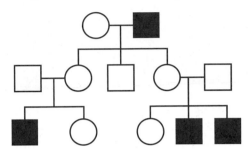

Phenotypes

○ female, dominant   ● female, recessive

□ male, dominant   ■ male, recessive

12 Using the diagram above, what can be determined about the genotypes of the first generation of offspring? SC.7.L.16.2

Ⓕ They are all homozygous recessive.

Ⓖ They are all homozygous dominant.

Ⓗ They are all heterozygous.

Ⓘ Two are heterozygous, and one is homozygous recessive.

13 If a pea plant with a homozygous dominant genotype for a trait is crossed with a pea plant with a heterozygous genotype for the same trait, what is the ratio of offspring? SC.7.L.16.2

Ⓐ 1 homozygous dominant: 2 heterozygous: 1 homozygous recessive

Ⓑ 2 homozygous dominant: 2 heterozygous: 0 homozygous recessive

Ⓒ 0 homozygous dominant: 2 heterozygous: 2 homozygous recessive

Ⓓ 1 homozygous dominant: 0 heterozygous: 3 homozygous recessive

NEED EXTRA HELP?

If You Missed Question...	1	2	3	4	5	6	7	8	9	10	11	12	13
Go to Lesson...	2	2	1	2	2	2	1	2	2	2	1	2	2

Benchmark Mini-Assessment   Chapter 12 • Lesson 1

Multiple Choice *Bubble the correct answer.*

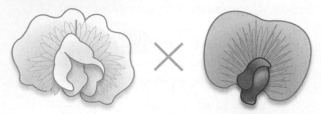

White   Purple

1. Look at the image above. What color flowers did Mendel discover were produced in the first generation? **SC.7.L.16.1**

   (A) The flowers were all blue.

   (B) The flowers were all pink.

   (C) The flowers were all purple.

   (D) The flowers were all white.

2. Which is NOT a reason that Mendel used pea plants for his experiments? **SC.7.L.16.1**

   (F) Pea plants do not self-pollinate.

   (G) Pea plants reproduce quickly.

   (H) Pea plants have many easily observed traits.

   (I) Pea plant reproduction could be controlled by Mendel.

Guinea Pig Fur Color		
Generation	White fur (number of offspring)	Black fur (number of offspring)
First	0	9
Second	6	19

3. Based on the table above, which statement is true? **SC.7.L.16.1**

   (A) Guinea pigs are not true-breeding for fur color.

   (B) The ratio of a hybrid cross in guinea pigs is about 2:1.

   (C) In the first generation, the trait for black fur is masked by the trait for white fur.

   (D) In guinea pigs, the trait for black fur is dominant, and the trait for white fur is recessive.

4. How did Mendel control pollination during his cross-pollination experiments involving the study of flower color in pea plants? **SC.7.L.16.1**

   (F) He allowed pollinators such as bees to pollinate the plant.

   (G) He removed the pistils from the plant being pollinated.

   (H) He removed the stamens from the plant being pollinated.

   (I) He transferred pollen from flower to flower on the same plant.

Copyright © Glencoe/McGraw-Hill, a division of The McGraw-Hill Companies, Inc.

Multiple Choice *Bubble the correct answer.*

1. Which Punnett square shows a cross between two heterozygous parents? **SC.7.L.16.2**

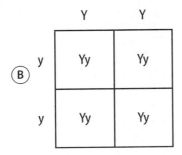

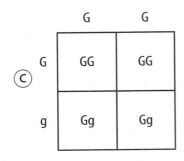

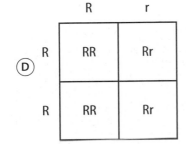

2. A cross between a plant with red flowers and a plant with white flowers produces offspring with pink flowers. What is this an example of? **SC.7.L.16.2**

(F) codominance

(G) dominance

(H) incomplete dominance

(I) polygenic inheritance

3. In one generation, two true-breeding plants produce offspring that look like one of the parents. In the second hybrid generation, the offspring have a ratio of 3:1 for the traits. Which statement is true of the parents of the first generation? **SC.7.L.16.2**

(A) Both were heterozygous for the dominant trait.

(B) Both were homozygous for the recessive trait.

(C) One was homozygous for the dominant trait and the other was homozygous for the recessive trait.

(D) One was homozygous for the dominant trait and the other was heterozygous for the dominant trait.

4. Which of these is a trait in humans that is determined by multiple alleles? **SC.7.L.16.1**

(F) height

(G) weight

(H) blood type

(I) skin color

Copyright © Glencoe/McGraw-Hill, a division of The McGraw-Hill Companies, Inc.

Multiple Choice *Bubble the correct answer.*

1. In the image above, what are *A, C, T,* and *G*? **SC.7.L.16.1**

   Ⓐ codons

   Ⓑ nucleotides

   Ⓒ amino acids

   Ⓓ nitrogen bases

2. How do mutations potentially cause harm? **SC.7.L.16.1**

   Ⓕ Mutations break down chromosomes.

   Ⓖ Mutations damage ribosomes.

   Ⓗ Mutations can cause a gene to code for a different protein than normal.

   Ⓘ Mutations prevent the formation of RNA, so translation does not take place.

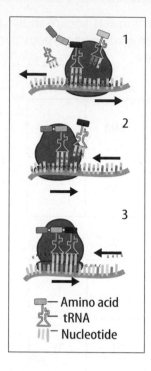

— Amino acid
— tRNA
— Nucleotide

3. In the image above, what is the correct order of the steps shown? **SC.7.L.16.1**

   Ⓐ 1, 2, 3

   Ⓑ 1, 3, 2

   Ⓒ 2, 3, 1

   Ⓓ 3, 2, 1

4. RNA carries the code for making **SC.7.L.16.1**

   Ⓕ acids.

   Ⓖ bases.

   Ⓗ DNA.

   Ⓘ proteins.

Copyright © Glencoe/McGraw-Hill, a division of The McGraw-Hill Companies, Inc.

Notes

Unit 4

EVOLUTION AND INTERDEPENDENCE OF ORGANISMS

350–341 B.C.
Greek philosopher Aristotle classifies organisms by grouping 500 species of animals into eight classes.

1735
Carl Linnaeus classifies nature within a hierarchy and divides life into three kingdoms: mineral, vegetable, and animal. He uses five ranks: class, order, genus, species and variety. Linnaeus's classification is the basis of modern biological classification.

1859
Charles Darwin publishes *On the Origin of Species,* in which he explains his theory of natural selection.

1866
German biologist Ernst Haeckel coins the term *ecology.*

1950

2000

1969
American ecologist Robert Whittaker is the first to propose a five-kingdom taxonomic classification of the world's biota. The five kingdoms are Animalia, Plantae, Fungi, Protista and Monera.

1973
Konrad Lorenz, Niko Tinbergen, and Karl von Frisch are jointly awarded the Nobel Prize for their studies in animal behavior.

1990
Carl Woese introduces the three-domain system that groups cellular life-forms into Archaea, Bacteria, and Eukaryote domains.

? Inquiry

Visit ConnectED for this unit's **STEM** activity.

History and Science

Nearly 50,000 years ago, a group of hunter-gatherers might have roamed through the forest searching for food among the lush plants. The plants and animals that lived in that environment provided the nutritional needs of these people. Humans adapted to the nutrients that the wild foods contained.

Many of the foods you eat today are very different from those eaten by hunter-gatherers. **Table 1** shows how some of these changes occurred.

Table 1 How Science Has Changed Foods

	What?	Advantages	Disadvantages
	Gathering Wild Foods— Foods found in nature were the diet of humans until farming began around 12,000 years ago.	Wild foods provided all the nutrients needed by the human body.	Finding wild foods is not reliable. People moved from place to place in search of food. Sometimes they didn't find food, so they went hungry or starved.
	Farming—People grew seeds from the plants they ate. If soil conditions were not ideal, farmers learned to add water or animal manure to improve plant growth.	Farming allowed more food to be grown in less space. Over time, people learned to breed plants for larger size or greater disease resistance.	Farming practices began to affect the nutrient content of foods. Crops became less nutrient-rich. People began to suffer from nutrient deficiencies and became more prone to disease.
	Hybridizing Plants— Gregor Mendel crossed a plant with genetic material from another, producing a hybrid. A hybrid is the offspring of genetically different organisms.	Hybridization produced new plant foods that combined the best qualities of two plants. The variety of plants available for food increased.	Hybrid crops are prone to disease because of their genetic similarity. Seeds from hybrids do not always grow into plants that produce food of the same quality as the hybrid.
	Genetically Modified (GM) Foods—Scientists remove or replace genes in plants. Removing genes that control flowering in spinach results in more leaves.	GM plants can increase crop yields, nutrient content, insect resistance, and shelf-life of foods. The lettuce shown here has been modified to produce insulin.	Inserted genes might spread to other plants, producing "superweeds." Allergies to GM foods might increase. The long-term effect on humans is unknown.

Active Reading

1. **Analyze** Identify two positive outcomes and two negative outcomes of how science has changed foods and food practices.

Positive Outcomes   **Negative Outcomes**

A Matter of Taste

In early history, food was eaten raw, just as it was found in nature. Cooking food probably occurred by accident. Someone might have accidentally dropped a root into a fire. When people ate the burnt root, it might have tasted better or been easier to chew. Over many generations, and with the influence of different cultures and their various ways to prepare food, the taste buds of people changed. People no longer enjoy as many raw foods.

Today, the taste buds of some people tempt them to eat high-calorie, low-nutrition, processed foods, as shown in **Figure 1.** These foods contain large amounts of calories, salt, and fat.

Figure 1 Processing foods increases convenience but removes nutrients and adds calories that could lead to obesity.

Active Reading **2. Assess** <u>Underline</u> the disadvantages of eating high-calorie, low-nutritional, processed foods.

In some parts of the world, people buy and prepare fresh fruits and vegetables every day, as shown in **Figure 2.** In general, these people have lower rates of obesity and fewer diseases that are common in people who eat more processed foods.

One scientist noted that people with a diet very different from their prehistoric ancestors are more susceptible to heart disease, cancer, diabetes, and other "diseases of civilization."

Active Reading **3. Distinguish** (Circle) the foods generally eaten by people with lower rates of obesity and fewer diseases than people who eat more processed foods.

Figure 2 People in China shop in markets where farmers sell fresh produce that comes directly from the farms.

Inquiry **LAB STATION** **Try It!** SC.7.N.2.1

MiniLab *What food would you design?* at connectED.mcgraw-hill.com

Apply It!

After you complete the lab, answer the questions below.

1. **Synthesize** Briefly summarize how the quality of food has changed over time.

2. **Evaluate** Discuss several things you can do to help insure you are eating healthy foods.

Notes

Tree Snails

A population of a tree snail species has different patterns in their shells. These slight differences in appearance among the individual members of the species are called variations. How do you think these variations get passed on from one generation of tree snails to the next? Circle the answer that best matches your thinking.

A. from parents to offspring

B. through the environment

C. from both parents to offspring and through the environment

Explain your thinking. Describe how variations are passed on from one generation to the next.

Megalodon

The Oceans' Largest Predator

Imagine you have an ancient ancestor who was four times your size. Imagine your ancestor also had 276 razor-sharp teeth to devour its food, which included large whales, dolphins, porpoises, and giant sea turtles. Both of those things would be true if you were a great white shark.

This ancient-ancestor shark is called a Megalodon shark. Megalodon sharks lived approximately 25 to 1.5 million years ago during the Cenozoic era. Based on fossils discovered off the coast of Florida and other locations around the world, Megalodon was the oceans' largest predator. It could grow to a length of 18 to 20 meters! At that size, it is the largest shark ever known.

How can scientists determine the size of this ancient species? Fossils of Megalodon teeth and vertebrae are the only remaining evidence of this giant shark.

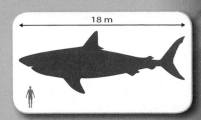

18 m

Scientists use the fossilized teeth to estimate how big Megalodon was, and what it looked like. By comparing the size, shape, and angle of the teeth in the mouth, scientists can determine which current species of sharks are related to Megalodon sharks. Scientists currently hypothesize that the Megalodon shark is a distant, ancient relative of the great white shark.

Scientists can determine the biological evolution of sharks by comparing fossils for similarities and putting them in a sequence to trace how sharks have changed over time. By studying how populations of related organisms have changed over time, scientists can better understand how new species evolve from other species.

This Megalodon shark exhibit, created by the Florida Museum of Natural History in Gainesville, Florida, shows the huge size of Megalodon shark jaws. These jaws are big enough to easily swallow a human.

Growth series of Megalodon 30 to 60 feet long

It's Your Turn

RESEARCH The only remaining fossils of Megalodon sharks are their teeth. Determine which fossilization process created these amazing fossils. Conduct research about the biology of Megalodon sharks and determine why other parts of their bodies never turned into fossils. Summarize your information and present it to your class.

Theory of Evolution by NATURAL SELECTION

ESSENTIAL QUESTIONS

 Who was Charles Darwin?

 How does Darwin's theory of evolution by natural selection explain how species change over time?

 How are adaptations evidence of natural selection?

Vocabulary

naturalist p. 519

variation p. 521

natural selection p. 522

adaptation p. 523

camouflage p. 524

mimicry p. 524

selective breeding p. 525

 Florida NGSSS

LA.7.2.2.3 The student will organize information to show understanding (e.g., representing main ideas within text through charting, mapping, paraphrasing, summarizing, or comparing/contrasting);

SC.7.L.15.2 Explore the scientific theory of evolution by recognizing and explaining ways in which genetic variation and environmental factors contribute to evolution by natural selection and diversity of organisms.

SC.7.N.1.1 Define a problem from the seventh grade curriculum, use appropriate reference materials to support scientific understanding, plan and carry out scientific investigation of various types, such as systematic observations or experiments, identify variables, collect and organize data, interpret data in charts, tables, and graphics, analyze information, make predictions, and defend conclusions.

 inquiry Launch Lab

SC.7.N.1.1, LA.6.2.2.3

20 minutes

Are there variations within your class?

All populations contain variations in some characteristics of their members.

1. Read and complete a lab safety form.

2. Use a **meterstick** to measure the length from your elbow to the tip of your middle finger in centimeters. Record the measurement.

3. Add your measurement to the class list.

4. Organize all of the measurements from shortest to longest.

5. Break the data into regular increments, such as 31–35 cm, 36–40 cm, and 41–45 cm. Count the number of measurements within each increment.

6. Construct a bar graph using the data. Label each axis and give your graph a title.

Think About This

1. What are the shortest and longest measurements?

2. How much do the shortest and longest lengths vary from each other?

3. **Key Concept** Describe how your results provide evidence of variations within your classroom population.

Inquiry Do you think these are exactly the same?

1. Look closely at these zebras. Are they all exactly the same? How are they different? What accounts for these differences? How do the stripes help these organisms survive in their environments?

_____

_____

_____

_____

_____

_____

Active Reading

2. Explain Who was Charles Darwin?

_____

_____

_____

_____

Charles Darwin

How many species of birds can you name? You might think of robins or chickens. Scientists estimate that about 10,000 species of birds live on Earth today. Each bird species has similar characteristics. Each has wings, feathers, and a beak. Scientists hypothesize that all birds evolved from an earlier, or ancestral, population of birdlike organisms. As this population evolved into different species, birds became different sizes and colors. They developed different songs and eating habits, but all retained similar bird characteristics.

How do birds and other species evolve? One scientist who worked to answer this question was Charles Darwin. Darwin was an English naturalist who, in the mid-1800s, developed a theory of how evolution works. _A_ **naturalist** _is a person who studies plants and animals by observing them._ Darwin spent many years observing plants and animals in their natural habitats before developing his theory. Recall that a theory is an explanation of the natural world that is well supported by evidence. Darwin was not the first to develop a theory of evolution, but his theory is the one best supported by evidence today.

Active Reading

LA.7.2.2.3,
SC.7.L.15.2,
SC.7.L.15.3

FOLDABLES

Make a small, four-door shutterfold book. Use it to investigate the who, what, when, and where of Charles Darwin, the Galápagos Islands, and the theory of evolution by natural selection.

Who? What?

When? Where?

 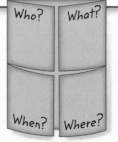

Intermediate Tortoise
- Shell shape is between dome and saddleback
- Can reach low and high vegetation

Saddleback Tortoise
- Large space between shell and neck
- Can reach high vegetation

 3. **Visual Check Infer** Which characteristics does the Domed Tortoise have? Fill in the bullets below.

Domed Tortoise

- _____
- _____

Figure 6 Each island in the Galápagos has a different environment. Tortoises look different depending on which island environment they inhabit.

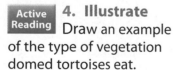 4. **Illustrate** Draw an example of the type of vegetation domed tortoises eat.

[ ]

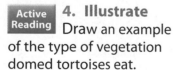 5. **Recall** <u>Underline</u> what made Darwin become curious about the organisms that lived on the Galápagos Islands.

Voyage of the *Beagle*

Darwin served as a naturalist on the HMS *Beagle,* a survey ship of the British navy. During his voyage around the world, Darwin observed and collected many plants and animals.

The Galápagos Islands

Darwin was especially interested in the organisms he saw on the Galápagos (guh LAH puh gus) Islands. These islands are located 1,000 km off the South American coast in the Pacific Ocean. Darwin saw that each island had a slightly different environment. Some were dry. Some were more humid. Others had mixed environments.

Tortoises Giant tortoises lived on many of the islands. When a resident told him that the tortoises on each island looked different, as shown in **Figure 6,** Darwin became curious.

Mockingbirds and Finches Darwin also became curious about the variety of mockingbirds and finches he saw and collected on the islands. Like the tortoises, different types of mockingbirds and finches lived in different island environments. Later, he was surprised to learn that many of these varieties were different enough to be separate species.

Darwin's Theory

Darwin realized there was a relationship between each species and the food sources of the island it lived on. Look again at **Figure 6.** You can see that tortoises with long necks lived on islands that had tall cacti. Their long necks enabled them to reach high to eat the cacti. The tortoises with short necks lived on islands that had plenty of short grass.

Common Ancestors

Darwin became convinced that all the tortoise species were related. He thought they all shared a common ancestor. He suspected that a storm had carried a small ancestral tortoise population to one of the islands from South America millions of years before. Eventually, the tortoises spread to the other islands. Their neck lengths and shell shapes changed to match their islands' food sources. How did this happen?

Variations

Darwin knew that individual members of a species exhibit slight differences, or variations. *A* **variation** *is a slight difference in an inherited trait of individual members of a species.* Even though the snail shells in **Figure 7** are not all exactly the same, they are all from snails of the same species. You can also see variations in the zebras in the photo at the beginning of this lesson. Variations arise naturally in populations. They occur in the offspring as a result of sexual reproduction. You might recall that variations are caused by random mutations, or changes, in genes. Mutations can lead to changes in phenotype. Recall that an organism's phenotype is all of the observable traits and characteristics of the organism. Genetic changes to phenotype can be passed on to future generations.

Figure 7 The variations among the shells of a species of tree snail occur naturally within the population.

Active Reading

6. Describe (Circle) three snail shells in the image above. Describe each variation below.

_____

_____

_____

_____

_____

_____

_____

_____

_____

_____

_____

_____

By comparing the anatomy of organisms and looking for homologous or analogous structures, scientists can determine if organisms had a common ancestor.

Some organisms have vestigial structures, suggesting that they descended from a species that used the structure for a purpose.

Pharyngeal pouches

Human

Scientists use evidence from developmental and molecular biology to help determine if organisms are related.

Inquiry SC.7.N.1.1, SC.N.1.5, SC.7.L.15.2, SC.7.L.15.3

LAB STATION **Try It!**

Inquiry *Model Adaptations in an Organism*

Use Vocabulary

1 Define *embryology* in your own words.

_____

_____

2 Distinguish between a homologous structure and an analogous structure.

_____

_____

3 Use the term *vestigial structure* in a complete sentence.

_____

Understand Key Concepts

4 Scientists use molecular biology to determine how two species are related by comparing the genes in one species to genes

  (A) in extinct species.   (C) in related species.
  (B) in human species.   (D) in related fossils.

5 **Explain** why vestigial structures in whale fossils support the theory of evolution. SC.7.L.15.1

_____

_____

Interpret Graphics

6 **Assess** Fill in the graphic organizer below to identify four areas of study that provide evidence for evolution.

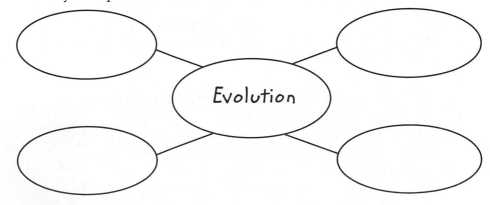

Evolution

Critical Thinking

7 **Predict** what a fossil that illustrates the evolution of a bird from a reptile might look like. SC.7.L.15.1

_____

_____

Think About It! Natural selection is a primary mechanism leading to change over time in organisms. Through natural selection, species adapt to changing environments.

Key Concepts Summary

Vocabulary

LESSON 1 Fossil Evidence of Evolution

- Fossils form in many ways, including mineral replacement, carbonization, and impressions in sediment.

- Scientists can learn the ages of fossils by techniques of relative-age dating and absolute-age dating.

- Though incomplete, the **fossil record** contains patterns suggesting the **biological evolution** of related species.

fossil record p. 509

mold p. 511

cast p. 511

trace fossil p. 511

geologic time scale p. 513

extinction p. 514

biological evolution p. 515

LESSON 2 Theory of Evolution by Natural Selection

- The 19th century **naturalist** Charles Darwin developed a theory of evolution that is still studied today.

- Darwin's theory of evolution by **natural selection** is the process by which populations with **variations** that help them survive in their environments live longer and reproduce more than those without beneficial variations. Over time, beneficial variations spread through populations, and new species that are adapted to their environments evolve.

- **Camouflage, mimicry,** and other **adaptations** are evidence of the close relationships between species and their changing environments.

naturalist p. 519

variation p. 521

natural selection p. 522

adaptation p. 523

camouflage p. 524

mimicry p. 524

selective breeding p. 525

LESSON 3 Biological Evidence of Evolution

- Fossils provide only one source of evidence of evolution. Additional evidence comes from living species, including studies in **comparative anatomy, embryology,** and molecular biology.

- Through evolution by natural selection, all of Earth's organisms are related. The more recently they share a common ancestor, the more closely they are related.

comparative anatomy p. 530

homologous structure p. 530

analogous structure p. 531

vestigial structure p. 531

embryology p. 532

FOLDABLES® Chapter Project

Assemble your lesson Foldables as shown to make a Chapter Project. Use the project to review what you have learned in this chapter.

Use Vocabulary

Distinguish between the following terms.

1 *absolute-age dating* and *relative-age dating*

_____

_____

2 *variations* and *adaptations*

_____

_____

3 *natural selection* and *selective breeding*

_____

_____

4 *homologous structure* and *analogous structure*

_____

_____

5 *vestigial structure* and *homologous structure*

_____

_____

Link Vocabulary and Key Concepts

Use vocabulary terms from the previous page to complete the concept map. **SC.7.L.15.1**

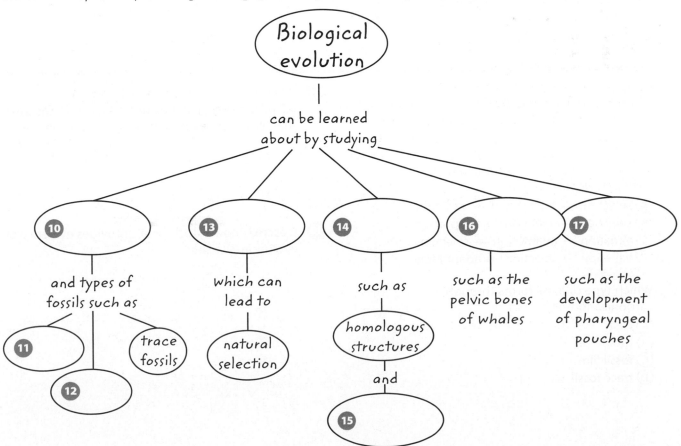

Fill in the correct answer choice.

🔑 Understand Key Concepts

1 Why do scientists think the fossil record is incomplete? SC.7.L.15.1

- (A) Fossils decompose over time.
- (B) The formation of fossils is rare.
- (C) Only organisms with hard parts become fossils.
- (D) There are no fossils before the Phanerozoic eon.

2 What do the arrows on the graph below represent? SC.7.L.15.3

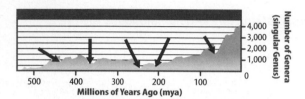

- (A) extinction events
- (B) meteorite impacts
- (C) changes in Earth's temperature
- (D) the evolution of a new species

3 What can scientists learn about fossils using techniques of absolute-age dating? SC.7.L.15.1

- (A) estimated ages of fossils in rock layers
- (B) precise ages of fossils in rock layers
- (C) causes of fossil disappearances in rock layers
- (D) structural similarities to other fossils in rock layers

4 Which is the sequence by which natural selection works? SC.7.L.15.2

- (A) selection → adaptation → variation
- (B) selection → variation → adaptation
- (C) variation → adaptation → selection
- (D) variation → selection → adaptation

5 Which type of fossil forms through carbonization? SC.7.L.15.1

- (A) cast
- (B) mold
- (C) fossil film
- (D) trace fossil

Critical Thinking

6 **Explain** the relationship between fossils and extinction events. SC.7.L.15.1, SC.7.L.15.3

7 **Infer** In 2004, a fossil of an organism that had fins and gills, but also lungs and wrists, was discovered. What might this fossil suggest about evolution? SC.7.L.15.1

8 **Summarize** Darwin's theory of natural selection using the Galápagos tortoises or finches as an example. SC.7.L.15.2

9 **Assess** how the determination that Earth is 4.6 billion years old provided support for the idea that all species evolved from a common ancestor. LA.7.2.2.3

10 **Describe** how cytochrome *c* provides evidence of evolution. SC.7.L.15.2

11 Explain why the discovery of genes was powerful support for Darwin's theory of natural selection. **SC.7.L.15.2**

_____

_____

_____

_____

12 Interpret Graphics The diagram below shows two different methods by which evolution by natural selection might proceed. Discuss how these two methods differ. **SC.7.L.15.2**

_____

_____

_____

_____

_____

_____

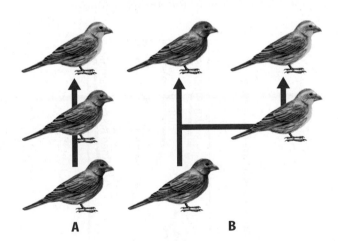

A　　　　　　　　**B**

Big Idea Review

14 How do species adapt to changing environments over time? Explain how evidence from the fossil record and from living species suggests that Earth's species are related. List each type of evidence and provide an example of each. **SC.7.L.15.1**

_____

_____

_____

_____

_____

_____

_____

_____

_____

15 The photo on the chapter opener shows an orchid that looks like a bee. How might this adaptation be evidence of evolution by natural selection? **SC.7.L.15.2**

_____

_____

_____

Math Skills　　MA.6.A.3.6

Use Scientific Notation

16 The earliest fossils appeared about 3,500,000,000 years ago. Express this number in scientific notation.

17 The oldest fossils visible to the unaided eye are about 565,000,000 years old. What is this time in scientific notation?

18 The oldest human fossils are about 1×10^4 years old. Express this as a whole number.

Writing in Science

13 Write a paragraph on a separate sheet of paper explaining how natural selection and selective breeding are related. Include a main idea, supporting details, and a concluding sentence. **SC.7.L.15.2**

Fill in the correct answer choice.

Multiple Choice

1 According to the theory of natural selection, why are some individuals more likely than others to survive and reproduce? SC.7.L.15.2

Ⓐ They do not acquire any adaptations.

Ⓑ They are better adapted to exist in their environment than others.

Ⓒ They acquire only harmful characteristics.

Ⓓ The environment randomly decides which organisms will reproduce.

2 What do homologous structures, vestigial structures, and fossils provide evidence of? SC.7.L.15.1

Ⓕ analogous structures

Ⓖ food choice

Ⓗ populations

Ⓘ evolution

Use the figure below to answer question 3.

Bat wing   Insect wing

3 The analogous structures shown above are not related. However, they both evolved through natural selection. What are they both examples of? SC.7.L.15.2

Ⓐ vestigial organs

Ⓑ homologous structures

Ⓒ adaptations

Ⓓ variations

4 What is an adaptation? SC.7.L.15.2

Ⓕ a body part that has lost its original function through evolution

Ⓖ a characteristic that better equips an organism to survive in its environment

Ⓗ a feature that appears briefly during early development

Ⓘ a slight difference among the individuals in a species

5 What causes variations to arise in a population? SC.7.L.15.2

Ⓐ changes in the environment

Ⓑ competition for limited resources

Ⓒ random mutations in genes

Ⓓ rapid population increases

Use the image below to answer question 6.

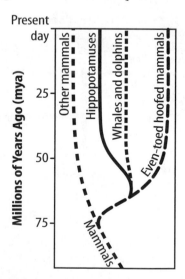

6 The image above shows that even-toed hoofed mammals and other mammals shared a common ancestor. When did this ancestor live? SC.7.L.15.1

Ⓕ 25–35 million years ago

Ⓖ 50–60 million years ago

Ⓗ 60–75 million years ago

Ⓘ 75 million years ago

7 What is an extinct species? SC.7.L.15.3

- Ⓐ a species that blends in easily with its environment
- Ⓑ a species that resembles another species
- Ⓒ a species that was unable to adapt and died out
- Ⓓ a species that is bred for specific characteristics

Use the figure below to answer question 8.

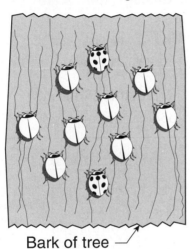

Bark of tree

8 What unique characteristic would the beetles shown above develop through biological adaptation if, over a period of years, the bark on the trees shown became spotted? SC.7.L.15.2

- Ⓕ The beetles would become spotted.
- Ⓖ The beetles would become plain.
- Ⓗ About half of the beetles would become spotted, and half would not.
- Ⓘ There would be no change.

9 When compared with other fossils, what structure indicates the ancestors of whales used to walk on land? SC.7.L.15.1

- Ⓐ pharyngeal pouches
- Ⓑ camouflage
- Ⓒ homologous wings
- Ⓓ vestigial pelvic bones

Use the figure below to answer question 10.

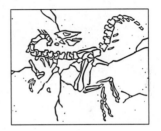

10 What method can scientists use to determine when the fossil above appeared on Earth? SC.7.L.15.1

- Ⓕ mimicry
- Ⓖ selective breeding
- Ⓗ relative-age dating
- Ⓘ biological evolution

11 Which is considered an important factor in natural selection? SC.7.L.15.2

- Ⓐ limited reproduction
- Ⓑ competition for resources
- Ⓒ no variations within a population
- Ⓓ plentiful food and other resources

NEED EXTRA HELP?

If You Missed Question...	1	2	3	4	5	6	7	8	9	10	11
Go to Lesson...	2	1,3	2	2	2	3	2	2	1,3	1	2

Multiple Choice *Bubble the correct answer.*

1. Which is least likely to appear in the fossil record? **SC.7.L.15.1**

   Ⓐ a bird's beak

   Ⓑ a dinosaur's teeth

   Ⓒ a mammal's skull

   Ⓓ a worm's body

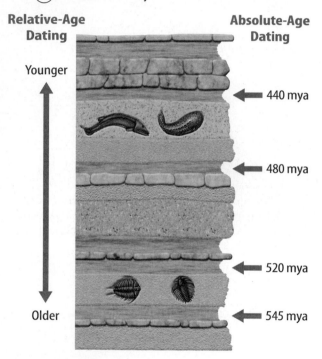

Relative-Age Dating

Younger

Older

Absolute-Age Dating

← 440 mya

← 480 mya

← 520 mya

← 545 mya

2. The dates shown in the diagram above refer to the ages of **SC.7.L.15.1**

   Ⓕ original material.

   Ⓖ trace fossils.

   Ⓗ igneous rock layers.

   Ⓘ sedimentary rock layers.

3. An organism dies, and its body leaves an impression in mud. Over time, the mud hardens into rock, and the impression becomes a fossil. Which kind of fossil was formed? **SC.7.L.15.1**

   Ⓐ cast

   Ⓑ mold

   Ⓒ original material

   Ⓓ trace fossil

The Geologic Time Scale

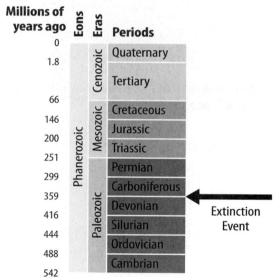

4. The extinction event shown in the chart above happened about how long ago? **SC.7.L.15.3**

   Ⓕ 350 years ago

   Ⓖ 350,000 years ago

   Ⓗ 350 million years ago

   Ⓘ 350 billion years ago

Copyright © Glencoe/McGraw-Hill, a division of The McGraw-Hill Companies, Inc.

Multiple Choice *Bubble the correct answer.*

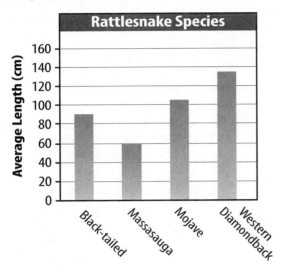

Rattlesnake Species

Average Length (cm)

Black-tailed | Massasauga | Mojave | Western Diamondback

1. Which type of information about rattlesnakes does the bar graph above show? **SC.7.L.15.2**

   (A) adaptation

   (B) evolution

   (C) mimicry

   (D) variation

2. A theory is an explanation that has **SC.7.L.15.2**

   (F) been proven beyond a doubt.

   (G) little evidence to support it.

   (H) much evidence to support it.

   (I) no evidence to support it.

3. The alligator snapping turtle has a worm-like structure on its tongue. When this structure wiggles, fish are attracted to it and swim into the turtle's mouth. This wormlike appendage is an example of **SC.7.L.15.2**

   (A) camouflage.

   (B) mimicry.

   (C) behavioral adaptation.

   (D) functional adaptation.

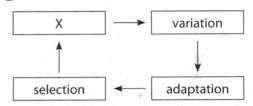

4. The diagram above shows the process of natural selection. What is X? **SC.7.L.15.2**

   (F) competition

   (G) evolution

   (H) mimicry

   (I) reproduction

Copyright © Glencoe/McGraw-Hill, a division of The McGraw-Hill Companies, Inc.

Multiple Choice *Bubble the correct answer.*

1. Frogs, robins, snakes, cows, and humans have tongues. The tongues of these animals are an example of which kind of structure? **SC.7.L.15.1**

   (A) analogous

   (B) homologous

   (C) pharyngeal

   (D) vestigial

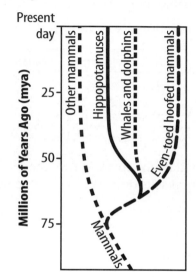

2. The diagram above shows how four species of mammals with a common ancestor diverged from one another through time. Which pair of animals is most closely related? **SC.7.L.15.2**

   (F) hippopotamuses and even-toed hoofed mammals

   (G) whales/dolphins and even-toed hoofed mammals

   (H) whales/dolphins and hippopotamuses

   (I) hippopotamuses and other mammals

3. Gene mutation is the source of **SC.7.L.15.2**

   (A) adaptation.

   (B) evolution.

   (C) natural selection.

   (D) variation.

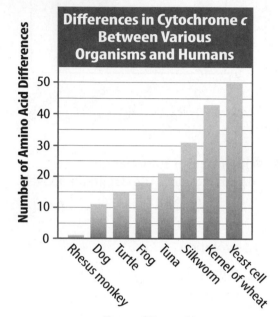

4. According to the information in the graph above, which organism is least like a human? **SC.7.L.15.1**

   (F) dog

   (G) frog

   (H) rhesus monkey

   (I) yeast cell

Copyright © Glencoe/McGraw-Hill, a division of The McGraw-Hill Companies, Inc.

Notes

Notes

What's the relationship?

Some organisms interact with other organisms over time and form relationships. Scientists describe these relationships as symbiotic. What do you think a symbiotic relationship is? Put an X next to all the examples that you think are symbiotic relationships.

A.__ Some microorganisms live inside a termite's digestive system and digest the wood that a termite eats.

B.__ A tick feeds on the blood of a dog.

C.__ A strangler fig vine grows around a tree and eventually kills the tree.

D.__ A cleaner shrimp feeds on the decaying food particles inside a fish's mouth.

E.__ Leafcutter ants provide a fungus with food (the leaves) and then eat the fungus.

F.__ A seed sticks to an animal's fur but eventually falls off the animal.

G.__ A wasp lays its eggs inside a caterpillar. When the eggs hatch, the new wasps feed on the caterpillars.

Explain your thinking. Describe the rule or reasoning you used to decide if a relationship between two organisms is symbiotic.

Interactions of
LIVING
THINGS

FLORIDA BIG IDEAS

1 The Practice of Science

17 Interdependence

How do living things interact with and depend on the other parts of an ecosysytem?

These prairie dogs live together in a network of burrows and tunnels. By cooperating with each other, they can survive more easily than each prairie dog can on its own.

1 What types of resources do you think the prairie dogs need in order to survive?

_____

_____

_____

2 How do you think living together helps them survive?

_____

_____

_____

3 How might these prairie dogs interact with each other and their environment?

_____

_____

_____

What do you think about interactions of living things?

Before you read, decide if you agree or disagree with each of these statements. As you read this chapter, see if you change your mind about any of the statements.

		AGREE	DISAGREE
1	An ecosystem contains both living and nonliving things.	☐	☐
2	All changes in an ecosystem occur over a long period of time.	☐	☐
3	Changes that occur in an ecosystem can cause populations to become larger or smaller.	☐	☐
4	Some organisms have relationships with other types of organisms that help them to survive.	☐	☐
5	Most of the energy used by organisms on Earth comes from the Sun.	☐	☐
6	Both nature and humans affect the environment.	☐	☐

There's More Online!
Video • Audio • Review • ⓘLab Station • WebQuest • Assessment • Concepts in Motion • Multilingual eGlossary

Ecosystems and BIOMES

Vocabulary

ecosystem p. 551

abiotic factor p. 552

biotic factor p. 553

population p. 553

community p. 553

biome p. 554

succession p. 555

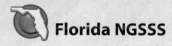 **Florida NGSSS**

LA.7.2.2.3 The student will organize information to show understanding (e.g., representing main ideas within text through charting, mapping, paraphrasing, summarizing, or comparing/contrasting);

SC.7.L.17.3 Describe and investigate various limiting factors in the local ecosystem and their impact on native populations, including food, shelter, water, space, disease, parasitism, predation, and nesting sites.

SC.7.N.1.3 Distinguish between an experiment (which must involve the identification and control of variables) and other forms of scientific investigation and explain that not all scientific knowledge is derived from experimentation.

inquiry Launch Lab SC.7.N.1.1

20 minutes

How do environments differ?

The different environments on Earth can vary greatly. Some organisms are more comfortable in one environment than another.

Procedure

1. Read and complete a lab safety form.

2. Your teacher will divide the class into three groups. Students in one group will wear **heavy coats**. Students in another group will hold a **bag of ice**. Students in the third group will have no change to their environment.

3. After five minutes, determine how comfortable you are in your environment.

Think About This

1. What types of environments were represented by the three groups?

2. **Key Concept** How did your environment change? What did you do to respond?

1. This Arctic tundra is a cold region with little rainfall and a short growing season. What kinds of organisms might live here? What might happen if the climate got warmer?

What are ecosysytems?

How are you similar to a wolf and a pine tree? All three of you are very different living things, also called organisms. All organisms have some characteristics in common. They all use energy and do certain things to survive. Organisms also interact with their environments. Ecology is the study of how organisms interact with each other and with their environments.

Every organism on Earth lives in an **ecosystem**—*the living and nonliving things in one place*. Different organisms depend on different parts of an ecosystem to survive. For example, the deer in **Figure 1** might eat leaves and drink water from a stream. Although leaves are alive and the water is not, the deer needs both to survive. A fish in the stream also needs water to survive, but it interacts differently with water than the deer does.

 2. Define What is an ecosystem?

`Active Reading`

Abiotic Factors

Water is just one example of a part of an ecosystem that was never alive. **Abiotic factors** *are the nonliving parts of an ecosystem*. Some important abiotic factors include water, light, temperature, atmosphere, and soil. The types and amounts of these factors available in an ecosystem help determine which organisms can live there.

Figure 1 This Key Deer lives in the Florida Keys.

3. NGSSS Check
Locate Which abiotic factors can you identify in this photo? SC.7.L.17.3

Active Reading

FOLDABLES® LA.7.2.2.3

Make a two-tab Fold-able as shown. Choose an ecosystem and describe the abiotic and biotic factors that might be in that ecosystem.

Ecosystem: _____

| Abiotic Factors | Biotic Factors |

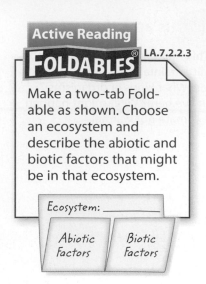

SCIENCE USE v. COMMON USE

atmosphere

Science Use the mix of gases surrounding a planet

Common Use a surrounding influence or feeling

Water

All organisms need water to live, but some need more water than others do. A cactus, like the one shown in **Figure 2**, grows in a desert, where it does not rain often. Ferns and vines live in a rain forest where it is very moist. The type of water in an ecosystem also helps determine which organisms can live there. Some organisms must live in saltwater environments, such as oceans, while others, like humans, must have freshwater to survive.

Light and Temperature

The amount of light available and the temperature of an ecosystem can also determine which organisms can live there. Some organisms, such as plants, require light energy for making food. Temperatures in ecosystems vary, and ecosystems with more sunlight generally have higher temperatures. Ferns thrive in a warm rain forest. A cactus survives in a desert environment that can be very hot during the day and very cold at night.

Atmosphere

Very few living things can survive in an ecosystem without oxygen. Earth's **atmosphere** contains oxygen gas as well as other gases, such as water vapor, carbon dioxide, and nitrogen, that organisms need.

Figure 2 Deserts and rain forests have different amounts of water.

Active Reading   **4. Describe** (Circle) one plant in each photo and describe how they differ from each other.

_____

_____

_____

Soil

Different ecosystems contain different amounts and types of nutrients, minerals, and rocks in the soil. Soil also can have different textures and hold different amounts of moisture. The depth of soil in an ecosystem can differ as well. All of these factors determine which organisms can live in an ecosystem. How do you think the soils in the environments in **Figure 2** differ?

Biotic Factors

You have read about the nonliving, or abiotic, parts of an ecosystem. These parts are important to the survival of living things. **Biotic factors** *are all of the living or once-living things in an ecosystem.* A parrot is one biotic factor in a rain forest, and so is a fallen tree. Biotic factors can be categorized and studied in several ways.

Populations

Think of the last time you saw a flock of birds. The birds you saw were part of a population. *A* **population** *is made up of all the members of one species that live in an area.* For example, all the gray squirrels living in a neighborhood are a population. Organisms in a population interact and compete for food, shelter, and mates.

Communities

Most ecosystems contain many populations, and these populations form a community. *A* **community** *is all the populations living in an ecosystem at the same time.* For example, populations of coral, fish, and sponges are part of a Florida coral reef community. These populations interact with each other in some way. Coral and sponges can become food for fish. They also can provide shelter for other organisms. Some corals even eat fish.

WORD ORIGIN

community
from Latin *communitatem,* means "fellowship"

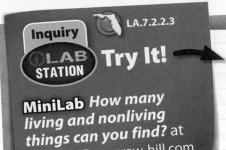

Inquiry **LAB STATION** **Try It!**   LA.7.2.2.3

MiniLab *How many living and nonliving things can you find?* at connectED.mcgraw-hill.com

Apply It!

After you complete the lab, answer the questions below.

1. Choose two types of living things that you found. In what ways might they interact?

2. What other living things also might be part of this community?

Biomes

The populations and communities that interact in a desert are very different from those that interact in an ocean. Deserts and oceans are different biomes. *A* **biome** *is a geographic area on Earth that contains ecosystems with similar biotic and abiotic features.* Biomes contain ecosystems, populations, and communities, as well as specific biotic and abiotic factors. As a result, biomes can be very different from each other. Some examples of Earth's major biomes are shown in **Figure 3.**

Active Reading

5. Define What is a biome?

_____

_____

_____

All biomes are part of the biosphere—the part of Earth that supports life. Earth's biomes can be described as either terrestrial or aquatic. *Terrestrial* means related to land, and *aquatic* means related to water. Terrestrial biomes include forests, deserts, tundra, and grasslands. Aquatic biomes include saltwater areas and freshwater areas. Biomes—like communities—can affect each other. For example, a beach ecosystem is part of both a terrestrial and an aquatic biome. Some organisms from each of these biomes interact in the beach ecosystem.

Figure 3 The Earth's biosphere includes all the different ecosystems.

Active Reading

6. Observe Record one observation about each biome.

Desert biome

Forest biome

Freshwater biome

Grassland biome

Tundra biome

Saltwater biome

What happens when environments change?

The photographs in **Figure 4** are of Mount St. Helens, an active volcano in the state of Washington. The ecosystem changed dramatically in a very short period of time because of the volcanic eruption. Over time, Earth's ecosystems, including Mount St. Helens, have undergone countless changes ranging from tiny to enormous.

Changes in the environment are caused by both natural processes and human actions. Some of these changes can occur rapidly, like the erupting volcano at Mount St. Helens. Other changes, such as the river flow that slowly carved into the land and created the Grand Canyon in Arizona, can take millions of years.

Response to Change

Sometimes changes have positive effects on ecosystems, such as greater rainfall that results in more plants growing. Other changes can have negative effects. A very dry season could cause plants to die, and animals might starve. Usually, a change in an ecosystem results in both positive and negative effects.

Succession

Over long periods of time, communities can change through succession until they are very different. **Succession** *is the gradual change from one community to another community in an area.* The environment around Mount St. Helens changed due to succession.

Active Reading 7. **Identify** Which biotic and abiotic factors changed after the Mount St. Helens eruption?

_____

_____

_____

_____

Figure 4 A volcanic eruption changed Mount St. Helens suddenly. Over many years, succession occured and Mount St. Helens changed again.

May 17, 1980

Active Reading 8. **Analyze** The top photo was taken just before the eruption of Mount St. Helens. When do you think the next two photos were taken? Fill in the blanks with your answers.

Lesson 1 • EXPLAIN 555

Biotic factors are the living parts of an ecosystem.

Earth's biosphere contains many different biomes.

Changes in a community can be very slow or very rapid.

Use Vocabulary

1 **Describe** the parts of an ecosystem.

_____

_____

2 **Define** *succession* using your own words.

_____

_____

3 **Distinguish** between *biome* and *biosphere*.

_____

_____

Understand Key Concepts 🔑

4 **Explain** the difference between biotic and abiotic factors.

_____

_____

5 Choose which describes a biotic factor.

(A) community   (C) temperature
(B) sunlight   (D) water

Interpret Graphics

6 **Compare** two different biomes. Write the name of each biome in the top row. Write the characteristics of each biome in the spaces below its name.

Critical Thinking

7 **Design** your own ecosystem. Tell how the abiotic and biotic factors affect the living things in your ecosystem. Which types of plants and animals live there? How do they interact?

_____

_____

_____

AMERICAN
MUSEUM ŏ
NATURAL
HISTORY

CAREERS
in SCIENCE

All for One,
One for All

▲ Gordon uses a device called a theodolite to measure locations of ant colonies. She tracks changes in a colony's behavior as it ages and grows.

If you've ever watched ants move single file across a sidewalk, you might have wondered how they know which way to go. This is a question that Dr. Deborah M. Gordon, an ecologist at Stanford University, might ask. She studies the behavior of red harvester ants.

Gordon studies the organization of ant colonies. A colony has one or more reproductive queens and many sterile workers living together. At any given time, ants might be working together on a specific task, such as building new tunnels, protecting the colony, or collecting food. However, no one ant in the colony directs the other ants. Gordon investigates how each ant within a colony takes on different tasks and how they work together as a group.

Ants communicate using chemicals. They release chemicals that other ants smell with their antennae. Each colony has its own unique odor, and only ants from that colony can recognize it. In addition, harvester ants have a chemical "vocabulary." They signal a specific task by communicating with a particular chemical.

To study ant communication, Gordon and her colleagues closely observe red harvester ants in their habitats and conduct experiments. Her team might isolate one of the communication chemicals and then place it in different locations or in the same location at different times of day. Gordon has learned that these ants can change tasks when they meet other ants. When one ant's antennae touch another ant, it can tell by the odor what task the other ant is doing. Gordon is studying how ants use encounters to interact with each other and their environments.

▲ The red harvester ant is one of only 50 ant species that have been well studied. Scientists have discovered about 10,000 ant species in all.

It's Your Turn

FIELD JOURNAL Find two animals to observe in their natural habitats. Record your observations in your Science Journal. How did they interact with the environment or with other animals? Share your results with classmates.

Populations and COMMUNITIES

ESSENTIAL QUESTIONS

How do individuals and groups of organisms interact?

How do daily activities impact the environment?

Vocabulary

limiting factor p. 561

biotic potential p. 561

carrying capacity p. 561

habitat p. 562

niche p. 562

symbiosis p. 562

Florida NGSSS

LA.7.2.2.3 The student will organize information to show understanding (e.g., representing main ideas within text through charting, mapping, paraphrasing, summarizing, or comparing/contrasting);

SC.7.L.17.2 Compare and contrast the relationships among organisms such as mutualism, predation, parasitism, competition, and commensalism.

SC.7.L.17.3 Describe and investigate various limiting factors in the local ecosystem and their impact on native populations, including food, shelter, water, space, disease, parasitism, predation, and nesting sites.

SC.7.N.1.1 Define a problem from the seventh grade curriculum, use appropriate reference materials to support scientific understanding, plan and carry out scientific investigation of various types, such as systematic observations or experiments, identify variables, collect and organize data, interpret data in charts, tables, and graphics, analyze information, make predictions, and defend conclusions.

SC.7.N.1.6 Explain that empirical evidence is the cumulative body of observations of a natural phenomenon on which scientific explanations are based.

Inquiry Launch Lab

SC.7.N.1.1

15 minutes

What is the density of your environment?

The different environments on Earth can vary greatly. Some organisms are more comfortable in one environment than another.

Procedure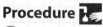

1. Use a **meterstick** to measure the length and width of the classroom.

2. Multiply the length by the width to find the area of the room in square meters.

3. Draw a map of your classroom, including the number and position of desks. Count the number of individuals in your class. Divide the area of the classroom by the number of individuals. Record how much space each person has.

Data and Observations

Think About This

1. What would happen if the number of students in your classroom doubled?

2. **Key Concept** How might your interactions with your classmates change if each person had less space?

Inquiry Community?

1. These leafcutter ants work together making homes and getting food. The ants in the population interact with each other as well as with other populations in the community. What other populations might the ants interact with?

_____

_____

_____

_____

_____

_____

Figure 5 Populations can live in very large areas, such as an ocean, or small areas, such as a lake.

Populations

Have you ever gone fishing at a lake? If so, recall that each individual fish in the lake is a member of a population. You read in Lesson 1 that a population is all the members of one species that live in an area. The area in which a population lives can be very large, such as the population of all fish in an ocean, as shown at the top of **Figure 5.** The area in which a population lives can also be small, like fish in a lake as shown at the bottom of **Figure 5.**

Recall that organisms respond to abiotic and biotic factors in their ecosystem. Sunlight, temperature, and water quality are examples of abiotic factors that affect the fish in the lake and in the ocean. Biotic factors, such as the plants they eat and other organisms that hunt them, also affect the fish. If any of these factors change, the fish population can also change.

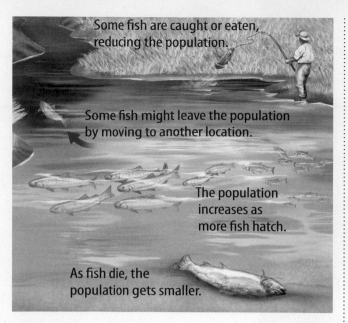

Some fish are caught or eaten, reducing the population.

Some fish might leave the population by moving to another location.

The population increases as more fish hatch.

As fish die, the population gets smaller.

Figure 6 The size of a fish population can change in several different ways.

Active Reading

2. Explain How do you think fish hatching affects the other populations in the community?

_____

_____

_____

_____

Population Size

Think about the fish in the lake. What might happen to the population if a large number of fish eggs hatched or a hundred people caught fish all at once?

Population size can increase or decrease, as shown in **Figure 6.** Populations can increase when individuals move into an area or when more individuals are born. Populations can decrease when individuals move away from an area or die. Sometimes the size of a population changes because the ecosystem changes. For example, if there is not much rainfall, a pond might shrink and some fish might die.

Population Density

Are the hallways in your school crowded or is there lots of room? The population density in the hallways reflects how crowded it is. Population density describes the number of organisms in the population relative to the amount of space available. A very dense population might have one fish for every few cubic meters of water. If a population is very dense, organisms might have a hard time finding enough resources to survive. They might not grow as large as individuals in less crowded conditions.

Inquiry

SC.7.N.1.1,
SC.7.N.1.3,
SC.7.E.6.6

LAB STATION **Try It!**

MiniLab *How does a fish population change?* at connectED.mcgraw-hill.com

Apply It! After you complete the lab, answer these questions.

1. What was the maximum size of the fish population?

2. What do you think would happen if the population stayed at that size for a long time?

Limiting Factors

What do you think keeps populations from becoming too large? **Limiting factors** *are factors that can limit the growth of a population.* The amount of available water, space, shelter, and food affects a population's size. With too few resources, some individuals cannot survive. Other factors, such as the availability of nesting sites, predation, competition, disease, and parasitism, also can limit how many individuals survive. Some limiting factors in a rabbit population are shown in **Figure 7.**

Biotic Potential Imagine a population of rabbits with a constant supply of food, an unlimited amount of land to live on, and no predators. The population would keep growing until it reached its biotic potential. **Biotic potential** *is the potential growth of a population if it could grow in perfect conditions with no limiting factors.* The population's rate of birth is the highest it can be, and its rate of death is the lowest it can be.

Carrying Capacity Almost no population reaches its biotic potential. Instead, it reaches its carrying capacity. **Carrying capacity** *is the largest number of individuals of one species that an ecosystem can support over time.* The limiting factors of an area determine the area's carrying capacity.

Overpopulation When a population's size grows beyond the ability of the area to support it overpopulation occurs. This often results in overcrowding, a lack of resources, and an unhealthy environment. For instance, waste from a population of trout in an overcrowded lake might build up faster than it can be broken down, making the population sick.

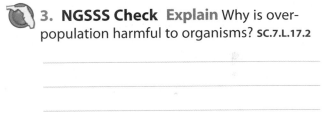

3. **NGSSS Check** **Explain** Why is overpopulation harmful to organisms? **SC.7.L.17.2**

_____

_____

_____

_____

Figure 7 The number of limiting factors in an area limits the growth of a population.

Active Reading 4. **Identify** Underline each limiting factor in the captions below.

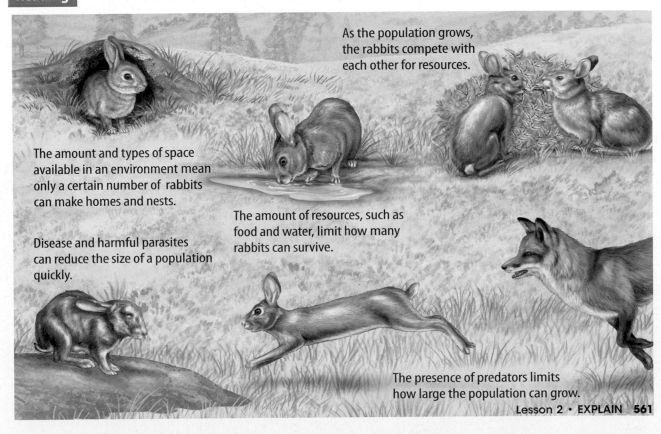

As the population grows, the rabbits compete with each other for resources.

The amount and types of space available in an environment mean only a certain number of rabbits can make homes and nests.

The amount of resources, such as food and water, limit how many rabbits can survive.

Disease and harmful parasites can reduce the size of a population quickly.

The presence of predators limits how large the population can grow.

A producer changes the energy available in the environment into food energy.

Consumers must use the energy and nutrients stored in other organisms for living and reproducing.

An energy pyramid shows how much food energy is available to organisms at each level of a community.

Use Vocabulary

1 **Describe the** process of chemosynthesis.

2 **Distinguish** between producers and consumers. SC.7.L.17.1

3 **Write** a sentence using the word *decomposer.* SC.7.L.17.1

Understand Key Concepts

4 **Explain** how energy from the Sun enters an ecosystem.

5 Which does NOT cycle through an environment?

(A) nitrogen   (C) sunlight

(B) oxygen   (D) water

Interpret Graphics

6 **Organize Information** Fill in the graphic organizer below for a food chain of herbivores, carnivores, and producers. SC.7.L.17.1

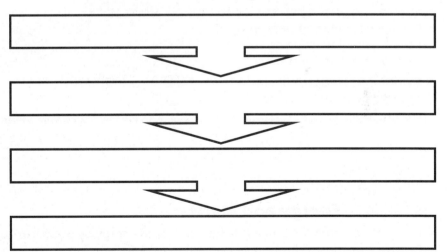

Critical Thinking

7 **Draw a** food web for a local habitat on a separate piece of paper. What are the producers and the consumers? Which are omnivores, carnivores, herbivores, and decomposers? Identify trophic levels where energy is reduced.

Math Skills

8 One liter of air contains about 0.21 L of oxygen (O_2). When filled, the human lungs hold about 6.0 L of air. How much O_2 is in the lungs when they are filled with air?

Chapter 14 Study Guide

Think About It! Plants and animals, including humans, interact with and depend upon each other and their environment to satisfy their basic needs. Energy flows from the Sun through producers to consumers.

🔑 Key Concepts Summary

Vocabulary

LESSON 1 Ecosystems and Biomes

- An **ecosystem** is made up of all the living and nonliving things in a location.
- **Biomes** are large regions that have specific typesof climate, physical characteristics, and organisms.
- One environment changes into another in a process called **succession.**

ecosystem p. 551
abiotic factor p. 552
biotic factor p. 553
population p. 553
community p. 553
biome p. 554
succession p. 555

LESSON 2 Populations and Communities

- Organisms must compete with each other to obtain resources, such as food, water, and living space.
- **Symbiotic relationships** include mutualism, parasitism, and commensalism.

limiting factor p. 561
biotic potential p. 561
carrying capacity p. 561
habitat p. 562
niche p. 562
symbiosis p. 562

LESSON 3 Energy and Matter

- Light energy from the Sun is changed into food energy by **producers.** Energy then moves through an ecosystem as organisms eat producers or other **consumers.**
- Energy movement can be modeled simply as a **food chain.** A **food web** models the movement of energy through many food chains in an ecosystem.

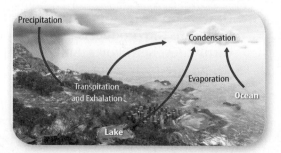

- Matter moves through ecosystems in cycles. Examples of matter cycles include the carbon, water, and oxygen cycles.

producer p. 568
consumer p. 569
food chain p. 570
food web p. 570
energy pyramid p. 571

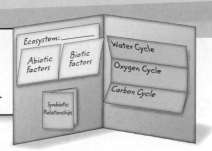

Active Reading

FOLDABLES® **Chapter Project**

Assemble your lesson Foldables as shown to make a Chapter Project. Use the project to review what you have learned in this chapter.

Use Vocabulary

1 A(n) _____ factor is any of the living parts of an ecosystem.

2 The biosphere contains all of Earth's
_____ .

3 The _____ of a population is the largest number of individuals that can survive in a location over time.

4 Each organism has a(n) _____ that includes the ways it survives, obtains food and shelter, and avoids danger in its habitat.

5 A(n) _____ cannot make food and must obtain energy from other organisms. **SC.7.L.17.1**

6 A(n) _____ models how much energy is available as it moves through an ecosystem. **SC.7.L.17.1**

Link Vocabulary and Key Concepts

Use vocabulary terms from the previous page to complete the concept map.

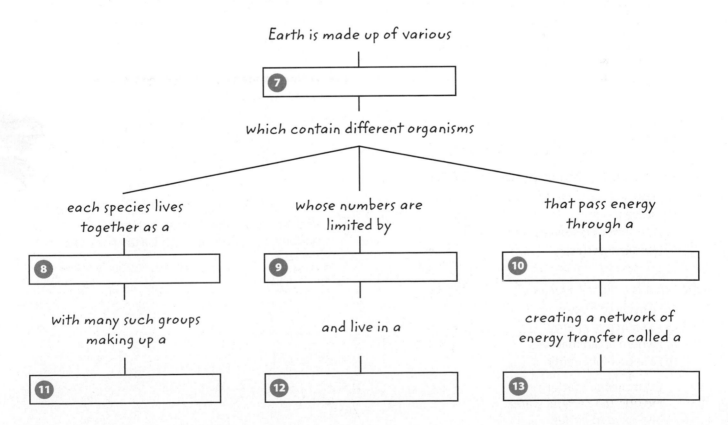

Earth is made up of various

7 [ ]

which contain different organisms

each species lives together as a

8 [ ]

with many such groups making up a

11 [ ]

whose numbers are limited by

9 [ ]

and live in a

12 [ ]

that pass energy through a

10 [ ]

creating a network of energy transfer called a

13 [ ]

Fill in the correct answer choice.

🔑 Understand Key Concepts

1 Which is NOT an abiotic factor? **LA.7.2.2.3**
- (A) atmosphere
- (B) prey
- (C) sunlight
- (D) temperature

2 Which biome is shown below? **LA.7.2.2.3**

- (A) desert
- (B) forest
- (C) grassland
- (D) tundra

3 Which term describes a slow change in an environment? **LA.7.2.2.3**
- (A) abiotic
- (B) biotic
- (C) regression
- (D) succession

4 Which is NOT a limiting factor? **LA.7.2.17.3**
- (A) competition
- (B) disease
- (C) amount of resources
- (D) biotic potential

5 What is a niche? **LA.7.2.2.3**
- (A) a cycle of matter
- (B) a source of energy
- (C) where an animal lives
- (D) a source of energy

Critical Thinking

6 **Contrast** aquatic and terrestrial ecosystems, and explain how they might interact. **LA.7.2.2.3**

_____

_____

_____

7 **List** some of the biotic and abiotic factors you respond to each day. Describe how they impact you. **SC.7.L.17.3**

_____

_____

_____

_____

8 **Consider** the value of the study of ecology. Why is it important or not important? **LA.7.2.2.3**

_____

_____

_____

9 **Differentiate** between a habitat and a niche. **LA.7.2.2.3**

_____

_____

_____

10 **Reflect** on whether human beings have a symbiotic relationship with Earth. **SC.7.L.17.2**

_____

_____

_____

11 **Describe** the symbiotic relationship shown below. **SC.7.L.17.2**

_____
_____
_____
_____

12 **Hypothesize** what might happen to a lake ecosystem if very little rain fell for many years. How would the lake community change? What would happen if the amount of rain greatly increased rather than decreased? **SC.7.L.17.3**

_____
_____
_____
_____
_____

13 **Construct** a food chain that includes you. **SC.7.L.17.1**

Writing in Science

14 **Visualize** succession in your community. What changes have occurred in your community over the last hundred years? What might occur over the next hundred? The next thousand? On a separate piece of paper, write a paragraph describing your predictions. **SC.7.N.1.6**

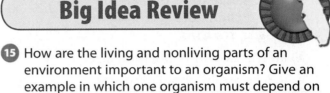
Big Idea Review

15 How are the living and nonliving parts of an environment important to an organism? Give an example in which one organism must depend on another organism in order to survive. **SC.7.L.17.2**

_____
_____
_____
_____

16 The photo at the beginning of the chapter shows a population of prairie dogs. How might living together as a population be helpful to the prairie dogs? How might living together be more difficult than living alone? **SC.7.L.17.2**

_____
_____
_____

Math Skills   MA.6.A.3.6

Use Proportion

17 A large tree releases about 0.31 kg of oxygen per day. A typical adult uses 0.84 kg of oxygen per day. How many trees would provide enough oxygen for the adult?

18 In one year, an acre of trees can provide enough oxygen for about 18 people. How many acres of trees would provide enough oxygen for the population of New York City—about 8,300,000?

Record your answers on the answer sheet provided by your teacher or on a sheet of paper.

Multiple Choice

1 Which change to a population would cause increased competition among its members? **SC.7.L.17.3**

  Ⓐ an increase in population size

  Ⓑ a decrease in population size

  Ⓒ an increase in predation on the population

  Ⓓ an increase of disease in the population

Use the table below to answer question 2.

Organism	Interaction
Cattails	make food using sunlight energy
Mice	eat plants
Fish	eat plants, insects, or other fish
Snakes	eat fish or mice

2 The organisms in the table above live in or around a lake. Which organism would receive the least amount of energy from the food it consumes? **SC.7.L.17.1**

  Ⓕ cattail

  Ⓖ mice

  Ⓗ fish

  Ⓘ snake

3 Which symbiotic relationship occurs when two species in a community benefit from the relationship? **SC.7.L.17.2**

  Ⓐ commensalism

  Ⓑ parasitism

  Ⓒ mutualism

  Ⓓ predation

4 Which type of consumer eats only producers? **SC.7.L.17.1**

  Ⓕ detritivores

  Ⓖ herbivores

  Ⓗ omnivores

  Ⓘ carnivores

Use the table below to answer question 5.

5 Which limiting factor does the diagram illustrate? **SC.7.L.17.1**

  Ⓐ carrying capacity

  Ⓑ overpopulation

  Ⓒ predation

  Ⓓ competition for resources

6 Which of the choices below is a decomposer? **SC.7.L.17.1**

  Ⓕ mushroom

  Ⓖ grass

  Ⓗ cricket

  Ⓘ fish

7 In which type of symbiotic relationship does one organism benefit and the other is neither helped nor harmed? **SC.7.L.17.2**

  Ⓐ competition

  Ⓑ commensalism

  Ⓒ mutualism

  Ⓓ parasitism

8 Which limiting factor is NOT affected by the amount of organisms in a given area? **SC.7.L.17.3**

Ⓕ foodwase

Ⓖ drought

Ⓗ shelter

Ⓘ disease

Use the diagram below to answer question 9.

9 Which level has the least amount of food energy? **SC.7.L.17.1**

Ⓐ birds

Ⓑ grasses

Ⓒ grasshoppers

Ⓓ coyote

10 Which is a symbiotic relationship? **SC.7.L.17.2**

Ⓕ a hawk eating a mouse

Ⓖ a fish laying eggs in a stream

Ⓗ a mistletoe plant living on a tree

Ⓘ a plant using sunlight and making food

11 Which category are consumers usually classified by? **SC.7.L.17.1**

Ⓐ appearance

Ⓑ diet

Ⓒ habitat

Ⓓ size

12 Which pair of organisms would most likely compete for food? **SC.7.L.17.2**

Ⓕ cow and chicken

Ⓖ snake and hawk

Ⓗ mushroom and shrub

Ⓘ brown bear and salmon

Use the diagram below to answer question 13.

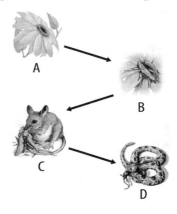

13 Which organism is a producer? **SC.7.L.17.1**

Ⓐ A

Ⓑ B

Ⓒ C

Ⓓ D

NEED EXTRA HELP?

If You Missed Question...	1	2	3	4	5	6	7	8	9	10	11	12	13
Go to Lesson...	2	3	2	3	2	3	2	2	3	2	3	2	3

Multiple Choice *Bubble the correct answer.*

1. Which is an abiotic factor that plays a role in a desert biome? **SC.7.L.17.1**

   (A)

   (B)

   (C)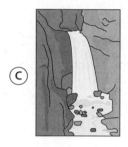

   (D)

2. In an open ocean biome, most life is found in the upper layer of water. Which abiotic factor likely plays the greatest role in this distribution of organisms? **SC.7.L.17.1**

   (F) precipitation

   (G) salinity

   (H) soil

   (I) sunlight

3. The image above is an example of a(n) **SC.7.L.17.2**

   (A) biosphere.

   (B) community.

   (C) ecosystem.

   (D) population.

4. A volcano on an island in the Pacific Ocean erupts after staying dormant for hundreds of years. What happens to the communities on the island? **SC.7.L.17.3**

   (F) The communities are all permanently destroyed.

   (G) The communities undergo succession and change over time.

   (H) The abiotic factors in the area change, but have little effect on communities.

   (I) The populations of many organisms in the communities increase.

Copyright © Glencoe/McGraw-Hill, a division of The McGraw-Hill Companies, Inc.

Multiple Choice *Bubble the correct answer.*

1. In the image above, which term describes the effect organism W is having on a population of R organisms? **SC.7.L.17.1**

   Ⓐ commensalism

   Ⓑ mutualism

   Ⓒ carrying capacity

   Ⓓ limiting factor

2. In a particular forest, the deer are smaller than normal and carry diseases. What might you conclude about the deer population? **SC.7.L.17.3**

   Ⓕ The deer population has grown beyond the forest's carrying capacity.

   Ⓖ The deer population has too many predators.

   Ⓗ The forest does not have any limiting factors that affect the growth of the deer population.

   Ⓘ The forest has the perfect conditions for the deer population, so it is reaching its biotic potential.

3. In the image above, which term describes the relationship between the two organisms? **SC.7.L.17.2**

   Ⓐ commensalism

   Ⓑ mutualism

   Ⓒ carrying capacity

   Ⓓ limiting factor

4. In which of these areas would you expect to find the largest population of rabbits? **SC.7.L.17.3**

   Ⓕ an area that has a large coyote population

   Ⓖ an area that has a large plant population

   Ⓗ an area that has a small amount of water

   Ⓘ an area that has a small insect population

Copyright © Glencoe/McGraw-Hill, a division of The McGraw-Hill Companies, Inc.

Multiple Choice *Bubble the correct answer.*

Use the image below to answer questions 1 and 2.

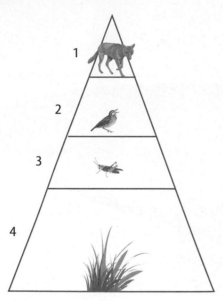

1. In which level of the energy pyramid would you expect to find the largest population? **SC.7.L.17.1**

   (A) 1

   (B) 2

   (C) 3

   (D) 4

2. In which level of the energy pyramid would you expect to find the smallest amount of energy? **SC.7.L.17.1**

   (F) 1

   (G) 2

   (H) 3

   (I) 4

Use the image below to answer questions 3 and 4.

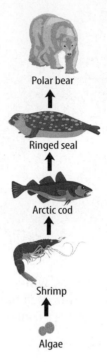

3. In the food chain, which organism is a producer? **SC.7.L.17.1**

   (A) algae

   (B) shrimp

   (C) arctic cod

   (D) polar bear

4. In the food chain, which organism is NOT a carnivore? **SC.7.L.17.1**

   (F) shrimp

   (G) arctic cod

   (H) polar bear

   (I) ringed seal

Copyright © Glencoe/McGraw-Hill, a division of The McGraw-Hill Companies, Inc.

Notes

Glossary/Glosario

g Multilingual eGlossary

A science multilingual glossary is available on the science Web site. The glossary includes the following languages.

Arabic	Hmong	Tagalog
Bengali	Korean	Urdu
Chinese	Portuguese	Vietnamese
English	Russian	
Haitian Creole	Spanish	

Cómo usar el glosario en español:
1. Busca el término en inglés que desees encontrar.
2. El término en español, junto con la definición, se encuentran en la columna de la derecha.

Pronunciation Key

Use the following key to help you sound out words in the glossary.

a . . . . b**a**ck (BAK)	**ew** . . . . f**oo**d (FEWD)
ay . . . . d**ay** (DAY)	**yoo** . . . . p**u**re (PYOOR)
ah . . . . f**ather** (FAH thur)	**yew** . . . . f**ew** (FYEW)
ow . . . . fl**ow**er (FLOW ur)	**uh** . . . . comm**a** (CAH muh)
ar . . . . **car** (CAR)	**u (+ con)** . . . . r**u**b (RUB)
e . . . . l**e**ss (LES)	**sh** . . . . **sh**elf (SHELF)
ee . . . . l**ea**f (LEEF)	**ch** . . . . na**t**ure (NAY chur)
ih . . . . tr**i**p (TRIHP)	**g** . . . . **g**ift (GIHFT)
i (i + com + e) **i**dea (i DEE uh)	**j** . . . . **g**em (JEM)
oh . . . . g**o** (GOH)	**ing** . . . . s**ing** (SING)
aw . . . . s**o**ft (SAWFT)	**zh** . . . . vi**si**on (VIH zhun)
or . . . . **or**bit (OR buht)	**k** . . . . ca**k**e (KAYK)
oy . . . . c**oi**n (COYN)	**s** . . . . **s**eed, **c**ent (SEED, SENT)
oo . . . . f**oo**t (FOOT)	**z** . . . . **z**one, rai**s**e (ZOHN, RAYZ)

English · A · Español

abiotic factor/Air Quality Index (AQI) | **factor abiótico/Índice de calidad del aire (ICA)**

abiotic factor (ay bi AH tihk • FAK tuhr): a non-living thing in an ecosystem.

absorption: the transfer of energy from a wave to the medium through which it travels.

accuracy: a description of how close a measurement is to an accepted or true value.

acid precipitation: precipitation that has a lower pH than that of normal rainwater (5.6).

adaptation (a dap TAY shun): an inherited trait that increases an organism's chance of surviving and reproducing in a particular environment.

Air Quality Index (AQI): a scale that ranks levels of ozone and other air pollutants.

factor abiótico: componente no vivo de un ecosistema.

absorción: transferencia de energía desde una onda hacia el medio a través del cual viaja.

exactitud: descripción de qué tan cerca está una medida a un valor aceptable.

precipitación ácida: precipitación que tiene un pH más bajo que el del agua de la lluvia normal (5.6).

adaptación: rasgo heredado que aumenta la oportunidad de un organismo de sobrevivir y reproducirse en su medioambiente.

Índice de calidad del aire (ICA): Escala que clasifica los niveles de ozono y de otros contaminantes del aire.

GLOSSARY/ GLOSARIO

allele (uh LEEL): a different form of a gene.

amplitude: the maximum distance a wave varies from its rest position.

analogous (uh NAH luh gus) structures: body parts that perform a similar function but differ in structure.

Archean (ar KEE un) eon: the time from about 4 to 2.5 million years ago.

asexual reproduction: a type of reproduction in which one parent organism produces offspring without meiosis and fertilization.

asthenosphere (as THE nuh sfir): the partially melted portion of the mantle below the lithosphere.

alelo: forma diferente de un gen.

amplitud: distancia máxima que varía una onda desde su posición de reposo.

estructuras análogas: partes del cuerpo que ejecutan una función similar pero tienen una estructura distinta.

eón arcaico: tiempo desde hace cerca de 4 a 2.5 millones de años.

reproducción asexual: tipo de reproducción en la cual un organismo parental produce crías sin mitosis ni fertilización.

astenosfera: porción parcialmente fundida del manto debajo de la litosfera.

B

basin: area of subsidence; region with low elevation.

biological evolution: the change over time in populations of related organisms.

biome: a geographic area on Earth that contains ecosystems with similar biotic and abiotic features.

biotic factor (bi AH tihk • FAK tuhr): a living or once-living thing in an ecosystem.

biotic potential: the potential growth of a population if it could grow in perfect conditions with no limiting factors.

budding: the process during which a new organism grows by mitosis and cell division on the body of its parent.

cuenca: área de hundimiento; región de elevación baja.

evolución biológica: cambio a través del tiempo en las poblaciones de organismos relacionados.

bioma: área geográfica en la Tierra que contiene ecosistemas con características bióticas y abióticas similares.

factor biótico: ser vivo o que una vez estuvo vivo en un ecosistema.

potencial biótico: crecimiento potencial de una población si puede crecer en condiciones perfectas sin factores limitantes.

germinación: proceso durante el cual un organismo nuevo crece por medio de mitosis y división celular en el cuerpo de su progenitor.

C

camouflage (KAM uh flahj): an adaptation that enables a species to blend in with its environment.

carrying capacity: the largest number of individuals of one species that an ecosystem can support over time.

cast: a fossil copy of an organism made when a mold of the organism is filled with sediment or mineral deposits.

camuflaje: adaptación que permite a las especies mezclarse con su medioambiente.

capacidad de carga: número mayor de individuos de una especie que un medioambiente puede mantener.

contramolde: copia fósil de un organismo compuesto en un molde de el organismo está lleno de sedimentos o los depósitos de minerales.

chemical energy: energy that is stored in and released from the bonds between atoms.

cinder cone: a small, steep-sided volcano that erupts gas-rich, basaltic lava.

cleavage: the breaking of a mineral along a smooth, flat surface.

cloning: a type of asexual reproduction performed in a laboratory that produces identical individuals from a cell or a cluster of cells taken from a multicellular organism.

closed system: a system that does not exchange matter or energy with the environment.

codominance: an inheritance pattern in which both alleles can be observed in a phenotype.

community: all the populations living in an ecosystem at the same time.

comparative anatomy: the study of similarities and differences among structures of living species.

composite volcano: a large, steep-sided volcano that results from explosive eruptions of andesitic and rhyolitic lavas along convergent plate boundaries.

compression: the squeezing force at a convergent boundary; region of a longitudinal wave where the particles of the medium are closest together.

conduction (kuhn DUK shun): the transfer of thermal energy due to collisions between particles.

constants: the factors in an experiment that remain the same.

consumer: an organism that cannot make its own food and gets energy by eating other organisms.

continental drift: Wegener's hypothesis which suggested that the continents are in constant motion on Earth's surface.

convection: the circulation of particles within a material caused by differences in thermal energy and density; the transfer of thermal energy by the movement of particles from one part of a material to another.

energía química: energía almacenada en y liberada por los enlaces entre los átomos.

cono de ceniza: volcán pequeño de lados empinados que expulsa lava rica en gas basáltico.

exfoliación: rompimiento de un mineral en láminas o superficies planas.

clonación: tipo de reproducción asexual realizada en un laboratorio que produce individuos idénticos a partir de una célula o grupo de células tomadas de un organismo pluricelular.

sistema cerrado: sistema que no intercambia materia o energía con el ambiente.

condominante: patrón heredado en el cual los dos alelos se observan en un fenotipo.

comunidad: todas las poblaciones que viven en un ecosistema al mismo tiempo.

anatomía comparativa: estudio de las similitudes y diferencias entre las estructuras de las especies vivas.

volcán compuesto: volcán grande de lados empinados producido por erupciones explosivas de lavas andesíticas y riolíticas a lo largo de límites convergentes.

compresión: tensión en un límite convergente; región de una onda longitudinal donde las partículas del medio están más cerca.

conducción: transferencia de energía térmica debido a colisiones entre partículas.

constantes: factores en un experimento que permanecen iguales.

consumidor: organismo que no elabora su propio alimento y obtiene energía comiendo otros organismos.

deriva continental: hipótesis de Wegener que sugirió que los continentes están en constante movimiento en la superficie de la Tierra.

convección: circulación de partículas en el interior de un material causada por diferencias en la energía térmica y la densidad; transferencia de energía térmica por el movimiento de partículas de una parte de la materia a otra.

convergent plate boundary: the boundary between two plates that move toward each other.

core: the dense metallic center of Earth.

cornea (KOR nee uh): a convex lens made of transparent tissue located on the outside of the eye.

crest: the highest point on a transverse wave.

critical thinking: comparing what you already know with information you are given in order to decide whether you agree with it.

crust: the brittle, rocky outer layer of Earth.

crystal structure: the orderly, repeating pattern of atoms in a crystal.

crystallization: the process by which atoms form a solid with an orderly, repeating pattern.

límite convergente de placas: límite entre dos placas que se acercan una hacia la otra.

núcleo: centro de la Tierra denso y metálico.

córnea: lente convexo hecho de tejido transparente, ubicado en la parte externa del ojo.

cresta: punto más alto en una onda transversal.

pensamiento crítico: comparación que se hace cuando se sabe algo acerca de información nueva, y se decide si se está o no de acuerdo con ella.

corteza: capa frágil y rocosa superficial de la Tierra.

estructura del cristal: patrón repetitivo y ordenado de los átomos en un cristal.

cristalización: proceso por el cual los átomos forman un sólido con un patrón ordenado y repetitivo.

D

deforestation: the removal of large areas of forests for human purposes.

density: the mass per unit volume of a substance.

dependent variable: the factor a scientist observes or measures during an experiment.

deposition: the laying down or settling of eroded material.

description: a spoken or written summary of an observation.

desertification: the development of desertlike conditions due to human activities and/or climate change.

diffraction: the change in direction of a wave when it travels by the edge of an object or through an opening.

diploid: a cell that has pairs of chromosomes.

divergent plate boundary: the boundary between two plates that move away from each other.

DNA: the abbreviation for deoxyribonucleic (dee AHK sih ri boh noo klee ihk) acid, an organism's genetic material.

deforestación: eliminación de grandes áreas de bosques con propósitos humanos.

densidad: cantidad de masa por unidad de volumen de una sustancia.

variable dependiente: factor que el científico observa o mide durante un experimento.

deposición: establecimiento o asentamiento de material erosionado.

descripción: resumen oral o escrito de una observación de.

desertificación: desarrollo de condiciones parecidas a las del desierto debido a actividades humanas y/o al cambio en el clima.

difracción: cambio en la dirección de una onda cuando ésta viaja por el borde de un objeto o a través de una abertura.

diploide: célula que tiene pares de cromosomas.

límite divergente de placas: límite entre dos placas que se alejan una de la otra.

ADN: abreviatura para ácido desoxirribonucleico, material genético de un organismo.

GLOSSARY/ GLOSARIO

dominant (DAH muh nunt) trait: a genetic factor that blocks another genetic factor.

rasgo dominante: factor genético que bloquea otro factor genético.

earthquake: vibrations caused by the rupture and sudden movement of rocks along a break or a crack in Earth's crust.

echo: a reflected sound wave.

ecosystem: all the living things and nonliving things in a given area.

egg: the female reproductive, or sex, cell; forms in an ovary.

electric energy: energy carried by an electric current.

electromagnetic wave: a transverse wave that can travel through empty space and through matter.

embryology (em bree AH luh jee): the science of the development of embryos from fertilization to birth.

energy: the ability to cause change.

energy pyramid: a model that shows the amount of energy available in each link of a food chain.

energy transfer: the process of moving energy from one object to another without changing form.

energy transformation: the conversion of one form of energy to another.

epicenter: the location on Earth's surface directly above an earthquake's focus.

explanation: an interpretation of observations.

extinct: all of the members of a species have died.

extinction (ihk STINGK shun): event that occurs when the last individual organism of a species dies.

extrusive rock: igneous rock that forms when volcanic material erupts, cools, and crystallizes on Earth's surface.

terremoto: vibraciones causadas por la ruptura y el movimiento repentino de las rocas en una fractura o grieta en la corteza de la Tierra.

eco: onda sonora reflejada.

ecosistema: todos los seres vivos y los componentes no vivos de un área dada.

óvulo: célula reproductiva femenina o sexual; forma en un ovario.

energía eléctrica: energía transportada por una corriente eléctrica.

onda electromagnética: onda transversal que puede viajar a través del espacio vacío y de la materia.

embriología: ciencia que trata el desarrollo de embriones desde la fertilización hasta el nacimiento.

energía: capacidad de ocasionar cambio.

pirámide energética: modelo que explica la cantidad de energía disponible en cada vínculo de una cadena alimentaria.

transferencia de energía: proceso por el cual se mueve energía de un objeto a otro sin cambiar de forma.

transformación de energía: conversión de una forma de energía a otra.

epicentro: lugar en la superficie de la Tierra justo encima del foco de un terremoto.

explicación: interpretación que se hace de las observaciones.

extinto: especie en la cual todos los miembros han muerto.

extinción: evento que ocurre cuando el último organismo individual de una especie muere.

roca extrusiva: roca ígnea que se forma cuando el material volcánico sale, se enfría y se cristaliza en la superficie de la Tierra.

fault: a crack or a fracture in Earth's lithosphere along which movement occurs.

falla: grieta o fractura en la litosfera de la Tierra en la cual ocurre el movimiento.

fault-block mountain: parallel ridge that forms where blocks of crust move up or down along faults.

fault zone: an area of many fractured pieces of crust along a large fault.

fertilization (fur tuh luh ZAY shun): a reproductive process in which a sperm joins with an egg.

fission: cell division that forms two genetically identical cells.

focus: a location inside Earth where seismic waves originate and rocks first move along a fault and from which seismic waves originate.

folded mountain: mountain made of layers of rocks that are folded.

foliation (foh lee AY shun): rock texture that forms when uneven pressures cause flat minerals to line up, giving the rock a layered appearance.

food chain: a model that shows how energy flows in an ecosystem through feeding relationships.

food web: a model of energy transfer that can show how the food chains in a community are interconnected.

fossil: the preserved remains or evidence of past living organisms.

fossil record: record of all the fossils ever discovered on Earth.

fracture: the breaking of a mineral along a rough or irregular surface.

frequency: the number of wavelengths that pass by a point each second.

montaña de bloques fallados: dorsal paralela que se forma donde los bloques de corteza se mueven hacia arriba o hacia abajo en las fallas.

zona de falla: área de muchos pedazos fracturados de corteza en una falla extensa.

fertilización: proceso reproductivo en el cual un espermatozoide se une con un óvulo.

fisión: división celular que forma dos células genéticamente idénticas.

foco: lugar en el interior de la Tierra donde se originan las ondas sísmicas, las cuales son producidas por el movimiento de las rocas a lo largo de un falla.

montaña plegada: montaña constituida de capas de rocas plegadas.

foliación: textura de la roca que se forma cuando presiones disparejas causan que los minerales planos se alineen, dándole a la roca una apariencia de capas.

cadena alimentaria: modelo que explica cómo la energía fluye en un ecosistema a través de relaciones alimentarias.

red alimentaria: modelo de transferencia de energía que explica cómo las cadenas alimentarias están interconectadas en una comunidad.

fósil: restos conservados o evidencia de organismos vivos del pasado.

registro fósil: registro de todos los fósiles descubiertos en la Tierra.

fractura: rompimiento de un mineral en una superficie desigual o irregular.

frecuencia: número de longitudes de onda que pasan por un punto cada segundo.

G

gene (JEEN): a section of DNA on a chromosome that has genetic information for one trait.

genetics: the study of how traits are passed from parents to offspring.

genotype (JEE nuh tipe): the alleles of all the genes on an organism's chromosomes; controls an organism's phenotype.

gen: parte del ADN en un cromosoma que contiene información genética para un rasgo.

genética: estudio de cómo los rasgos pasan de los padres a los hijos.

genotipo: de los alelos de todos los genes en los cromosomas de un organismo, los controles de fenotipo de un organismo.

geologic time scale: a chart that divides Earth's history into different time units based on changes in the rocks and fossils.

geosphere: the solid part of Earth.

global warming: an increase in the average temperature of Earth's surface.

grain: an individual particle in a rock.

gravity: an attractive force that exists between all objects that have mass.

greenhouse effect: the natural process that occurs when certain gases in the atmosphere absorb and reradiate thermal energy from the Sun.

escala de tiempo geológico: tabla que divide la historia de la Tierra en diferentes unidades de tiempo, basado en los cambios en las rocas y fósiles.

geosfera: parte sólida de la Tierra.

calentamiento global: incremento en la temperatura promedio de la superficie de la Tierra.

grano: partícula individual de una roca.

gravedad: fuerza de atracción que existe entre todos los objetos que tienen masa.

efecto invernadero: proceso natural que ocurre cuando ciertos gases en la atmósfera absorben y vuelven a irradiar la energía térmica del Sol.

(H)

habitat: the place within an ecosystem where an organism lives; provides the biotic and abiotic factors an organism needs to survive and reproduce.

Hadean (hay DEE un) eon: the first 640 million years of Earth history.

half-life: the time required for half of the amount of a radioactive parent element to decay into a stable daughter element.

haploid: a cell that has only one chromosome from each pair.

heat: the movement of thermal energy from a region of higher temperature to a region of lower temperature.

heredity (huh REH duh tee): the passing of traits from parents to offspring.

heterozygous (he tuh roh ZI gus): a genotype in which the two alleles of a gene are different.

homologous (huh MAH luh gus) chromosomes: pairs of chromosomes that have genes for the same traits arranged in the same order.

homologous (huh MAH luh gus) structures: body parts of organisms that are similar in structure and position but different in function.

homozygous (hoh muh ZI gus): a genotype in which the two alleles of a gene are the same.

hot spot: a location where volcanoes form far from plate boundaries.

hábitat: lugar en un ecosistema donde vive un organismo; proporciona los factores bióticos y abióticos que un organismo necesita para vivir y reproducirse.

eón hadeano: los primeros 640 millones de años de la historia de la Tierra.

vida media: tiempo requerido para que la mitad de cierta cantidad de un elemento radiactivo se desintegre en otro elemento estable.

haploide: célula que tiene solamente un cromosoma de cada par.

calor: movimiento de energía térmica desde una región de alta temperatura a una región de baja temperatura.

herencia: paso de rasgos de los padres a los hijos.

heterocigoto: genotipo en el cual los dos alelos de un gen son diferentes.

cromosomas homólogos: pares de cromosomas que tienen genes de iguales rasgos dispuestos en el mismo orden.

estructuras homólogas: partes del cuerpo de los organismos que son similares en estructura y posición pero diferentes en función.

homocigoto: genotipo en el cual los dos alelos de un gen son iguales.

punto caliente: lugar lejos de los límites de las placas donde se forman volcanes.

hypothesis: a possible explanation for an observation that can be tested by scientific investigations.

hipótesis: explicación posible de una observación que se puede probar por medio de investigaciones científicas.

incomplete dominance: an inheritance pattern in which an offspring's phenotype is a combination of the parents' phenotypes.

independent variable: the factor that is changed by the investigator to observe how it affects a dependent variable.

inference: a logical explanation of an observation that is drawn from prior knowledge or experience.

interference: occurs when waves overlap and combine to form a new wave.

International System of Units (SI): the internationally accepted system of measurement.

intrusive rock: igneous rock that forms as magma cools underground.

iris: the colored part of the eye.

isostasy (i SAHS tuh see): the equilibrium between continental crust and the denser mantle below it.

dominancia incompleta: patrón heredado en el cual el fenotipo de un hijo es una combinación de los fenotipos de los padres.

variable independiente: factor que el investigador cambia para observar cómo afecta la variable dependiente.

inferencia: explicación lógica de una observación que se extrae de un conocimiento previo o experiencia.

interferencia: ocurre cuando las ondas coinciden y combinan para forma una onda nueva.

Sistema Internacional de Unidades (SI): sistema de medidas aceptado internacionalmente.

roca intrusiva: roca ígnea que se forma cuando el magma se enfría bajo el suelo.

iris: parte coloreada del ojo.

isostasia: equilibrio entre la corteza continental y el manto más denso debajo de la corteza.

K

kinetic energy: energy due to motion.

energía cinética: energía debida al movimiento.

L

landform: a topographic feature formed by processes that shape Earth's surface.

lava: magma that erupts onto Earth's surface.

law of conservation of energy: law that states that energy can be transformed from one form to another, but it cannot be created or destroyed.

law of reflection: law that states that when a wave is reflected from a surface, the angle of reflection is equal to the angle of incidence.

lens: a transparent object with at least one curved side that causes light to change direction.

accidente geográfico: característica topográfica formada por procesos que moldean la superficie de la Tierra.

lava: magma que sale a la superficie de la Tierra.

ley de la conservación de la energía: ley que plantea que la energía puede transformarse de una forma a otra, pero no puede crearse ni destruirse.

ley de la reflexión: ley que establece que cuando una onda se refleja desde una superficie, el ángulo de reflexión es igual al ángulo de incidencia.

lente: objeto transparente que tiene, al menos, un lado curvo que hace que la luz cambie de dirección.

GLOSSARY/
GLOSARIO

light ray: represents a narrow beam of light that travels in a straight line.

light source: something that emits light.

limiting factor: a factor that can limit the growth of a population.

lithification: the process through which sediment turns into rock.

lithosphere (LIH thuh sfihr): the rigid outermost layer of Earth that includes the uppermost mantle and crust.

longitudinal (lahn juh TEWD nul) wave: a wave in which the disturbance is parallel to the direction the wave travels.

luster: the way a mineral reflects or absorbs light at its surface.

rayo de luz: haz de luz angosto que viaja en línea recta.

fuente lumínica: algo que emite luz.

factor limitante: factor que limita el crecimiento de una población.

litificación: proceso mediante el cual el sedimento se vuelve roca.

litosfera: capa rígida más externa de la Tierra que incluye el manto superior y la corteza.

onda longitudinal: onda en la que la perturbación es paralela a la dirección en que viaja la onda.

brillo: forma en que un mineral refleja o absorbe la luz en su superficie.

M

magma: molten rock stored beneath Earth's surface.

magnetic reversal: an event that causes a magnetic field to reverse direction.

magnetosphere: the outer part of Earth's magnetic field that interacts with charged particles.

mantle: the thick middle layer in the solid part of Earth.

mechanical energy: sum of the potential energy and the kinetic energy in a system.

mechanical wave: a wave that can travel only through matter.

medium: a material in which a wave travels.

meiosis: a process in which one diploid cell divides to make four haploid sex cells.

mid-ocean ridge: long, narrow mountain range on the ocean floor; formed by magma at divergent plate boundaries.

mimicry (MIH mih kree): an adaptation in which one species looks like another species.

mineral: a naturally occurring, inorganic solid that has a crystal structure and a definite chemical composition.

mirror: any reflecting surface that forms an image by regular reflection.

magma: roca fundida depositada bajo la superficie de la Tierra.

inversión magnética: evento que causa que un campo magnético invierta su dirección.

magnetosfera: parte externa del campo magnético de la Tierra que interactúa con partículas cargadas.

manto: capa delgada central de la parte sólida de la Tierra.

energía mecánica: suma de la energía potencial y de la energía cinética en un sistema.

onda mecánica: onda que puede viajar sólo a través de la materia.

medio: material en el cual viaja una onda.

meiosis: proceso en el cual una célula diploide se divide para constituir cuatro células sexuales haploides.

dorsal oceánica: cordillera larga y angosta en el lecho del océano, formada por magma en los límites de las placas divergentes.

mimetismo: una adaptación en el cual una especie se parece a otra especie.

mineral: sólido inorgánico de origen natural que tiene estructura de cristal y composición química definida.

espejo: cualquier superficie reflectora que forma una imagen por reflexión común.

GLOSSARY/
GLOSARIO

mold: the impression of an organism in a rock.

mountain: landform with high relief and high elevation.

mutation (myew TAY shun): a permanent change in the sequence of DNA, or the nucleotides, in a gene or a chromosome.

molde: impresión de un organismo en una roca.

montaña: accidente geográfico de alto relieve y elevación alta.

mutación: cambio permanente en la secuencia de ADN, de los nucleótidos, en un gen o en un cromosoma.

N

natural selection: the process by which organisms with variations that help them survive in their environment live longer, compete better, and reproduce more than those that do not have the variations.

naturalist: a person who studies plants and animals by observing them.

niche (NICH): the way a species interacts with abiotic and biotic factors to obtain food, find shelter, and fulfill other needs.

nonpoint-source pollution: pollution from several widespread sources that cannot be traced back to a single location.

nonrenewable energy resource: an energy resource that is available in limited amounts or that is used faster than it can be replaced in nature.

normal polarity: when magnetized objects, such as compass needles, orient themselves to point north.

nuclear energy: energy stored in and released from the nucleus of an atom

nucleotide (NEW klee uh tide): a molecule made of a nitrogen base, a sugar, and a phosphate group.

selección natural: proceso por el cual los organismos con variaciones que las ayudan a sobrevivir en sus medioambientes viven más, compiten mejor y se reproducen más que aquellas que no tienen esas variaciones.

naturalista: persona que estudia las plantas y los animales por medio de la observación.

nicho: forma como una especie interactúa con los factores abióticos y bióticos para obtener alimento, encontrar refugio y satisfacer otras necesidades.

contaminación de fuente no puntual: contaminación de varias fuentes apartadas que no se pueden rastrear hasta una sola ubicación.

recurso energético no renovable: recurso energético disponible en cantidades limitadas o que se usa más rápido de lo que se repone en la naturaleza.

polaridad normal: ocurre cuando los objetos magnetizados, tales como las agujas de la brújula, se orientan a sí mismas para apuntar al norte.

energía nuclear: energía almacenada en y liberada por el núcleo de un átomo.

nucelótido: molécula constituida de una base de nitrógeno, azúcar y un grupo de fosfato.

O

observation: the act of using one or more of your senses to gather information and take note of what occurs.

ocean trench: a deep, underwater trough created by one plate subducting under another plate at a convergent plate boundary.

observación: acción de usar uno o más sentidos para reunir información y tomar notar de lo que ocurre.

fosa oceánica: depresión profunda debajo del agua formada por una placa que se desliza debajo de otra placa, en un límite de placas convergentes.

GLOSSARY/ GLOSARIO

opaque: a material through which light does not pass.

open system: a system that exchanges matter or energy with the environment.

ore: a deposit of minerals that is large enough to be mined for a profit.

opaco: material por el que no pasa la luz.

sistema abierto: sistema que intercambia materia o energía con el ambiente.

mena: depósito de minerales suficientemente grandes como para ser explotados con un beneficio.

P

Pangaea (pan JEE uh): name given to a supercontinent that began to break apart approximately 200 million years ago.

particulate matter: the mix of both solid and liquid particles in the air.

Phanerozoic (FAN uh ruh ZO ihk) eon: the time in Earth's history from 542 million years ago to the present.

phenotype (FEE nuh tipe): how a trait appears or is expressed.

photochemical smog: air pollution that forms from the interaction between chemicals in the air and sunlight.

pitch: the perception of how high or low a sound is; related to the frequency of a sound wave.

plain: landform with low relief and low elevation.

plate tectonics: theory that Earth's surface is broken into large, rigid pieces that move with respect to each other.

plateau: an area with low relief and high elevation.

point-source pollution: pollution from a single source that can be identified.

polygenic inheritance: an inheritance pattern in which multiple genes determine the phenotype of a trait.

population: all the organisms of the same species that live in the same area at the same time.

potential (puh TEN chul) energy: stored energy due to the interactions between objects or particles.

precision: a description of how similar or close measurements are to each other.

Pangea: nombre dado a un supercontinente que empezó a separarse hace aproximadamente 200 millones de años.

partículas en suspensión: mezcla de partículas sólidas y líquidas en el aire.

eón fanerozoico: tiempo en la historia de la Tierra desde hace 542 millones de años hasta el presente.

fenotipo: forma como aparece o se expresa un rasgo.

smog fotoquímico: polución del aire que se forma de la interacción entre los químicos en el aire y la luz solar.

tono: percepción de qué tan alto o bajo es el sonido; relacionado con la frecuencia de la onda sonora.

plano: accidente geográfico de bajo relieve y baja elevación.

tectónica de placas: teoría que afirma que la superficie de la Tierra está divida en piezas enormes y rígidas que se mueven una con respecto a la otra.

meseta: área de bajo relieve y alta elevación.

contaminación de fuente puntual: contaminación de una sola fuente que se puede identificar.

herencia poligénica: patrón de herencia en el cual genes múltiples determinan el fenotipo de un rasgo.

población: todos los organismos de la misma especie que viven en la misma área al mismo tiempo.

energía potencial: energía almacenada debido a las interacciones entre objetos o partículas.

precisión: sescripción de qué tan similar o cercana están las mediciones una de otra.

GLOSSARY/
GLOSARIO

prediction: a statement of what will happen next in a sequence of events.

primary wave (also P-wave): a type of seismic wave which causes particles in the ground to move in a push-pull motion similar to a coiled spring.

principle of superposition: principle that states that in rock layers that have not been folded or deformed, the oldest rock layers are on the bottom.

producer: an organism that uses an outside energy source, such as the Sun, and produces its own food.

Proterozoic (PROH ter oh zoh ihk) eon: the time from 2.5 to 0.542 billion years ago.

protocontinents: small, early continents.

Punnett square: a model that is used to show the probability of all possible genotypes and phenotypes of offspring.

pupil: an opening into the interior of the eye at the center of the iris.

predicción: afirmación de lo que ocurrirá a continuación en una secuencia de eventos.

onda primaria (también, onda P): tipo de onda sísmica que causa un movimiento de atracción y repulsión en las partículas del suelo, similar a un resorte.

principio de superposición: principio que establece que en las capas de las rocas que no se han doblado ni deformado, las rocas más viejas se encuentran en la parte inferior.

productor: organismo que usa una fuente de energía externa, como el Sol, para elaborar su propio alimento.

eón proterozoico: tiempo desde hace 2.5 a 0.542 mil millones de años.

protocontinentes: continentes pequeños y antiguos.

cuadro de Punnett: modelo que se utiliza para demostrar la probabilidad de que todos los genotipos y fenotipos posibles de cría.

pupila: abertura en el interior del ojo y en el centro del iris.

R

radiant energy: energy carried by an electromagnetic wave.

radiation: the transfer of thermal energy by electromagnetic waves.

radioactive decay: the process by which an unstable element naturally changes into another element that is stable.

rarefaction: region of a longitudinal wave where the particles of the medium are farthest apart.

recessive (rih SE sihv) trait: a genetic factor that is blocked by the presence of a dominant factor.

reclamation: a process in which mined land must be recovered with soil and replanted with vegetation.

reflection: the bouncing of a wave off a surface.

reforestation: process of planting trees to replace trees that have been cut or burned down.

energía radiante: energía que transporta una onda electromagnética.

radiación: transferencia de energía térmica por ondas electromagnéticas.

desintegración radiactiva: proceso mediante el cual un elemento inestable cambia naturalmente en otro elemento que es estable.

rarefacción: región de una onda longitudinal donde las partículas del medio están más alejadas.

rasgo recesivo: factor genético boqueado por la presencia de un factor dominante.

recuperación: proceso por el cual las tierras explotadas se deben recubrir con suelo y se deben replantar con vegetación.

reflexión: rebote de una onda desde una superficie.

reforestación: proceso de siembra de árboles para reemplazar los árboles que se han cortado o quemado.

GLOSSARY/ GLOSARIO

refraction: the change in direction of a wave as it changes speed in moving from one medium to another.

regeneration: a type of asexual reproduction that occurs when an offspring grows from a piece of its parent.

renewable energy resource: an energy resource that is replaced as fast as, or faster than, it is used.

replication: the process of copying a DNA molecule to make another DNA molecule.

retina (RET nuh): an area at the back of the eye that includes special light-sensitive cells—rod cells and cone cells.

reversed polarity: when magnetized objects reverse direction and orient themselves to point south.

ridge push: the process that results when magma rises at a mid-ocean ridge and pushes oceanic plates in two different directions away from the ridge.

RNA: ribonucleic acid, a type of nucleic acid that carries the code for making proteins from the nucleus to the cytoplasm.

rock: a naturally occurring solid composed of minerals, rock fragments, and sometimes other materials such as organic matter.

rock cycle: the series of processes that change one type of rock into another type of rock.

refracción: cambio en la dirección de una onda a medida que cambia de velocidad al moverse de un medio a otro.

regeneración: tipo de reproducción asexual que ocurre cuando un organismo se origina de una parte de su progenitor.

recurso energético renovable: recurso energético que se repone tan rápido, o más rápido, de lo que se consume.

replicación: proceso por el cual se copia una molécula de ADN para hacer otra molécula de ADN.

retina: área en la parte posterior del ojo que incluye especiales sensibles a la luz—bastones y conos.

polaridad inversa: ocurre cuando los objetos magnetizados invierten la dirección y se orientan a sí mismos para apuntar al sur.

empuje de dorsal: proceso que resulta cuando el magma se levanta en la dorsal oceánica y empuja las placas oceánicas en dos direcciones diferentes, lejos de la dorsal.

ARN: ácido ribonucleico, un tipo de ácido nucléico que contiene el código para hacer proteínas del núcleo para el citoplasma.

roca: sólido de origen natural compuesto de minerales, acumulación de fragmentos y algunas veces de otros materiales como materia orgánica.

ciclo geológico: series de procesos que cambian un tipo de roca en otro tipo de roca.

(S)

science: the investigation and exploration of natural events and of the new information that results from those investigations.

scientific law: a rule that describes a pattern in nature.

scientific theory: an explanation of observations or events that is based on knowledge gained from many observations and investigations.

seafloor spreading: the process by which new oceanic crust forms along a mid-ocean ridge and older oceanic crust moves away from the ridge.

ciencia: la investigación y exploración de los eventos naturales y de la información nueva que es el resultado de estas investigaciones.

ley científica: regla que describe un patrón dado en la naturaleza.

teoría científica: explicación de observaciones o eventos con base en conocimiento obtenido de muchas observaciones e investigaciones.

expansión del lecho marino: proceso mediante el cual se forma corteza oceánica nueva en la dorsal oceánica, y la corteza oceánica vieja se aleja de la dorsal.

secondary wave (also S-wave): a type of seismic wave that causes particles to move at right angles relative to the direction the wave travels.

sediment: rock material that forms when rocks are broken down into smaller pieces or dissolved in water as rocks erode.

seismic wave: energy that travels as vibrations on and in Earth.

seismogram: a graphical illustration of seismic waves.

seismologist (size MAH luh just): scientist that studies earthquakes.

seismometer (size MAH muh ter): an instrument that measures and records ground motion and can be used to determine the distance seismic waves travel.

selective breeding: the selection and breeding of organisms for desired traits.

sexual reproduction: type of reproduction in which the genetic material from two different cells—a sperm and an egg—combine, producing an offspring.

shear: parallel forces acting in opposite directions at a transform boundary.

shield volcano: a large volcano with gentle slopes of basaltic lavas, common along divergent plate boundaries and oceanic hotspots.

significant digits: the number of digits in a measurement that are known with a certain degree of reliability.

slab pull: the process that results when a dense oceanic plate sinks beneath a more buoyant plate along a subduction zone, pulling the rest of the plate that trails behind it.

sound energy: energy carried by sound waves.

sound wave: a longitudinal wave that can travel only through matter.

sperm: a male reproductive, or sex, cell; forms in a testis.

onda secundaria (también, onda S): tipo de onda sísmica que causa que las partículas se muevan en ángulos rectos respecto a la dirección en que la onda viaja.

sedimento: material rocoso que se forma cuando las rocas se rompen en piezas pequeñas o cuando se disuelven en agua al erosionarse.

onda sísmica: energía que viaja en forma de vibraciones por encima y dentro de la Tierra.

sismograma: ilustración gráfica de las ondas sísmicas.

sismólogo: científico que estudia los terremotos.

sismómetro: instrumento que mide y registra el movimiento del suelo y que determina la distancia de las ondas sísmicas.

cría selectiva: selección y la cría de organismos para las características deseadas.

reproducción sexual: tipo de reproducción en la cual el material genético de dos células diferentes de un espermatozoide y un óvulo se combinan, produciendo una cría.

cizalla: fuerzas paralelas que actúan en direcciones opuestas en un límite transformante.

volcán escudo: volcán grande con ligeras pendientes de lavas basálticas, común a lo largo de los límites de placas divergentes y puntos calientes oceánicos.

cifras significativas: número de dígitos que se conoce con cierto grado de fiabilidad en una medida.

convergencia de placas: proceso que resulta cuando una placa oceánica densa se hunde debajo de una placa flotante en una zona de subducción, arrastrando el resto de la placa detrás suyo.

energía sonora: Energía que transportan las ondas sonoras.

onda sonora: onda longitudinal que sólo viaja a través de la materia.

esperma: célula reproductora masculina o sexual; forma en un testículo.

sphere: a ball shape with all points on the surface at an equal distance from the center.

strain: a change in the shape of rock caused by stress.

streak: the color of a mineral's powder.

subduction: the process that occurs when one tectonic plate moves under another tectonic plate.

subsidence: the downward vertical motion of Earth's surface.

succession: the gradual change from one community to another community in an area.

surface wave: a type of seismic wave that causes particles in the ground to move up and down in a rolling motion.

symbiosis: a close, long-term relationship between two species that usually involves an exchange of food or energy.

esfera: figura de bola cuyos puntos en la superficie están ubicados a una distancia igual del centro.

deformación: cambio en la forma de una roca causado por la presión.

raya: color del polvo de un mineral.

subducción: proceso que ocurre cuando una placa tectónica se mueve debajo de otra placa tectónica.

hundimiento: movimiento vertical hacia abajo de la superficie de la Tierra.

sucesión: cambio gradual de una comunidad a otra comunidad en un área.

onda superficial: tipo de onda sísmica que causa un movimiento de rodamiento hacia arriba y hacia debajo de las partícula en el suelo.

simbiosis: relación estrecha a largo plazo entre dos especies que generalmente involucra intercambio de alimento o energía.

technology: the practical use of scientific knowledge, especially for industrial or commercial use.

temperature: the measure of the average kinetic energy of the particles in a material.

tension: the pulling force at a divergent boundary.

texture: a rock's grain size and the way the grains fit together.

thermal conductor: a material through which thermal energy flows quickly.

thermal energy: the sum of the kinetic energy and the potential energy of the particles that make up an object.

thermal insulator: a material through which thermal energy flows slowly.

trace fossil: the preserved evidence of the activity of an organism.

transcription: the process of making mRNA from DNA.

transform fault: fault that forms where tectonic plates slide horizontally past each other.

transform plate boundary: the boundary between two plates that slide past each other.

tecnología: uso práctico del conocimiento científico, especialmente para uso industrial o comercial.

temperatura: medida de la energía cinética promedio de las partículas de un material.

tensión: fuerza de tracción en un límite divergente.

textura: tamaño del grano de una roca y la forma como los granos encajan.

conductor térmico: material en el cual la energía térmica se mueve con rapidez.

energía térmica: suma de la energía cinética y potencial de las partículas que componen un objeto.

aislante térmico: material a través del cual la energía térmica fluye con lentitud.

traza fósil: evidencia conservada de la actividad de un organismo.

transcripción: proceso por el cual se hace mARN de ADN.

falla transformante: falla que se forma donde las placas tectónicas se deslizan horizontalmente una con respecto a la otra.

límite de placas transcurrente: límite entre dos placas que se deslizan una con respecto a la otra.

GLOSSARY/
GLOSARIO

translation: the process of making a protein from RNA.

translucent: a material that allows most of the light that strikes it to pass through, but through which objects appear blurry.

transmission: the passage of light through an object.

transparent: a material that allows almost all of the light striking it to pass through, and through which objects can be seen clearly.

transverse wave: a wave in which the disturbance is perpendicular to the direction the wave travels.

trough: the lowest point on a transverse wave.

traslación: proceso por el cual se hacen proteínas a partir de ARN.

translúcido: material que permite el paso de la mayor cantidad de luz que lo toca, pero a través del cual los objetos se ven borrosos.

transmisión: paso de la luz a través de un objeto.

transparente: material que permite el paso de la mayor cantidad de luz que lo toca, y a través del cual los objetos pueden verse con nitidez.

onda transversal: onda en la que la perturbación es perpendicular a la dirección en que viaja la onda.

seno: punto más bajo en una onda transversal.

U

uplift: the process that moves large bodies of Earth materials to higher elevations.

uplifted mountain: mountain that forms when large regions rise vertically with very little deformation.

urban sprawl: the development of land for houses and other buildings near a city.

levantamiento: proceso por el cual se mueven grandes cuerpos de materiales de la Tierra hacia elevaciones mayores.

montaña elevada: montaña que se forma cuando grandes regiones se levantan verticalmente, con muy poca deformación.

expansión urbana: urbanización de tierra para viviendas y otras construcciones cerca de la ciudad.

V

vaporization: the change in state from a liquid to a gas.

variable: any factor that can have more than one value.

variation (ver ee AY shun): a slight difference in an inherited trait among individual members of a species.

vegetative reproduction: a form of asexual reproduction in which offspring grow from a part of a parent plant.

vestigial (veh STIH jee ul) structure: body part that has lost its original function through evolution.

viscosity (vihs KAW sih tee): a measurement of a liquid's resistance to flow.

volcanic arc: a curved line of volcanoes that forms parallel to a plate boundary.

vaporización: cambio de estado líquido a gaseoso.

variable: cualquier factor que tenga más de un valor.

variación: ligera diferencia en un rasgo hereditario entre los miembros individuales de una especie.

reproducción vegetativa: forma de reproducción asexual en la cual el organismo se origina a partir de una planta parental.

estructura vestigial: parte del cuerpo que a través de la evolución perdió la función original.

viscosidad: medida de la resistencia de un líquido a fluir.

arco volcánico: línea curva de volcanes que se forman paralelos al límite de una placa.

GLOSSARY/ GLOSARIO

volcanic ash: tiny particles of pulverized volcanic rock and glass.

volcano: a vent in Earth's crust through which molten rock flows.

ceniza volcánica: partículas diminutas de roca y vidrio volcánicos pulverizados.

volcán: abertura en la corteza terrestre por donde fluye la roca derretida.

wave: a disturbance that transfers energy from one place to another without transferring matter.

wavelength: the distance between one point on a wave and the nearest point just like it.

work: the amount of energy used as a force moves an object over a distance.

onda: perturbación que transfiere energía de un lugar a otro sin transferir materia.

longitud de onda: distancia entre un punto de una onda y el punto más cercano similar al primero.

trabajo: cantidad de energía usada como fuerza que mueve un objeto a cierta distancia.

zygote (ZI goht): the new cell that forms when a sperm cell fertilizes an egg cell.

zigoto: célula nueva que se forma cuando un espermatozoide fecunda un óvulo.

Index

Abiotic factors

Italic numbers = illustration/photo **Bold numbers** = vocabulary term
lab = indicates entry is used in a lab on this page

Chemical energy

INDEX

INDEX

INDEX

9 The diagram below shows the leg bone structures of horses and their ancestors. The drawings of bone structures are based on fossil evidence. Fossils provide evidence for biological evolution.

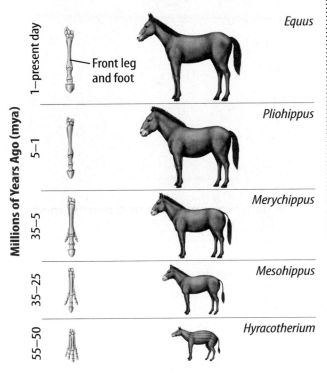

Which conclusion about the evolution of horses does fossil evidence support? **SC.7.L.15.1**

- (A) All of the horse-like animals shown are now extinct.
- (B) In the future, horses will be larger than they are today.
- (C) Some ancestors of modern horses did not have hooves.
- (D) Over millions of years, horses have evolved shorter hair.

10 This Punnett square shows a cross between two heterozygous parents.

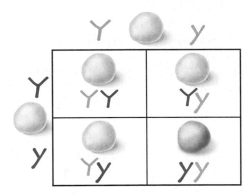

What is the probability that an offspring from this cross will have a heterozygous genotype? **SC.7.L.16.2**

- (F) 25 percent
- (G) 50 percent
- (H) 75 percent
- (I) 100 percent

11 Mitosis and meiosis are both processes that divide genetic material into new nuclei. These processes have many similarities. They also have important differences. Which can be a result of mitosis but not of meiosis? **SC.7.L.16.3**

- (A) the distribution of egg or sperm cells
- (B) a new organism that is identical to the parent cell
- (C) cells that contain half the genetic material that the parent cell contains
- (D) four cells that are all different from each other and from the parent cell

Copyright © Glencoe/McGraw-Hill, a division of The McGraw-Hill Companies, Inc.

NGSSS for Science Benchmark Practice continued

12 Mutualism, predation, parasitism, and commensalism are four types of symbiotic relationships. What is one way that mutualism is different from commensalism? **SC.7.L.17.2**

(F) Both organisms benefit from a mutualistic relationship, while both organisms are harmed in a commensal relationship.

(G) Both organisms are affected by a mutualistic relationship, while only one organism is affected by a commensal relationship.

(H) Mutualism can occur only between plants, while commensalism can occur between both plants and animals.

(I) Mutualistic relationships do not result in the death of either organism, while commensal relationships result in the death of at least one of the organisms.

13 The illustration below shows a desert food web.

If a disease resulted in the death of most of the rattlesnakes in this food web, which population of organisms would decrease most? **SC.7.L.17.1**

(A) birds

(B) coyotes

(C) lizards

(D) rodents

Copyright © Glencoe/McGraw-Hill, a division of The McGraw-Hill Companies, Inc.

NGSSS for Science Benchmark Practice *continued*

Copyright © Glencoe/McGraw-Hill, a division of The McGraw-Hill Companies, Inc.

14 A geologist studying rocks in Florida's central panhandle notices that many of the rocks contain fossils of marine organisms. He concludes that the rock from this area must have formed at the bottom of the ocean. What processes likely produced this rock? **SC.7.E.6.2**

(F) deposition of sediment and lithification

(G) extrusion of lava and crystallization

(H) formation of igneous rock and uplift

(I) intrusion of magma and crystallization

15 Diego made the drawing below to show a cross-section of rock that has four different rock layers.

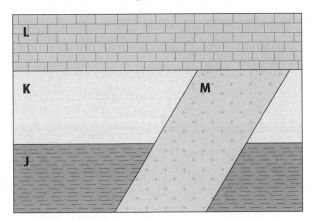

How does the law of superposition support Diego's conclusion that layer *K* is younger than layer *J*? **SC.7.E.6.3**

(A) Letter *K* comes after letter *J*.

(B) Layer *K* is on top of layer *J*.

(C) Layer *M* does not go through layer *L*.

(D) Layer *M* pushed through layer *K* first.

16 Pam finds this map in an encyclopedia article about tectonic plates.

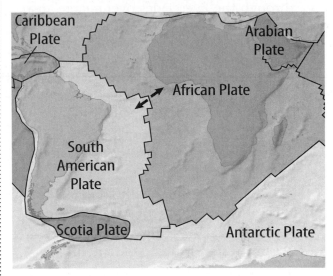

What is happening at the boundary between the South American Plate and the African Plate? **SC.7.E.6.5**

(F) A hot spot is causing an island chain to form.

(G) New oceanic crust is forming as magma rises.

(H) Volcanoes are forming along the edges of the continents.

(I) Large, deformed mountain ranges are forming where continental crust collides.

NGSSS for Science Benchmark Practice *continued*

17 The Everglades is a wetland ecosystem in southern Florida. Water flows slowly through the Everglades over a large area as if it were a very wide, shallow river. In some parts of the area, people have built neighborhoods, shopping malls, or farms that have disrupted this flow of water. What is the most likely effect of this development on the Everglades ecosystem? **SC.7.E.6.6**

Ⓐ improved water quality upstream of the development

Ⓑ improved water quality downstream of the development

Ⓒ areas that are either too dry or too wet upstream of the development

Ⓓ areas that are either too dry or too wet downstream of the development

18 Stress on rocks can cause Earth's surface to change shape and produce different kinds of landforms. The diagram below shows how one type of stress can affect rocks.

What kind of landform would this type of stress create? **SC.7.E.6.7**

Ⓕ eroded canyons

Ⓖ folded mountains

Ⓗ glacial moraine

Ⓘ karst topography

19 Akira makes a chart showing different types of light. At the top of the chart, she writes "Electromagnetic Waves." Under that category, she divides the chart into smaller categories. She labels one of these smaller categories "Visible Light from the Sun." Which falls within this smaller category? **SC.7.P.10.1**

Ⓐ black light

Ⓑ blue light

Ⓒ infrared light

Ⓓ ultraviolet light

20 An architect designs an office building. She wants bright, natural light to fill the front lobby. However, she does not want it to get too hot in the late afternoon. Which material would be best for the windows in the front lobby? **SC.7.P.10.2**

Ⓕ a material that absorbs all light

Ⓖ a material that transmits all light

Ⓗ a material that transmits most light but reflects the rest of it

Ⓘ a material that absorbs most light but refracts the rest of it

Copyright © Glencoe/McGraw-Hill, a division of The McGraw-Hill Companies, Inc.

NGSSS for Science Benchmark Practice continued

21 Two scientists study the effects of noise from motorboats on whales. One scientist places a detector above the surface of the water in the open air. The other places a detector underwater. Why does the underwater detector detect sounds before the open-air detector does? **SC.7.P.10.3**

(A) Sound travels faster through air than through water.

(B) Sound travels faster through water than through air.

(C) The underwater detector is farther away from the sounds.

(D) The detector located in the open air is farther away from the sounds.

22 Latisha steadily heated a pan of ice on a stove for 26 minutes while measuring its temperature. She made the graph below using the data she collected.

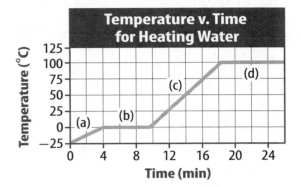

What could Latisha's friend conclude by looking at the graph? **SC.7.P.11.1**

(F) Areas **a** and **c** of the graph show when the water changed state.

(G) Areas **b** and **d** of the graph show when the water changed state.

(H) Areas **b** and **d** of the graph show when Latisha turned the stove off.

(I) Area **d** of the graph shows when there was no more water left in the pan.

Copyright © Glencoe/McGraw-Hill, a division of The McGraw-Hill Companies, Inc.

NGSSS for Science Benchmark Practice continued

23 Chris placed two fresh batteries in a flashlight. Then he switched on the flashlight and the bulb lit up. Which series of energy transformations took place? **SC.7.P.11.2**

Ⓐ chemical energy→electrical energy→ light energy

Ⓑ electrical energy→light energy→ chemical energy

Ⓒ light energy→electrical energy→light energy

Ⓓ nuclear energy→electrical energy→ light energy

24 Hannah placed her left hand in a bowl of cold water and her right hand in a bowl of warm water. She waited a few minutes while her left hand cooled down and her right hand warmed up. Then she pulled her hands out of each bowl of water and folded them together. Which statement best describes how heat flows between her hands now? **SC.7.P.11.4**

Ⓕ Heat flows from her left hand to her right hand until they are both the same temperature.

Ⓖ Heat flows from her right hand to her left hand until they are both the same temperature.

Ⓗ Heat flows from her left hand to her right hand until the right hand becomes warmer than the left hand.

Ⓘ Heat flows from her right hand to her left hand until the left hand becomes warmer than the right hand.

25 The diagram below shows different layers that make up the geosphere.

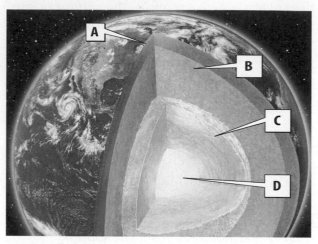

How are layers *C* and *D* alike and different? **SC.7.E.6.1**

Ⓐ They are both solid and part of Earth's core, but layer *C* is denser than layer *D*.

Ⓑ They are both part of Earth's core, but layer *C* is liquid while layer *D* is solid.

Ⓒ They are both solid, but layer *C* is part of Earth's mantle while layer *D* is part of Earth's core.

Ⓓ They are both liquid, but layer *C* is part of Earth's mantle while layer *D* is part of Earth's core.

Copyright © Glencoe/McGraw-Hill, a division of The McGraw-Hill Companies, Inc.

low are levels C and D alike and